KW-482-545

12.15
MCC

Mechanisms of Cognitive Development: Behavioral and Neural Perspectives

Edited by

James L. McClelland
Robert S. Siegler
Carnegie Mellon University

WITHDRAWN
from
STIRLING UNIVERSITY LIBRARY

  LAWRENCE ERLBAUM ASSOCIATES, PUBLISHERS
2001   Mahwah, New Jersey   London

Copyright © 2001 by Lawrence Erlbaum Associates, Inc.
All rights reserved. No part of the book may be reproduced in
any form, by photostat, microfilm, retrieval system, or any other
means, without the prior written permission of the publisher.

Lawrence Erlbaum Associates, Inc., Publishers
10 Industrial Avenue
Mahwah, New Jersey 07430-2262

Library of Congress Cataloging-in-Publication Data

Mechanisms of cognitive development : behavioral and neural perspectives /
edited by James L. McClelland, Robert S. Siegler.
     p.   cm. — (Carnegie Mellon symposia on cognition)
     Papers presented at the 29th Carnegie Mellon Symposia on Cognition
Mechanisms of Cognitive Development: Behavioral and Neural Perspectives,
held at Carnegie Mellon University, October 9–11, 1998.
     Includes bibliographical references and index.
     ISBN 0-8058-3275-0 (cloth: alk. paper) — ISBN 0-8058-3276-9
(pbk. : alk. paper)
     1. Cognition—Congresses.   I. McClelland, James L.   II. Siegler, Robert S.
III. Carnegie Symposium on Cognition (29th: 1998 : Carnegie Mellon
University)   IV. Series.
BF11.M432   2000
153—dc21   00-033163
                                                               CIP

Books published by Lawrence Erlbaum Associates are printed on acid-free paper,
and their bindings are chosen for strength and durability.

Printed in the United States of America
10 9 8 7 6 5 4 3 2 1

3555981000

University of Stirling Library, FK9 4LA
Tel. 01786 - 467220

LONG LOAN

Please RETURN or RENEW
no later than the last date stamped below
Subject to recall if requested

WITHDRAWN
from
STIRLING UNIVERSITY LIBRARY

Carnegie Mellon Symposia
on Cognition

David Klahr, Series Editor

Anderson: Cognitive Skills and Their Acquisition

Carroll/Payne: Cognition and Social Behavior

Carver/Klahr: Cognition and Instruction: Twenty-Five Years of Progress

Clark/Fiske: Affect and Cognition

Cohen/Schooler: Scientific Approaches to Consciousness

Cole: Perception and Production of Fluent Speech

Farah/Ratcliff: The Neuropsychology of High-Level Vision: Collected Tutorial Essays

Granrud: Visual Perception and Cognition in Infancy

Gregg: Knowledge and Cognition

Just/Carpenter: Cognitive Processes in Comprehension

Klahr: Cognition and Instruction

Klahr/Kotovsky: Complex Information Processing: The Impact of Herbert A. Simon

Lau/Sears: Political Cognition

MacWhinney: The Emergence of Language

MacWhinney: Mechanisms of Language Acquisition

McClelland/Siegler: Mechanisms of Cognitive Development: Behavioral and Neural Perspectives

Reder: Implicit Memory and Metacognition

Siegler: Children's Thinking: What Develops?

Sophian: Origins of Cognitive Skills

Steier/Mitchell: Mind Matters: A Tribute to Allen Newell

VanLehn: Architectures for Intelligence

Contents

Preface

This symposium carries on a tradition that began in the Psychology Department at Carnegie Mellon University more than 30 years ago. Each meeting has brought together a small group of leading scientists to explore an issue at the forefront of the study of human cognition or a related area of psychology.

The subject of this symposium can be stated as a simple question: What brings about the emergence of intelligence over the course of early life, and what sustains its further development? This question lay at the heart of the work of Jean Piaget, the 20th century's preeminent developmentalist, and it continues to be fundamental to our efforts to comprehend how we come to represent and reason about the world. Yet, the question has proven devilishly difficult to answer. Piaget himself used the descriptive biological metaphors of assimilation and accommodation to evoke intuitions about the nature of the process. Unfortunately, it has proven very difficult to build from Piaget's descriptions more precise accounts of how these change processes work. In the end, Piaget's primary legacy remains his remarkable and insightful observations about the nature of children's abilities at different points in time, rather than his account of how we get from here to there.

Now, at the beginning of the 21st century, there have been decades of additional research, and our understanding of children's thinking has deepened considerably. Clever experimentalists have devised techniques for demonstrating that infants and young children possess a variety of implicit understandings of basic concepts such as objects, time, space, number, and causality. Research on cognitive development has been broadened to include a wide range of domains, including informal physics, biology, and psychology. Until recently, however, relatively little progress had been made toward understanding what causes cognitive abilities to change over time.

This symposium celebrates the fact that this limited understanding is beginning to grow rapidly, and it brings together many of the researchers who are helping it to grow. The diverse papers in the symposium are united by the goal of determining how experience causes changes in how we think and act, and why it leads to change in some cases but fails miserably to lead to change in others. Before the symposium, all speakers were asked to think about four questions; the issues raised by these questions provide recurrent themes in the papers in this volume. The questions were:

1. Why do cognitive abilities emerge when they do during development?
2. What are the sources of developmental and individual differences, and of developmental anomalies in learning?
3. What happens in the brain when people learn?
4. How can experiences be ordered and timed to optimize learning?

As indicated by the title of this volume, the symposium also celebrates a second important trend in the study of cognitive development: The growing convergence of behavioral and neural approaches to understanding cognitive change. Several factors underlie this trend. One is our growing knowledge and understanding of the biology of brain development. Every day, molecular and developmental neurobiologists gain additional insights into the underlying physiological processes that lead to the emergence of the brain from its embryonic precursors. Among other things, we have learned that the brain is far from fully formed at birth. A considerable body of research now documents the dramatic impact of postnatal experience on the outcome of brain development, and suggests there is an intimate relationship between the structural and functional consequences of experience. New techniques allow us to examine the activity of the brain, and to see how this activity changes with experience. In animals, we may record from many neurons simultaneously to detect how their responses may be altered by experience. In humans, new functional brain imaging techniques allow us to visualize, albeit at a somewhat coarser grain, the consequences of experience for brain activity. Psychologists interested in cognitive change have rushed to embrace these methods, and several contributors to this symposium, among them Haier, Carpenter, Just, Minshew, Keller, Cherkassy, and Roth exemplify this trend.

A second key factor that is promoting increased understanding of change mechanisms at both behavioral and neural levels are new methods for exploring cognitive change. One such approach is the microgenetic method, which provides a child (or other experimental subject) with a great deal of focused experience, and densely samples the changing competence as the changes are occurring. Microgenetic methods are being used by both psychologists and neuroscientists, as is evident in the work of Goldin-Meadow, Siegler, McClelland, Merzenich, Thelen, and Kuhn in this volume. Such methods allow us to observe the often surprising paths that lead to changes in both behavior and brain activity.

A third factor that is promoting understanding of cognitive change is the emergence of computational models that characterize cognitive and behavioral change in terms of underlying neurobiological processes. These models hold promise of providing a synthetic demonstration of how experience can lead to physical changes in the brain, that in turn impact perception, cognition, and action. Such models are prominent in the work of Johnson and deHaan, and McClelland in this volume.

With these trends in view, one begins to feel that we are reaching the point where the distinction between neuroscientific and psychological approaches may be disappearing. Nowhere is this more evident than in the fact that it is not really possible to separate our speakers into those who represent behavioral and those who represent neural approaches. Robbie Case and Michael Mueller, by background psychologists, explore the role of neural processes in the emergence of higher cognitive processes. Michael Merzenich, by background a neuroscientist, describes behavioral studies of remediation of deficits in language perception. Many other examples from the list of speakers could be cited. Clearly, for many of the participants, the neural and behavioral perspectives mentioned in the title of our symposium are already well integrated.

Change occurs over different time spans, and the types of issues that can be investigated and the methods that can be used reflect these varying time spans. The first six chapters in the volume, which correspond to the papers presented on the first day of the conference, focus on changes that occur in response to particular experiences over relatively short spans of time: minutes, days, weeks, or months, rather than years. Repeated sampling of changing competence over these relatively short time spans allows evaluation of detailed and specific proposals concerning change mechanisms. A number of the studies that are reported do not involve children; rather, they involve college undergraduates, nonhuman primates, foreign adults learning English, and artificial neural networks. The basic assumption is that cognitive change in all of these types of learners shares a lot in common. Understanding cognitive changes in these diverse types of learners also provides a basis for distilling what is unique, and what is not, about children's cognitive development.

The next five chapters in the volume, which correspond to the papers presented on the second day of the conference, focus on change over longer time scales. Many of these talks and chapters, like the earlier ones, offer quite specific mechanistic accounts of the changes that we see over these longer scales. The accounts of the long-term changes tend to place greater emphasis on age-related biological changes and their interactions with new experiences, as contributors to cognitive growth. The final four papers, presented on the third day of the conference, focus on developmental disabilities. The existence of these disorders reminds us that there are large individual differences at the neural level, just as there are at the behavioral level. No explanation of cognitive development can be complete without an understanding of how the neural mechanisms enable development to succeed, as it normally does. Thus, Neville and Bavelier examine developmental plasticity in deaf and blind children, Galiburda and Rosen examine it in children with dyslexia, and Casey examines it in children with frontal lobe damage. The study of how development can go awry provides a contrast with the normal situation, and brings out clearly the need to complement an exploration of the role of experience with an understanding of how the complex machinery of the brain enables the normal process of development.

In addition to the speakers, a large number of other people and institutions made vital contributions to the conference, and we would like to thank them. Several groups provided generous support that helped defray the cost of the conference: The National Science Foundation, the National Institute of Mental Health, the National Institute of Child Health and Human Development, and the Psychology Department of Carnegie Mellon University. In addition, the NSF, in granting our request for funding, made a very constructive and welcome stipulation. They required us to raise additional money, so that we could bring a group of junior scientists—graduate students, postdoctoral students, and assistant professors—to Pittsburgh to participate in the Symposium. NIMH and NICHD came through with the extra funds. With the cooperation of matching funds from many of the fellows' home institutions, we were able to award 21 fellowships to such scientists, who came from as far away as The Netherlands and Great Britain. These junior scientists enriched the scientific discourse during the conference, adding excitement, new ideas, and promise for the future. We also would like to thank Mary Anne Cowden and Barb Dorney for their invaluable assistance in organizing the meeting and to recognize Barb Dorney for her invaluable help in preparing the volume as well. Their efforts help make this a memorable symposium for all who participated.

—*James L. McClelland*
—*Robert S. Siegler*

Part I

STUDIES OF THE MICROGENESIS OF COGNITIVE CHANGE

BEHAVIORAL APPROACHES

Chapter 1

Giving the Mind a Hand: The Role of Gesture in Cognitive Change

Susan Goldin-Meadow
University of Chicago

Many generations of developmental psychologists have documented the changes children exhibit as they are transformed from less knowledgeable into more knowledgeable members of their cultures. Documenting the changes is comparatively easy. It is accounting for these changes that continues to challenge developmentalists.

Even in domains where scholars are relatively comfortable granting the child some amount of innate knowledge (e.g., language, number, cf. Gelman & Williams, 1998), the problem of learning and change is not solved. Children need some way of cashing in on the knowledge they bring to the learning situation. For example, if we were to grant children a predisposition to develop communication systems equipped with noun-like and verb-like units (cf. Goldin-Meadow, Butcher, Mylander, & Dodge, 1994), children are still faced with the task of identifying nouns and verbs in the particular language they are learning. Thus, by anybody's account of development, we need to identify mechanisms of change.

Some mechanisms of change may be specialized according to domain, with different mechanisms for domains in which children come to the learning situation with more or less innate structure (core as opposed to noncore domains in Gelman and Williams', 1998, terms). Mechanisms of change need not be specialized, however. The same processes could function in areas that children are more or less prepared to learn. For example, structure mapping—the tendency to find and map inputs to our existing mental structures (Gentner, 1989)—has been proposed as a mechanism to explain learning in both core and noncore domains (Gelman & Williams, 1998).

Here, I consider a set of mechanisms that, in principle, could apply to all domains. The versatility of these mechanisms comes in large part from the fact that the central ingredient is the spontaneous gesture that accompanies talk. Gesture is a pervasive phenomenon, found across cultures (Feyereisen & de Lannoy, 1991), ages (McNeill, 1992), and a wide variety of tasks (Iverson & Goldin-Meadow, 1998a). Even congenitally blind children who have never seen others gesture move their hands spontaneously as they speak (Iverson & Goldin-Meadow, 1997; 1998b). In our previous work, we showed that learners' gestures are distinctive at moments of transition in their acquisition of a task (Goldin-Meadow, 1997; Goldin-Meadow, Alibali, & Church, 1993), thus associating gesture with periods of greatest change.

Identifying the mechanisms responsible for change is perhaps most easily accomplished at moments when change is imminent. Indeed, an entirely new type of study—the microgenetic study (Siegler & Crowley, 1991)—has grown up in large part to allow researchers to focus on the small steps learners take in their acquisition of a task, particularly the steps just prior to progress (e.g., Karmiloff-Smith, 1992; Kuhn, Garcia-Mila, Zohar, & Andersen, 1995; Siegler & Jenkins, 1989). The goal of microgenetic studies is to examine the learner, not just prior to and after instruction as in traditional training studies, but throughout the learning process. Such studies promise to offer insight into the period of transition itself.

The results of previously conducted microgenetic studies provide hints that transitional moments may not be completely captured in children's verbalizations. For example, Siegler found that prior to the acquisition of a new strategy, children become less articulate in speech (Siegler & Jenkins, 1989), and may behave as if they know a strategy although they have no verbal insight into the strategy (Siegler & Stern, 1998). Indeed, there is evidence that, prior to the acquisition of a task, children can demonstrate knowledge through nonverbal means that is not at all evident in their more explicit behaviors. For example, Siegler (1976) found that, although both 5- and 8-year-olds used a weight-only rule in solving a series of balance-scale problems, the 8-year-olds produced head movements that indicated they were also aware of the weights' distance from the fulcrum; the 5-year-olds gave no such evidence and made significantly less progress on the task. As a second example, Clements and Perner (1994) found that, through eye glances, children provided some awareness of the correct answer to a theory-of-mind task, an answer that they gave no evidence of in their speech. These findings suggest that nonverbal indices might go beyond speech in offering insight into moments of change and, perhaps then, into the processes that underlie change.

Our previous work has confirmed these suspicions—gesture, particularly when considered in relation to the speech it accompanies, can be used to identify when a child is in a transitional state. Specifically, children whose gestures often convey different information from their speech about a task, when given instruction, are likely to make progress on that task. In other words, these children are ready to learn the task and, in this sense, are in transition (Church & Goldin-Meadow, 1986; Perry, Church, & Goldin-Meadow, 1988). In this chapter, I ask whether gesture might not only index children who are in transition, but might also play a role in the transition process itself. In other words, I ask whether gesture might itself be involved in mechanisms of change.

The chapter is divided into three parts. In Part One, I review our previous work providing evidence that gesture is indeed a reliable index of children in transition, and that its use during periods of transition is associated with improved learning. Having shown that gesture is correlated with change, I then go on to ask whether it causes change. I consider two different, though not mutually exclusive, possibilities.

In Part Two, I explore the significance of the fact that gesture is "out there," occurring routinely in naturalistic talk of all sorts. Gesture, when interpreted in relation to speech, can signal that the speaker is in transition and open to new input. If communication partners are able to read the signals contained in a child's gestures, they may then be able to alter their interactions with the child accordingly. Gesture, by influencing the input children receive from others, would then be part of the process of change itself.

In Part Three, I consider whether gesture might not be more directly involved in learning, influencing the learners themselves. Gesture externalizes ideas differently and therefore may draw on different resources than speech. Conveying an idea across modalities may, in the end, require less effort than conveying the idea within speech alone; that is, gesture may serve as a cognitive prop, freeing up cognitive effort that can be used on other tasks. If so, using gesture may actually ease the learner's processing burden and, in this way, function as part of the mechanism of change.

Gesture Is Associated with Learning

Gesture-Speech Mismatch Indexes Openness to Learning.   I begin by describing the basic phenomenon. Consider a child asked to justify why he has just said that spreading out a row of checkers has altered the number of checkers in the row (the standard Piagetian number task). Children often argue that the number in the spread-out row is different "because you spread them out," and they often accompany this justification with a matching gesture—a spreading-out motion indicating how the checkers were moved. However, at times, children will produce this same verbal explanation, but accompany it with a very different gestural explanation—for example, a movement pairing each of the checkers in Row 1 with the checkers in Row 2, thus demonstrating, albeit silently, the fact that the checkers in the two rows can be put in one-to-one correspondence. Responses of this sort have been called gesture–speech "mismatches"—instances where gesture conveys information that is different from the information conveyed in the accompanying speech (Church & Goldin-Meadow, 1986).

Children often produce a large number of gesture–speech mismatches in their explanations of Piagetian conservation tasks. What is particularly striking, however, is that the children who produce many mismatches on the conservation task, when given instruction in the task, are significantly more likely to profit from the instruction than children who produce few mismatches (Church & Goldin-Meadow, 1986). Thus, producing different information in gesture on the conservation task is a signal that the child is ready to learn that task.

We have found this same phenomenon in other tasks. For example, when asked to solve addition problems of the following sort, $3 + 5 + 4 =$ _ $+ 4$, fourth-grade children frequently solve the problems incorrectly and offer incorrect explanations for those solutions—the same incorrect explanation in both speech and gesture. Children often say "I added the 3, the 5, the 4, and the 4, and got 16, and put it in the blank," while pointing at the 3, the 5, the 4 on the left side of the equation, the 4 on the right side, and then the blank (an "add all of the numbers in the problem" explanation). However, some children produce the same verbal explanation, but along with it produce a very different gestural explanation—for example, point at the 3, the 5, and the blank, the two numbers on the left side of the equation that can be added to arrive at the correct sum to put in the blank (a "grouping" explanation). Again, we find that it is those children who produce many gesture–speech mismatches who are significantly more likely to profit from instruction in mathematical equivalence than those children who produce few mismatches (Perry et al., 1988).

Gesture–Speech Mismatch Is a Step in the Learning Process.   The studies described thus far have all been traditional—a child's knowledge is assessed at pretest, instruction is given, and the child's knowledge is assessed again at posttest. These studies show that gesture–speech mismatch is associated with a propensity to learn, but they do not in any way shed light on the path of learning. To do so, we conducted a microgenetic study, assessing children repeatedly as they were exposed to instruction in mathematical equivalence (Alibali & Goldin-Meadow, 1993).

Children were given instruction in problems of the $3 + 5 + 4 =$ _ $+ 4$ variety, and were assessed three times over the course of the training period. The children began the study at different levels of understanding of mathematical equivalence (although none could solve the problems correctly), and made different gains as a result of instruction. Indeed, some made no progress at all, and a small number regressed. However, the interesting point is that the vast majority of the 35 children who gestured and made progress on the task did so following the same path: (a) Children began by producing the same explanation in both gesture and speech and that explanation was incorrect (Matching Incorrect). (b) The children then produced explanations in gesture that were different from their explanations in speech (Mismatching); the mismatching explanations were either both incorrect, or one (typically gesture) was correct and the other incorrect. (c) Finally, the children returned to producing the same explanation in both gesture and speech, but now the explanation was correct (Matching Correct). Over the course of the study, 11 children traversed the first two steps of the path, 15 traversed the last two steps, and 3 traversed all three steps, accounting for 83% of the 35 children (a number significantly higher than that expected by chance, $p < .001$, binomial test). Only 6 (17%) of the children who gestured and progressed on the task did so by skipping the mismatching step.

In addition to suggesting that gesture–speech mismatch can be a stepping-stone on the way toward mastery of a task, the findings also provide evidence that, when gesture–speech mismatch is a step on a child's path to mastery, learning is deeper and more robust. We compared the children's ability to generalize to multiplication what they had learned on addition problems. This generalization was measured on a posttest immediately following the training, and on a follow-up test 2 weeks after training. We focused on two groups: Children who progressed to a Matching Correct state by passing through a Mismatching state, and children who progressed to a Matching Correct state by skipping the Mismatching state. We found that children who arrived at the Matching Correct state by skipping the Mismatching state were significantly less likely to generalize their knowledge on the posttest, and significantly less likely to maintain their gains on the follow-up test 2-weeks later, than children who reached the Matching Correct state by going through a Mismatching state (Fig. 1.1; Alibali & Goldin-Meadow, 1993). These findings provide the first hint that gesture, when taken in relation to speech, may not only reflect learning, but also contribute to it.

Gesture Conveys Implicit Knowledge. I have shown thus far that gesture, when considered in relation to the speech it accompanies, can index the stability of a child's cognitive state. Can gesture also tell us something about the knowledge that contributes to that state? The information that a child conveys in gesture is a good candidate for implicit knowledge—

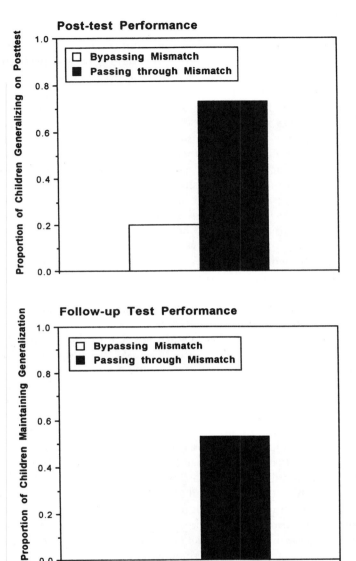

FIG. 1.1. Proportion of children who successfully generalized their mathematical equivalence training to the posttest immediately following the instruction (top graph), and who maintained that progress 2-weeks later on a follow-up test (bottom graph). All of the children had achieved a Matching Correct state by the end of training. Children who arrived at the Matching Correct state by passing through a Mismatching state (black bars) were significantly more likely to generalize their training and maintain that progress than children who arrived at the Matching Correct state by skipping a Mismatching state (white bars).

knowledge that at some level the child *has*, but is not able to articulate (Berry & Broadbent, 1984; Clements & Perner, 1994; Karmiloff-Smith, 1992; Nisbett & Wilson, 1977; Reber, 1993; Stanley, Mathews, Buss, & Kotler-Cope, 1989). There are two steps involved in showing that gesture reflects implicit knowledge.

The first is to demonstrate that the knowledge children convey in their gestures is not accessible to speech. To make this case, we examined the entire set of responses that fourth-grade children produced when asked to explain their solutions to the mathematical equivalence task. Looking only at children who gestured on at least some problems, we determined which types of problem-solving procedures were expressed (a) in gesture and never in speech, (b) in speech and never in gesture, or (c) in both gesture and speech. A procedure need not have been produced in gesture and speech on the same problem in order to find its way into the third category; it was sufficient for the child to produce the procedure in gesture on one problem and that same procedure in speech on another problem. Interestingly, we found that very few of the children's procedures were expressed uniquely in speech, that is, without some representation in gesture even if on another problem (Fig. 1.2). The children's procedures were either expressed in both gesture and speech, or uniquely in gesture

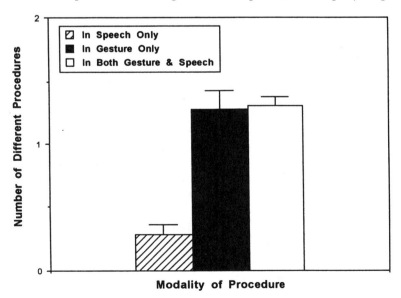

FIG. 1.2. Modalities in which a procedure was conveyed by a child over a series of six mathematical equivalence procedures. Very few procedures were conveyed uniquely in speech. Procedures were either expressed in both gesture and speech, or uniquely in gesture. Bars reflect standard errors.

(Goldin-Meadow, Alibali, & Church, 1993). Thus, most of the information these children possessed about the math problems was accessible to gesture, and some of that information was accessible *only* to gesture.

The second step is to demonstrate that children actually have the knowledge that we impute to them on the basis of their spontaneous gestures. We have assumed that when a string of points is produced along with a spoken problem-solving procedure, those points are conveying a problem-solving procedure. But the pointing gestures could be doing nothing more than directing the listener's attention to the numbers in the problem. There is evidence, however, that the string of gestures, when taken together, does reflect a procedure in its own right. The evidence comes from cases where the child's gestures do *not* match speech. For example, consider a child responding to the same problem described earlier, $3 + 5 + 4 = \_ + 4$. The child articulated an "add the numbers up to the equal sign" procedure in speech, "I added the 3, the 5, and 4, and put 12 in the blank." In gesture, however, the child produced a string of points that appeared to reflect a completely different procedure—point at the 3, the 5, the left 4, the right 4, and the blank, an "add all the numbers in the problem" procedure.

When later asked by a separate experimenter on a separate rating task which answers would be acceptable solutions to this problem (children, on the whole, were happy to accept more than one answer for a single problem), the child, of course, accepted 12—the answer generated by the "add the numbers up to the equal sign" procedure that she had conveyed in speech. More interestingly, when asked whether 16 would also be an acceptable answer—the answer generated by the "add all the numbers" procedure that she conveyed in gesture but not in speech—the child was significantly more likely to say "yes" than she was to answers generated by procedures that she had not produced in either gesture or speech (Garber, Alibali, & Goldin-Meadow, 1998). In other words, the child recognized the solution generated by the procedure that she had produced *only* in her gestures. Thus, strings of gestures can indeed reveal knowledge about problem-solving procedures. Children may not be able to express in speech the problem-solving knowledge that they convey in gesture, but they are able to recognize it on other, gesture-independent tasks. In this sense, the knowledge conveyed uniquely in gesture is implicit.

Gesture as a Mechanism of Learning Through Its Effect on Others

If gesture makes a child's implicit knowledge "visible" to a communication partner, the partner may be able to use that information to alter the way he or she interacts with the child. In this way, gesture could be part of the learning process itself. This hypothetical mechanism can, of course, only work if communication partners are able to interpret the gestures children produce.

Reading Preselected Gestures Off of Videotape. We have taken several steps to explore this hypothesis. We began by asking whether adults, not trained in gesture-coding, could observe children on videotape and glean substantive information from the gestures the children produced. Examples for the videotape were chosen so that half of the children produced gestures that conveyed the same information as their speech (matching explanations), and half produced gestures that conveyed different information from their speech (mismatching explantions). If the adults were able to report the information that the children produced only in gesture and not in speech, that is, on mismatching explanations, we would then have evidence that they can read child gesture.

We conducted two studies, one asking adults to describe what the child on the videotape knew about the conservation task (Goldin-Meadow, Wein, & Chang, 1992), and another asking adults to describe what the child knew about the mathematical equivalence problem (Alibali, Flevares, & Goldin-Meadow, 1997). In both studies, adults were given free rein in their responses: After each videotaped task, the tape was stopped, and the adult was asked to assess the child's understanding of the task. In both studies, we found that the adults were able to glean substantive information from the children's gestures, information that was not displayed anywhere in the children's speech. For example, a child responding to a conservation number task on the tape said that the rows of checkers were different " 'cause you moved 'em," but indicated some understanding of one-to-one correspondence in gesture (he moved his pointing hand from each checker in one row to the corresponding checker in the other row). When describing this child's reasoning, adults often commented that the child not only noticed that the checkers had been moved, but that the checkers matched up with one another. Comparable findings have been reported for other tasks (a narrative task, McNeill, Cassell, & McCullough, 1994) and even for child observers (Kelly & Church, 1997, 1998).

Reading Spontaneously Produced Gesture Online. These findings indicate that adults can read a child's gestures when those gestures are carefully chosen by the experimenter and presented twice on videotape. However, the mechanism for cognitive change that I propose requires that gesture be read 'online' in the give-and-take of naturalistic interaction. Our next step then was to ask adults to observe children "live" (Goldin-Meadow & Sandhofer, 1999). We asked adults to observe children participating in a series of Piagetian conservation tasks. In order for the adults to be able to assess each child's knowledge online, we gave them a checklist for each task the child performed. The list contained all of the explanations children typically give on tasks of this sort. The adults' job was to check off as many explanations as they thought the child had conveyed.

We first validated this checklist on two groups of adults asked to observe a videotape of children participating in Piagetian conservation tasks (Video Groups 1 and 2). We then asked one of these groups of adults to observe live, a series of children participating in six Piagetian tasks (Live Group 2). The children observed by these adults were randomly chosen from their classes. We found that the adults checked explanations that the children conveyed *only* in their gesture from 32% to 44% of the time in response to both the preselected videotaped gestures and the spontaneously produced online gestures.

These results suggest that the adults were able to read the children's gestures (albeit not all of the time). However, it is possible that the adults checked the explanations that the children conveyed uniquely in gesture, not because they actually read the children's gestures, but because these are the explanations that readily come to an adult's mind on tasks of this sort. To explore this possibility, we first established how often an adult checked a given explanation (e.g., one-to-one correspondence) when that explanation was *not* produced on a particular number task. In other words, we established a base-rate for how often adults erroneously checked one-to-one correspondence on this number task. We then compared this figure to how often adults checked one-to-one correspondence when it *was* conveyed uniquely in gesture on that same number task. We found, in both the video groups and the live group, that adults were significantly more likely to check an explanation when it was produced uniquely in gesture than when that same explanation was not produced at all on the same task (Fig. 1.3). Interestingly, there was no difference between the video and live groups in how much information the adults gleaned from gesture. Adults, then, are able to glean substantive information from the gestures children produce, even if those gestures are unedited and fleeting.

Does gesture affect the observer's ability to extract information from the *speech* it accompanies? The videotaped examples were selected so that we could examine whether listeners could abstract information from gesture. However, the design of the study also allowed us to explore whether gesture affects how accurately its accompanying speech is interpreted. In six of the examples, gesture conveyed the same explanation as the speech it accompanied; in the other six, gesture conveyed a different explanation. We found that the adults correctly checked off explanations conveyed in speech significantly more often when they were accompanied by a matching gesture than by a mismatching gesture.

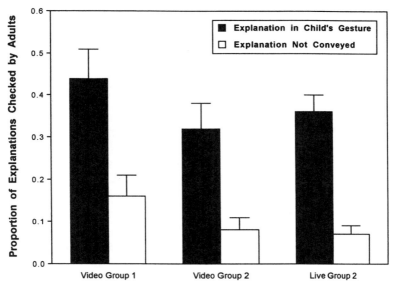

FIG. 1.3. Proportion of explanations identified on a checklist by adults who viewed a series of children participating in a conservation task on a preselected videotape (Video Groups 1 and 2) or in a live situation (Live Group 2). In all three groups, adults were significantly more likely to check an explanation when it was conveyed uniquely in a child's gestures versus when that same explanation was not conveyed at all by the child. Bars reflect standard errors.

To be certain that the differences between the two types of explanations (speech with a matching gesture vs. speech with a mismatching gesture) were attributable to the presence of gesture and not to differences in the speech itself, we presented an additional group of adults with only the audio portion of the videotape (i.e., the picture was turned off). As expected, the differences that we hypothesized to be due to gesture disappeared—the adults were equally likely to correctly check "yes" for the six explanations of each type when there was no picture and therefore no gesture to affect the interpretation of speech.

These results suggest that adults' ability to receive a message in speech is affected by the gestures that accompany that speech. However, it is not clear from these data whether a matching gesture *improves* the adult's ability to recognize an accompanying spoken explanation, or whether a mismatching gesture *diminishes* the adult's ability to recognize an accompanying spoken explanation. The adults in the naturalistic task observed some children producing explanations that contained speech and no gesture at all, a necessary ingredient to explore this hypothesis. Although the adults correctly identified spoken explanations *more* often when those explanations were accompanied by a matching gesture (88%) than when they were accompanied by no gesture at all (82%), this difference was not statistically significant. In contrast, the adults correctly identified spoken explanations significantly *less* often when those explanations were accompanied by a mismatching gesture (70%) than when they were accompanied by no gesture at all (82%). Thus, at the moment, we have no evidence that gesture improves a listener's ability to recognize a message produced in speech if gesture conveys the same message (although such an effect would be difficult to see in these data because accuracy was so high; see Krauss, Morrel-Samuels, & Colasante (1991) for whom ceiling effects were not a problem and who also found that matching gesture did not improve comprehension of a message). However, we do have evidence that gesture diminishes a listener's ability to recognize a spoken message if gesture conveys a different message (see also Kelly & Church, 1998).

Responding to Gesture in a Naturalistic Teaching Situation. The findings described thus far set the stage for gesture's participation in the learning process. Observing gesture as a third-party to a conversation is different from observing it as a participant in the conversation. If adults alter the way they interact with a child on the basis of the child's gestures, those adults are going to have to be able to read gesture as they interact with the child. The next study suggests that adults are able to do so.

In an ongoing study Melissa Singer, San Kim, and I asked eight teachers to interact individually with from 5 to 7 children. The teachers' task was to teach each child how to solve the mathematical equivalence problems. To acquaint the teachers with their pupils, before each tutorial, the teacher was requested to observe the child interacting with the experimenter, who asked the child to solve six mathematical equivalence problems and explain how he or she arrived at those solutions. The teacher was given five problems to use in instructing each child. In a previous study designed to explore whether teachers use gesture in math tutorials (Goldin-Meadow, Kim, & Singer, 1999), we discovered that teachers tend to do most of the talking during the tutorials. In order to be able to observe teacher responses to child gestures, we needed to increase student participation in the tutorials. We therefore asked that the teachers request the children to both solve each of the five problems and explain their solutions during the tutorial.

We have thus far coded three of the eight teachers' interactions with their students. The children did indeed produce gestures during the tutorials, and they conveyed problem-solving procedures with those gestures. On average, 39% of the children's turns contained gesture, 21% in which gesture conveyed the same information as their speech, and 18% in which gesture conveyed different information from their speech. We assumed conservatively that the teacher understood a child's gestures when that teacher reiterated on the very next turn the problem-solving procedure that the child had expressed uniquely in gesture. We found that, as in our studies of more constrained situations, the teachers reiterated procedures children expressed uniquely in gesture approximately 30% of the time. Moreover, the children's gestures had an impact on whether their words were heeded: Teachers reiterated the child's problem-solving procedures less often when those procedures were accompanied by child-gesture conveying a different procedure than when the procedures were accompanied by child-gesture conveying the same procedure or no gesture at all (this pattern was not determined by the correctness of the child's procedures; e.g., the pattern held even when we examined teacher reiterations following correct procedures only). Thus, even in a naturalistic situation in which the adult is an active participant, the adult is able to glean substantive, and unique, information from children's gestures.

Using Gesture to Alter Input to the Child. Our final step in exploring the hypothesis that gesture provides a mechanism by which children can shape the input they receive from adults is to show that adults alter their input to the child as a function of the information they glean from that child's gestures. At the moment, the evidence for this hypothesis is purely anecdotal. At times in our math tutorials, the teachers did not reiterate the problem-solving procedure found exclusively in the child's gesture, but rather used it as a stepping-stone for their next move. For example, for the problem $5 + 3 + 4 = \_ + 4$, the child began by pointing simultaneously at the left 4 with her left hand and the right 4 with her right hand. Rather than reiterate the notion that there are equal addends on each side of the equation, the teacher's next move was to articulate the grouping procedure in both speech and gesture—"you can solve the problem by adding the 5 and the 3 and putting the sum in the blank," accompanied by a V-shaped point at the 5 and 3. Note that the grouping procedure works in this problem because there are equal addends, one on each side of the equation, that can be canceled. The fact that the child demonstrated some awareness of equal addends in gesture seemed to give the teacher license to introduce grouping, a procedure that the child then picked up on in her next turn and continued to use throughout the interaction. Our future analyses will explore how often interactions of this sort occur in the math tutorials.

It is important to point out that adults need not be aware of the fact that have been influenced by the child's gesture. Indeed, the adult may get it wrong and still be able to provide useful input to the child. Consider, for example, the following teacher who participated in the study conducted by Alibali et al. (1997). The child said he solved the problem $5 + 6 + 7 =$

_ + 7 by adding the 5, 6, and 7 (an "add numbers up to the equal sign" procedure), while pointing only at the 5 and 6 (a "grouping" procedure). After observing this child, the teacher said that the child did *not* understand the grouping procedure: "What I'm picking up now is [the child's] inability to realize that these (indicates 5 and 6) are meant to represent the same number . . . there isn't a connection being made by the fact that the 7 on this side of the equal sign (indicates left side) is supposed to also be the same as this 7 on this side of the equal side (indicates right side), which would, you know, once you made that connection it should be fairly clear that the 5 and 6 belong in the box." Note that, at some level the teacher was incorrect—the child did indeed have an understanding, however implicit, of the grouping procedure that he expressed only in gesture. It is possible that the teacher chose the grouping procedure to highlight as the one the child did not know because she detected the procedure in the child's gestures. The fact that the teacher did not *explicitly* recognize the child's grasp of this procedure may not matter if, in instructing the child, the teacher were to focus on what she thought the child needed most—input about the grouping procedure. Instruction about grouping might be especially effective for this particular child because it might help him to transform or "redescribe" his emerging knowledge into a problem-solving procedure that he could apply and articulate in speech (cf. Karmiloff-Smith, 1992).

To summarize thus far, gesture routinely accompanies speech and appears to provide an undercurrent of conversation that participants may or may not explicitly notice but detect nonetheless. Teachers do pick up on at least some of the information children convey uniquely in gesture. Moreover, gestural communication is not a one-way street. Teachers produce gestures of their own, many of which express information that is different from the information they express in speech—and children pick up on those gestures, even if they convey (unintended) incorrect procedures (Goldin-Meadow, Kim, & Singer, 1999). Thus, gesture is an inevitable part of conversation, received as well as produced by the child. Adult gesture can therefore be a source of usable input for the child.

However, as I have shown here, gesture also allows the child to signal (perhaps without intending to do so) that he or she is on the cusp of insight. In this sense, gesture is comparable to a pheromone—a signal to a member of the species that the child is ready for input. The signal may be taken as a general one—a "teach me–I'm ready" announcement that elicits instruction, any sort of instruction, from a communication partner. Or, the signal may be taken as a specific call for input of a certain sort. If, in fact, gesture does pinpoint those areas in which the child is ready to learn, gesture may be functioning as an externalized index of the child's *proximal zone* (Vygotsky, 1978)—the range of skills on which a child can make progess if given appropriate assistance. As such, gesture may provide a concrete, externalized mechanism by which adults can calibrate their input to the child's most pressing needs.

Gesture as a Mechanism of Learning Through Its Effect on the Learner

I turn now to the second type of learning mechanism in which gesture may play a role. In addition to signaling the learner's cognitive state to others, gesture may function in some beneficial way for the learner him or herself. There are indeed a number of hints that gesture is there when good things happen. For example, Fisher and Brennan (personal communication, June 3, 1998) found that better recall was associated with gesture. Children observed a Red Cross lecture/demonstration, which they were asked to recall after a week had passed. Although accuracy of recall was generally high (80%), it was much higher (99%) when the children gestured along with their recalled responses.

Gesture was not a manipulated variable in Fisher and Brennan's study—it arose as a serendipitous finding. In contrast, Iverson (personal communication, June, 1998) deliberately manipulated gesture to determine its effects on recall. Adults were shown a cartoon and asked to retell the story immediately after viewing it. During the immediate retelling, half of the adults were told to keep their hands still on the arms of the chair, and half were given no particular instructions. The second group gestured in retelling the cartoon, and the first obviously did not. There were no differences between the groups in the number of story details that were recalled during the immediate retelling. Both groups were then asked to retell the cartoon again 1 week later, and this time none of the adults was restricted in their movements. Interestingly, the group that was initally allowed to gesture recalled more details about the cartoon than the group that was initially prevented from gesturing. Gesturing during the first retelling appeared to enhance the likelihood that the information would be retained and recalled during the second retelling. Although it is not yet clear what role gesture is playing in memory (e.g., at what point in the memory process does gesture make its contribution?), these studies do strongly suggest that gesture plays a beneficial role in recall or, at the least, that gesture is an important encoding factor that can affect memory if it is (or is not) replicated at the time of recall.

There is, in addition, some suggestion from our previous work that gesturing is associated with learning (Alibali & Goldin-Meadow, 1993). Several of the children in our microgenetic study of mathematical equivalence failed to gesture during the study. These children did less well on the posttest and the follow-up test than the children who gestured throughout the study, although the difference between groups did not quite reach statistical significance. Thus, when required to generalize what they had learned during instruction and retain that understanding over a 2-week period, children who gestured during the process showed a tendency to outperform children who did not gesture.

Finally, in a conservation training study, Church (1999) found that across-modality variability (that is, the number of gesture–speech mismatches the child produced at the start of the study) was a significantly better predictor of learning than within-modality variability (the number of different strategies the child produced in speech, either within a trial or across trials). Thus, it is not only the number of different strategies that matters in predicting learning, but whether those strategies are produced in gesture. Why might gesture be associated with learning?

Gesture Is Where Children Experiment. One possibility is that gesture is the place where learners experiment. Recall that in our microgenetic study (Alibali & Goldin-Meadow, 1993), children who gestured and progressed to a correct understanding of mathematical equivalence by going through a mismatching state (one in which they produced gestures that conveyed different information from their speech) did better on both a posttest and follow-up test than children who gestured but progressed to the same correct state without going through a period of gesture–speech mismatch.

To better understand this phenomenon, we examined the modality in which children produced each of the procedures in their repertoires prior to instruction (Alibali & Goldin-Meadow, 1993). Children who gestured and produced many matching explanations were identical to children who gestured and produced many mismatching explanations in two respects: Both groups expressed very few procedures *only* in speech (that is, without also expressing that procedure in gesture at some time over the set of six problems), and both groups expressed a relatively large, and equal, number of procedures in both speech and gesture (not necessarily in both modalities on the same problem, but across six problems). Where the matching and mismatching children differed was in procedures accessible to gesture: The mismatching children expressed a significantly larger number of procedures *only* in gesture compared to the matching children. Overall, the mismatching children expressed a wider variety of procedures than the matching children—and all of that variety resided in gesture. Thus, the variability that many theorists consider essential to developmental progress (e.g., Siegler, 1994; Thelen, 1989) is indeed present in these children—in their gestures.

This phenomenon is not only found across children but within-child as well (Alibali & Goldin-Meadow, 1993). Children who, with instruction, progressed from an incorrect matching state to a mismatching state significantly *increased* the number of procedures they expressed uniquely in gesture. Conversely, children who progressed from a mismatching state to a correct matching state significantly *decreased* the number of procedures they expressed uniquely in gesture. Thus, children have the largest number of different procedures in their repertoires when they are in a mismatching state, and the influx of new procedures is found uniquely in gesture.

In an attempt to observe the smallest steps children make when learning a task, Alibali (1994) gave children minimal instruction in mathematical equivalence. Not surprisingly, she found that, at best, children made only minimal progress; indeed, some appeared to make no progress at all, and some even regressed. Predictably, children who progressed from an incorrect matching state to a mismatching state *increased* the total number of different procedures in their repertoires; children who regressed from a mismatching state to an incorrect matching state *decreased* their total number of procedures; children who remained in an incorrect matching state or a mismatching state retained the same number of different procedures, a *low* number for the incorrect matchers, a *high* number for the mismatchers.

Alibali (1994; Goldin-Meadow & Alibali, 1995) then observed the way in which the children's repertoires changed over the course of instruction. A large number of children in all four groups were found to *maintain* at least one procedure over the study, suggesting that change may be more gradual than abrupt (cf. Alibali, 1999; Kuhn & Pearsall, 1998; Siegler & Chen, 1998). What about *abandoning* old procedures or *generating* new ones? As we might expect, children who progessed from an incorrect matching state to a mismatching state, not only maintained old procedures, but generated new ones (thus enlarging their repertoires). Children who regressed from a mismatching state to an incorrect matching state abandoned old procedures but did not generate new ones (thus shrinking their repertoires).

The interesting contrast comes from children who remained in the same state throughout the study. Children who remained in an incorrect matching state, predictably, neither abandoned old procedures nor generated new ones—they maintained the same number of procedures in their repertoires by not changing those repertoires at all. In contrast, children who remained in a mismatching state maintained the same number of procedures in their repertoires by *continuously revamping those repertoires*, generating new procedures while abandoning old ones. Thus, children in a mismatching state not only had a large number of different procedures in their repertoires, but those procedures were continuously changing, providing the kind of variability that may be necessary for change. Important for the argument I am making here, all of this "experimentation" with new ideas took place in gesture. Many of the newly generated procedures were incorrect, and were quickly abandoned. Gesture thus appears to be a place where children can air ideas that may not, in themselves, be all that sound but may be able to serve as stepping-stones for progress nonetheless.

Why experiment in gesture? One might imagine that gesture would be an ideal place to try out untried ideas, simply because there is essentially no social constriction on the gestures people produce (aside from the often rude gestural "emblems," which are conventional, frequently produced without any speech, and qualitatively different from the gestures we are considering here, cf. Ekman & Friesen, 1969). Or, perhaps gesture is the ideal place for experimentation because the ideas themselves are easier to express in the manual modality. This may be particularly true for domains such as mathematics, which lend themselves to visual thinking (Hadamard, 1945).

Gesture May Ease the Cognitive Burden.   Another possiblility, and one we are currently exploring (Goldin-Meadow, Nusbaum, Kelly, & Wagner, under review) is that gesturing itself can reduce cognitive effort, perhaps in the same way that writing a problem down can reduce the effort needed to solve the problem (see Alibali & DiRusso, 1999, who make this very argument with respect to pointing and learning to count). If gesture does serve as a kind of cognitive prop, the effort saved as the result of gesturing could be allocated toward working out new ideas that could, in turn, lead to progress in the task. We tested this hypothesis by asking whether gesturing on a task frees up effort that can then be used on another task. We gave fourth-grade children two tasks to perform simultaneously—(a) explain their solutions to a mathematical equivalence problem, and (b) recall a list of words (either a short list containing a single word, or a longer list containing three words). On each trial, children first solved the mathematical equivalence problem. After solving the problem, the child was given the list of words to be recalled, and was asked to explain how he or she had solved the math problem. After completing the explanation, the child was then asked to recall the list of words.

The children were asked to do these tasks under two conditions—one in which they were told to hold their hands completely still within the handprint drawn on a sheet of paper that we supplied (the gesture-prevented condition), and a second in which they were told that they could use their hands freely (the gesture-permitted condition). Our goal was to observe every child gesturing and not gesturing. A priori, we might expect gesturing to *increase* cognitive load simply because the gesturer must plan and execute communication in two modalities. If so, we would expect the children to remember *fewer* words when they gestured than when they did not gesture. Alternatively, gesturing might *decrease* cognitive load by increasing resources available to the child, for example, by shifting the burden from verbal to spatial memory. If so, we would expect the children to remember *more* words when they gestured than when they did not gesture.

Children did follow our instructions in the gesture-prevented condition—none produced any gestures on these problems. However, and perhaps not surprisingly, children did not gesture on every single problem in the gesture-permitted condition either. Indeed, some children did not gesture on any problems at all; we call these six children the No-Gesturers and set them aside for the moment. Turning to the 10 Gesturers, we ignored the condition (gesture-prevented vs. gesture-permitted) to which a problem had been assigned, and categorized each problem according to whether the child had actually gestured when explaining that problem. We then calculated the proportion of correctly recalled one- and three-word lists following problems on which the child gestured versus problems on which the child did not gesture.

Figure 1.4 displays the proportion of word lists correctly recalled following gesture problems versus no-gesture problems. Not surpisingly, children correctly recalled significantly more one-word lists than three-word lists ($F(1,9) = 19.33$, $p < .01$). If gesturing on the explanation task frees up space in working memory (perhaps by shifting some of the load from verbal working memory to spatial working memory), then we would expect children to be able to remember more words on the recall task when they gesture on the explanation task than when they do not—but only when their memories are taxed, that is, only on the three-word lists. In other words, we would expect an interaction between word list length and presence of gesture which, in fact, we found ($F(1,9) = 10.73$, $p < .01$). Children were essentially at ceiling in recalling the one-word lists, whether or not those lists followed gesture versus no-gesture problems. The crucial comparison involves the three-word lists, which were designed to tax the children's memory skills. Here we see that the children recalled significantly more three-word lists following problems on which they gestured than following problems on which they failed to gesture ($p < .01$, Newman-Keuls). These findings suggest that the act of gesturing may have eased the child's cognitive burden in the explanation task, freeing effort up for the word-recall task.

It is important to note that it is not just having the opportunity to gesture that is associated with improved recall, it is necessary for the child to actually do the gesturing for memory to be affected. Thus, if we ignore whether the child actually gestured and compare word-recall on the problems originally assigned to the gesture-permitted versus gesture-prevented conditions, the effect disappears. The effect is also not there for nongesturers, who do not produce gestures on any of their trials and thus obviously cannot reap the benefits of gesturing.

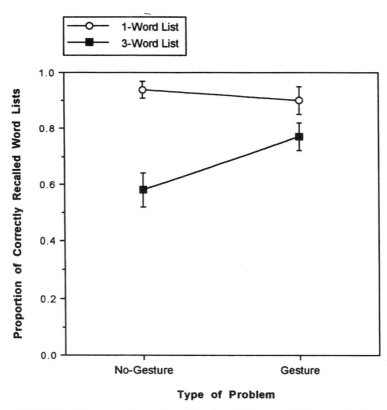

FIG. 1.4. The proportion of word lists that children recalled correctly following problems on which they gestured versus problems on which they did not gesture. When their memories were taxed (that is, on the three-word lists), the children recalled a significantly higher proportion of word lists following Gesture problems than No-Gesture problems, suggesting that the act of gesturing may have freed up cognitive effort that could then be used on the recall task. Bars reflect standard errors.

It is possible, however, that our gesture-prevented condition actually imposes a discomfort burden on the child. If so, this burden may be decreasing recall in the gesture-prevented condition relative to the gesture-permitted condition (as opposed to gesture enhancing recall in the gesture-permitted condition relative to the gesture-prevented condition). This possibility is unlikely for two reasons. First, when we divide problems in the gesture-permitted condition into those on which the child actually gestured versus those on which the child did not gesture (all of the problems were of equal difficulty), we find the same pattern seen in Fig. 1.4— and these particular no-gesture problems were not experimentally created and thus not obviously subject to the discomfort concern. Note that the children whom we included in this analysis were not nongesturers, but merely gesturers who did not use gesture on every problem. Importantly, these children were equally successful (or unsuccessful) at solving problems on which they gestured and problems on which they did not gesture.

Second, we have begun running children in a discomfort control condition and the results from a single child are promising. This child, who also participated in the hands-still study (and gestured on every problem in the gesture-permitted condition), was asked to do precisely the same task but this time she was to keep her feet completely still within a set of footprints stenciled on a sheet of paper. If discomfort is impeding recall, the results of the feet-still study should look just like the results of the hands-still study. However, if gesturing actually improves memory, then the results should be unaffected by holding one's feet still (unless, of course, keeping one's feet still affects gesturing which, for this child, was not the case—she gestured on every problem in the feet-still study). The results for the child, who appeared to find it just as disconcerting to keep her feet still as to

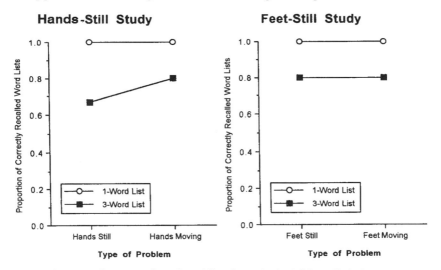

FIG. 1.5. The proportion of word lists that a single child recalled after explaining math problems under two different conditions: (a) holding her hands still when explaining half of the math problems (Hands-Still study, left graph), and (b) holding her feet still when explaining half of the problems (Feet-Still study, right graph). The child spontaneously gestured on all of the problems in the Feet-Still study, and on all of the problems in the Hands-Moving condition of the Hands-Still study. She recalled a lower proportion of three-word lists when she was prevented from gesturing in the Hands-Still condition of the first study than in the Hands-Moving condition. This relatively poor performance was not due merely to the discomfort of a keeping a body part still, as the same decrement was not found in the Feet-Still study.

keep her hands still, are presented in Fig. 1.5. The patterns suggest that holding one's feet still does not affect memory, but that holding one's hands still—which prevents gesturing—does. Thus, if children gesture on a task, they appear to have more cognitive effort left over for doing other things than if they do not gesture on the task. Gesturing can ease the child's cognitive burden.

What might gesturing be doing to ease the child's cognitive burden? The act of conveying information, whatever that information might be, in a second modality may make the task easier for the child. In other words, using two modalities rather than one may be the key, independent of the type of information that is expressed. However, the message itself may matter. When gesture conveys the same information as speech, it could be making the task easier for the child by adding redundancy. On the other hand, when gesture conveys different information from speech, it could be making the task easier by providing a vehicle that allows the child to express thoughts that he or she cannot yet express in speech. Future analyses of how the gestured explanations the children produced in the hands-still study affect their recall of the word lists will hopefully allow us to explore these possibilities.

Gesture's Role in Cognitive Change

Gesture is implicated in cognitive change. It is, of course, possible that gesture is nothing more than an epiphenomenon of change, associated with it but not in any way central to its causes. However, evidence is mounting that gesture may be involved in the process of change itself, communicating silent aspects of the learner's cognitive state to potential agents of change, or helping more directly to ease the learner's cognitive burden.

If gesture is causally involved in change, its effect is likely to be widespread. Gesture has been found to express substantive information, often information that differs from the information expressed in the speech it accompanies, in a variety of tasks and over a large age range: in toddlers going through a vocabulary spurt (Gershkoff-Stowe & Smith, 1997); preschoolers explaining a game (Evans & Rubin, 1979); elementary school children explaining mathematical equations (Perry et al., 1988) and seasonal change (Crowder & Newman, 1993); children and adults discussing moral dilemmas (Church, Schonert-Reichl, et al., 1995); adolescents explaining Piagetian bending-rods tasks (Stone, Webb, & Mahootian, 1991); and adults explaining gears (Perry & Elder, 1996; Schwartz & Black, 1996) and problems involving constant change (Alibali, Bassok, et al., 1995, 1999). Some of the tasks on which gesture has been found might be considered to be core tasks in Gelman and Williams' (1998) terms, others are more likely to be noncore. The fact that gesture is found in both types of tasks may mean that it has the potential to be involved in innately driven as well as non-innately driven learning, that is, to be a general mechanism of change.

Previous studies have shown that asking children to explain their responses to a problem has a beneficial effect on learning (e.g., Chi, Bassok, Lewis, Reimann, & Glaser, 1989; Siegler, 1997). Being forced to come up with an explanation encourages learners to articulate their, perhaps previously unexamined, presuppositions and to put their ill-formed ideas into words. This self-examination may be what is important in eliciting explanations from learners. Yet, requesting explanations may also be effective because it elicits gesture from the learner. Gesture offers the opportunity to explain in another modality, one that has very different representational demands and possibilities than does speech.

McNeill (1992) argued that gesture and speech form complementary components of a single, integrated system, with each modality best suited to expressing its own set of meanings. Gesture reflects a global–synthetic image. It is idiosyncratic and constructed at the moment of speaking—it does not belong to a conventional code. In contrast, speech reflects a linear-segmented, hierarchical linguistic structure, utilizing a grammatical pattern that embodies the language's standards of form and drawing on an agreed-on lexicon of words. Consider, for example, a speaker describing the coastline of the east coast of the United States. One well-formed gesture can do much more to convey the nuances of the coastline to a listener than even the best-chosen set of words (cf. Huttenlocher, 1976). Gesture thus allows speakers to convey thoughts that may not easily fit into the categorical system that their conventional language offers (Goldin-Meadow & McNeill, 1999). Taken together, gesture and speech offer the possibility of constructing multiple representations of a single task, and these multiple perspectives may prove useful, particularly in learning complex tasks.

In addition, because gesture is not regulated by an acknowledged codified system, the notions that are expressed in this modality can easily go unchallenged. Rarely are speakers criticized for their gestures, while the same message expressed in speech may well elicit comment and disapproval. Not only are the notions conveyed in gesture likely to go unchallenged by others, but they are also likely to go unchallenged by the self. A speaker can sneak in an idea, perhaps an ill-formed one, in gesture that does not cohere well with the set of ideas expressed in speech. Gesture may be an ideal place to try out inchoate, untamed, and innovative ideas simply because those ideas do not have to fit. Much experimentation may take place, and remain, in gesture, never reaching the conventionally shared spoken system; but the experimenation itself may be useful. Indeed, it may be that gesture is the place where we can expect to see children's worst guesses about how a task works.

Gesture is pervasive, routinely accompanying speech of all varieties. It is, however, not subjected to the same standards of approval as speech simply because it is not an explicit representational system in the same way speech is. On the other hand, gesture is symbolic in its own right. Gesturing about a procedure is not the same thing as enacting that procedure. Gesture reflects implicit knowledge that is at least one step removed from actually performing a procedure. Expressing knowledge in gesture may therefore represent an important step in the redescription process that Karmiloff-Smith (1992) describes, a process culminating in explicit awareness. As such, gesture has a unique status and may play a unique role in learning.

In sum, previous work has established that gesture is associated with learning. It can index moments of cognitive instability and reflect thoughts not yet found in speech. Here, I have raised the possibility that gesture might do more than just reflect learning—it might be involved in the learning process itself. I have considered two non-mutually exclusive possibilities. First, gesture could play a role in the learning process by displaying, for all to see, the learner's newest, and perhaps undigested, thoughts. Parents, teachers, and peers would then have the opportunity to react to those unspoken thoughts and provide the learner with the input necessary for future steps. Second, gesture could play a role in the learning process more directly by providing another representational format, one that would allow the learner to explore, perhaps with less effort, ideas that may be difficult to think through in a verbal format. Thus gesture has the potential to contribute to cognitive change, directly by influencing the learner and indirectly by influencing the learning environment.

ACKNOWLEDGMENTS

This research was supported by NIH grant R01 HD18617 and by a grant from the Spencer Foundation.

REFERENCES

Alibali, M. W. (1994). Processes of cognitive change revealed in gesture and speech. Unpublished doctoral dissertation. University of Chicago.

Alibali, M. W. (1999). How children change their minds: Strategy change can be gradual or abrupt. *Developmental Psychology, 35*, 127–145.

Alibali, M. W., Bassok, M., Olseth, K. L., Syc, S. E., & Goldin-Meadow, S. (1995). Gestures reveal mental models of discrete and continuous change. In J. D. Moore & J. F. Lehman

(Eds.), *Proceedings of the Seventeenth Annual Conference of the Cognitive Science Society*, (pp. 391–396). Hillsdale, NJ: Lawrence Erlbaum Associates.

Alibali, M. W., Bassok, M., Solomon, K. O., Syc, S. E., & Goldin-Meadow, S. (1999). Illuminating mental representations through speech and gesture. *Psychological Science, 10*, 327–333.

Alibali, M. W., & DiRusso, A. A. (1999). The function of gesture in learning to count: More than keeping track. *Cognitive Development, 14*, 37–56.

Alibali, M. W., Flevares, L., & Goldin-Meadow, S. (1997). Assessing knowledge conveyed in gesture: Do teachers have the upper hand? *Journal of Educational Psychology, 89*, 183–193.

Alibali, M. W. & Goldin-Meadow, S. (1993). Gesture-speech mismatch and mechanisms of learning: What the hands reveal about a child's state of mind. *Cognitive Psychology, 25*, 468–523.

Berry, D. C., & Broadbent, D. E. (1984). On the relationship between task performance and associated verbalizable knowledge. *Quarterly Journal of Experimental Psychology, 36A*, 209–231.

Chi, M. T. H., Bassok, M., Lewis, M. W., Reimann, P., & Glaser, R. (1989). Self-explanations: How students study and use examples in learning to solve problems. *Cognitive Science, 13*, 145–182.

Church, R. B. (1999). Using gesture and speech to capture transitions in learning. *Cognitive Development, 14*, 313–342.

Church, R. B., & Goldin-Meadow, S. (1986). The mismatch between gesture and speech as an index of transitional knowledge. *Cognition, 23*, 43–71.

Church, R. B., Schonert-Reichl, K., Goodman, N., Kelly, S. D., & Ayman-Nolley, S. (1995). The role of gesture and speech communication as reflections of cognitive understanding. *Journal of Contemporary Legal Issues, 6*, 123–154.

Clements, W. A., & Perner, J. (1994). Implicit understanding of belief. *Cognitive Development, 9*, 377–395.

Crowder, E. M., & Newman, D. (1993). Telling what they know: The role of gesture and language in children's science explanations. *Pragmatics and Cognition, 1*, 341–376.

Ekman, P., & Friesen, W. V. (1969). The repertoire of non-verbal behavior: Categories, origins, usage and coding. *Semiotica, 1*, 49–98.

Evans, M. A., & Rubin, K. H. (1979). Hand gestures as a communicative mode in school-aged children. *The Journal of Genetic Psychology, 135*, 189–196.

Feyereisen, P., & de Lannoy, J.-D. (1991). *Gestures and speech: Psychological investigations*. Cambridge: Cambridge University Press.

Garber, P., Alibali, M. W., & Goldin-Meadow, S. (1998). Knowledge conveyed in gesture is not tied to the hands. *Child Development, 69*, 75–84.

Gelman, R., & Williams, E. (1998). Enabling constraints on cognitive development: In D. Kuhn & R. S. Siegler (Eds.), *Cognition, perception and language*, Vol. 2 (pp. 575–630), W. Damon (Ed.), *Handbook of child development*, 5th edition. New York: Wiley.

Gentner, D. (1989). Structure mapping: A theoretical framework for analogy. *Cognitive Science, 7*, 155–170.

Gershkoff-Stowe, L., & Smith, L. B. (1997). A curvilinear trend in naming errors as a function of early vocabulary growth. *Cognitive Psychology, 34*, 37–71.

Goldin-Meadow, S. (1997). When gesture and words speak differently. *Current Directions in Psychological Science, 6*, 138–143.

Goldin-Meadow, S. & Alibali, M. W. (1995). Mechanisms of transition: Learning with a helping hand. In D. Medin (Ed.), *The Psychology of learning and motivation* (Vol. 33; pp. 115–157). New York: Academic Press.

Goldin-Meadow, S., Alibali, M. W., & Church, R. B. (1993). Transitions in concept acquisition: Using the hand to read the mind. *Psychological Review*, 279–297.

Goldin-Meadow, S., Butcher, C., Mylander, C., & Dodge, M. (1994). Nouns and verbs in a self-styled gesture system: What's in a name? *Cognitive Psychology, 27,* 259–319.

Goldin-Meadow, S., Kim, S., & Singer, M. (1999). What the teacher's hands tell the student's mind about math. *Journal of Educational Psychology, 91,* 720–730.

Goldin-Meadow, S., & McNeill, D. (1999). The role of gesture and mimetic representation in making language the province of speech. In Michael C. Corballis & Stephen Lea (Eds.), *The descent of mind* (pp. 155–172). Oxford: Oxford University Press.

Goldin-Meadow, S., Nusbaum, H., Kelly, S., & Wagner, S. (under review). Gesturing helps you remember. Manuscript submitted for publication.

Goldin-Meadow, S., & Sandhofer, C. (1999). Gestures convey substantive information about a child's thoughts to ordinary listeners. *Developmental Science, 2,* 67–74.

Goldin-Meadow, S., Wein, D., & Chang, C. (1992). Assessing knowledge through gesture: Using children's hands to read their minds. *Cognition and Instruction, 9(3),* 201–219.

Hadamard, J. (1945). *The psychology of invention in the mathematical field.* New York, NY: Dover.

Huttenlocher, J. (1976). Language and intelligence. In L. B. Resnick (ed.), *The nature of intelligence* (pp. 261–281). Hillsdale, NJ: Lawrence Erlbaum Associates.

Iverson, J., & Goldin-Meadow, S. (1997). What's communication got to do with it: Gesture in blind children. *Developmental Psychology, 33,* 453–467.

Iverson, J. M., & Goldin-Meadow, S. (Eds.). (1998a). *The nature and functions of gesture in children's communications,* in the *New Directions for Child Development* series, No. 79, San Francisco: Jossey-Bass.

Iverson, J. M., & Goldin-Meadow, S. (1998b). Why people gesture as they speak. *Nature, 396,* 228.

Karmiloff-Smith, A. (1992). *Beyond modularity: A developmental perspective on cognitive science.* Cambridge, MA: MIT Press.

Kelly, S., & Church, R. B. (1997). Can children detect conceptual information conveyed through other children's non-verbal behaviors? *Cognition and Instruction, 15,* 107–134.

Kelly, S., & Church, R. B. (1998). A comparison between children's and adults' ability to detect conceptual information conveyed through nonverbal behaviors. *Child Development, 69,* 85–93.

Krauss, R. M., Morrel-Samuels, P., & Colasante, C. (1991). Do conversational hand gestures communicate? *Journal of Personality and Social Psychology, 61,* 743–754.

Kuhn, D., Garcia-Mila, M., Zohar, A., & Andersen, C. (1995). Strategies of knowledge acquisition. *Monographs of the Society for Research in Child Development, 60,* Serial No. 245.

Kuhn, D., & Pearsall, S. (1998). Relations between metastrategic knowledge and strategic performance. *Cognitive Development, 13,* 227–247.

McNeill, D. (1992). *Hand and mind.* Chicago: University of Chicago Press.

McNeill, D., Cassell, J., & McCullough, K.-E. (1994). Communicative effects of speech-mismatched gestures. *Research on Language and Social Interaction, 27,* 223–237.

Nisbett, R. E., & Wilson, T. D. (1977). Telling more than we know: Verbal reports on mental processes. *Psychological Review, 84,* 231–259.

Perry, M., Church, R. B., & Goldin-Meadow, S. (1988). Transitional knowledge in the acquisition of concepts. *Cognitive Development, 3,* 359–400.

Perry, M., & Elder, A. D. (1996). Knowledge in transition: Adults' developing understanding of a principle of physical causality. *Cognitive Development, 12,* 131–157.

Reber, A. S. (1993). *Implicit learning: An essay on the cognitive unconscious.* New York: Oxford University Press.

Schwartz, D. L., & Black, J. B. (1996). Shuttling between depictive models and abstract rules: Induction and fallback. *Cognitive Science, 20,* 457–497.

Siegler, R. S. (1976). Three aspects of cognitive development. *Cognitive Psychology, 8,* 481–520.

Siegler, R. S. (1994). Cognitive variability: A key to understanding cognitive development. *Current Directions in Psychological Science, 3,* 1–5.

Siegler, R. S. (1997, April). A microgenetic study of self-explanation. Paper presented at the biennial meeting of the Society for Research of Child Development, Washington, DC.

Siegler, R. S., & Chen, Z. (1998). Developmental differences in rule learning: A microgenetic analysis. *Cognitive Psychology, 30,* 273–310.

Siegler, R. S., & Crowley, K. (1991). The microgenetic method: A direct means for studying cognitive development. *American Psychologist, 46*(6), 606–620.

Siegler, R. S., & Jenkins, E. (1989). *How children discover new strategies.* Hillsdale, NJ: Lawrence Erlbaum Associates.

Siegler, R. S., & Stern, E. (1998). Conscious and unconscious strategy discoveries: A microgenetic analysis. *Journal of Experimental Psychology: General, 127,* 377–398.

Stanley, W., Mathews, R., Buss, R., & Kotler-Cope, S. (1989). Insight without awareness: On the interaction of verbalization, instruction, and practice in a simulated process control task. *Quarterly Journal of Experimental Psychology: Human Experimental Psychology, 41*(3A), 553–577.

Stone, A., Webb, R., & Mahootian, S. (1992). The generality of gesture-speech mismatch as an index of transitional knowledge: Evidence from a control-of-variables task. *Cognitive Development, 6,* 301–313.

Thelen, E. (1989). Self-organization in developmental processes: Can systems approaches work? In M. Gunnar & E. Thelen (Eds.), *Systems and development: The Minnesota Symposium on Child Psychology* (pp. 77–117). Hillsdale, NJ: Lawrence Erlbaum Associates.

Vygotsky, L. S. (1978). *Mind in society.* Cambridge, MA: Harvard University Press.

Chapter 2

Children's Discoveries and Brain-Damaged Patients' Rediscoveries

Robert S. Siegler
Carnegie Mellon University

Among developmental psychologists, there is nearly universal agreement that how change occurs is a central issue facing the discipline. The growth that occurs from infancy to childhood to adolescence is so profound that it simply demands explanation.

Change also occurs throughout adulthood, but the changes tend to be less dramatic. Because most changes during adulthood are gradual, and because many of the changes depend on the particulars of individual experience, interest in change among those who study adult cognition is highly variable. Some researchers are interested, others are not. Reflecting this variable interest, and also reflecting the fact that it is more difficult to study change than to study steady states, cognitive and neuropsychological research has focused far more on static states than on change.

Yet, cognitive change is pervasive at all points in life. This is true for adults with brain damage as well as for others. The cognitive system is rarely frozen after brain injury (Finger & Stein, 1982; Will & Kelche, 1992). Instead, both brain and behavior continue to change as people adapt to new circumstances, both physiological and environmental. Understanding change after brain damage therefore seems critical for understanding the relation between behavioral and neural functioning.

Consistent with this view, a small but increasingly numerous group of neuropsychologists and neuroscientists are focusing on recovery of function after brain damage (e.g., Finger, LeVere, Almli, & Stein, 1988a; Rose & Johnson, 1992). Research in this area spans multiple levels of analysis, including molecular studies of how neural mechanisms respond to trauma, animal studies of variables that are associated with greater recovery of function, studies of human patients that focus on behavioral compensation and reacquisition of functioning, and evaluation of alternative rehabilitation techniques. It is increasingly focusing on fine-grained analyses of neural mechanisms, among them axonal sprouting, synaptogenesis, and dendritic branching (Kolb, 1992; Stein & Glasier, 1992). However, as Kolb (1992) noted:

> There needs to be a concerted effort to develop a behavioral methodology that will allow us to make better correlations with increasingly detailed measures of chemistry, anatomy, and physiology of the brain. . . . There needs to be a careful determination of how the behavior has been changed initially by the injury and how it has evolved over time. (p. 182)

The central argument of this chapter is that the microgenetic method, originally developed to study cognitive change in typical children, provides the type of behavioral methodology called for in Kolb's statement. To make this argument, I first describe a series of studies on the development of numerical skills and understanding in typical children. These studies culminate in a microgenetic study of strategy discovery and a computational model of the discovery process. Then, an application of the microgenetic method to examining recovery of numerical skills and understanding in a brain-damaged adult is described. Together, the studies illustrate how microgenetic methods can produce a deeper understanding of cognitive change.

NORMATIVE DEVELOPMENT OF CHILDREN'S ARITHMETIC

From age 4 onward, children use a variety of strategies to solve single-digit arithmetic problems. The most common strategies are listed and illustrated in Table 2.1. Probably the best known strategy is counting from one (the *sum strategy*). For example, 4-year-olds often solve small number problems such as 3 + 4 by putting up 3 fingers, putting up 4 fingers, and counting the 7 raised fingers. Another common strategy involves counting from the larger addend (the *min strategy*). On 3 + 4, a child using this strategy would either count "4, 5, 6, 7" or "5, 6, 7." A third common strategy is *decomposition*, which involves dividing the problem into two or more simpler ones. A child presented 3 + 4 might think, "3 + 3 = 6, 4 is one more than 3, 6 + 1 is 7." Yet, a fourth common strategy is *retrieval*, in which a child posed 3 + 4 would recall that the answer was 7.

TABLE 2.1
Young Children's Addition Strategies

Strategy	Typical Use of Strategy to Solve 3 + 5
Sum	Put up 3 fingers while saying "1, 2, 3," put up 5 fingers while saying "1, 2, 3, 4, 5," count all fingers by saying "1, 2, 3, 4, 5, 6, 7, 8."
Min	Say "5, 6, 7, 8," or "6, 7, 8," perhaps simultaneously putting up one finger on each count beyond 5.
Short-cut sum	Say "1, 2, 3, 4, 5, 6, 7, 8," perhaps simultaneously putting up one finger on each count.
Retrieval	Say an answer and explain it by saying "I just knew it."
Guessing	Say an answer and explain it by saying "I guessed."
Decomposition	Say "3 + 5 is like 4 + 4, so it's 8."

Individual children typically use several of these strategies to solve simple arithmetic problems. For example, the large majority of 4- to 7-year-olds used at least three of the strategies to solve single-digit addition problems (Geary & Burlingham-Dubree, 1989; Siegler, 1987), and the large majority of 7- and 9-year-olds use at least three strategies to solve the converse subtraction problems (Siegler, 1989). Particularly striking, when given the same problem twice, on occasions a week apart, about a third of children used different strategies on the two occasions (Siegler & Shrager, 1984). Use of varied strategies on these single-digit problems also persists into adulthood. College students in Canada, the United States, and Sweden have been found to use strategies other than retrieval to solve 15% to 30% of single-digit addition and multiplication problems (Geary & Wiley, 1991; Ladd, 1987; LeFevre, Bisanz, et al., 1996; LeFevre, Sadesky, et al., 1996; Svenson, 1985). Moreover, use of multiple arithmetic strategies is seen in a wide range of societies including China (Geary, Fan, & Bow-Thomas, 1992), France (Lemaire & Siegler, 1995), and Norway (Ostad, 1998). Thus, strategic variability is a widespread phenomenon in arithmetic as it is in many other domains (Siegler, 1996).

Evidence supporting these claims that individual children use varied strategies comes from several types of data: direct observation, self-reports, and chronometric analyses. In most studies, strategy use is initially inferred from a combination of observation of overt behavior and immediately retrospective answers to the question, "How did you solve that problem?" The strategy classifications allow separation of trials according to the strategy that appears to have been used. Then, the solution times believed to have been generated by each strategy are analyzed to determine if the predictors suggested by the inherent nature of the strategy account for the solution time pattern on those trials. For example, Siegler (1987) reported that on trials on which children were classified as using the min strategy, the size of the smaller addend was the best predictor of median solution time on each problem, accounting for 86% of the variance in the times on the 45 problems examined in that study. This made sense in that when children used the min strategy, the main source of differences among solution times on different problems should be the number of counts upward from the larger addend to the sum, that is, the size of the smaller addend. Similarly, the best predictor of solution times on trials on which children were classified as using the sum strategy was the sum of the addends. Again this made sense, in that the number of counts required to execute the sum strategy was the number indicated by the sum. Thus, behavioral observations and self-reports of arithmetic strategies yield valid strategy classifications.

Use of multiple strategies has important consequences for cognitive functioning. It allows people to solve problems more effectively than if they consistently used a single strategy. For example, when college students were allowed to choose between mental arithmetic or a calculator to solve multiplication problems, their performance was faster and more accurate than when they had to always use one strategy or always use the other (Siegler & Lemaire, 1997). They used mental arithmetic more often when they could solve the problem faster that way and used the calculator more often when they could solve the problem faster that way.

Analyses of the strategy use of younger and older children indicate that development of arithmetic skills involves several distinct types of changes (Lemaire & Siegler, 1995). One type of change involves discovery of new strategies. For example, most children discover the min strategy during first grade. A second type of change involves shifting distributions of strategies that are already used. From kindergarten through third grade, children decrease their use of the sum strategy, first increase and then decrease their use of the min strategy, and steadily increase their use of retrieval. A third type of change involves increasingly effective use of existing strategies. Execution of each individual strategy becomes progressively faster and more accurate from first grade through college age. Yet, a fourth type of change is increasingly adaptive choices among strategies. As children gain increasing experience with arithmetic problems, they choose increasingly often the types of strategy that work best on each problem (Lemaire & Siegler, 1995).

CHILDREN'S STRATEGY DISCOVERIES

How do children acquire new arithmetic strategies? Parents and teachers directly instruct them in some procedures, such as the sum strategy. However, very few children are instructed in other procedures, such as the min strategy. Children appear to discover these strategies in the course of their problem solving experience (Groen & Resnick, 1977). The question is: How do they make these discoveries?

Microgenetic methods are particularly well suited to examining the discovery process. As noted by Siegler and Crowley (1991), the method is characterized by three key properties:

1. Observations span the period during which the change of interest occurs;
2. The density of observations is high relative to the rate of change of the phenomenon;
3. Observed behavior is analyzed intensively with the goal of inferring the processes that gave rise to both quantitative and qualitative aspects of the change.

Most methods for studying children's thinking assess change only indirectly. Both of the two main methods, cross-sectional and longitudinal, require investigators to infer the process of change from observations of performance at widely spaced times. For example, children's understanding of a given task might be observed at ages 5, 7, and 9 years. The long time span between observations makes it difficult to infer the specifics of the change process. In contrast, microgenetic methods provide a dense sampling of changes as they occur. The greater density of observations during the period of change provides a much more detailed portrayal of the changes than do traditional longitudinal or cross-sectional methods, and thus reduce the amount of speculation needed to infer the underlying process that produced the changes.

Microgenetic methods that include assessments of strategy use on each trial are particularly advantageous for studying the process of strategy discovery. The reason is that they allow identification of the trial on which each child first used the new approach. Identifying the trial on which a new approach was first used allows examination of the nature of the discovery (for example, whether the child became excited at figuring out the new approach, whether the child was even aware of having used a new approach, and whether he or she could explain why the new approach was advantageous). Knowing exactly when the new approach was first used also allows examination of performance just before the discovery: What types of problems immediately preceded the discovery, whether the child was having difficulty solving the preceding problems, whether the child was taking an unusually long period of time to solve those problems, and so on. Moreover, knowing when the discovery was made allows examination of performance just after the discovery: How consistently the child used the new strategy on the same type of problem, how broadly the child generalized the new strategy to other types of problems, how efficiently

ecuted the new strategy, and how all of these dimensions of
e change as the child gains experience with the new approach.
crogenetic studies provide a kind of microscope for examining
performance just before, during, and after discoveries are made.

The logic of studying cognitive change through use of microgenetic
methods is the same as the logic that underlies studying cognitive performance through use of high-temporal-resolution neuro-imagining
methods. Many crucial aspects of cognitive performance and change are
present only briefly. Without fine-grained methods, important parts of the
process cannot be detected. In the case of strategy discovery, these parts
often involve brief-lived transition strategies that are crucial to discovery
of more enduring strategies but that themselves are used only for a short
time. They also can involve brief-lived unconscious use of new strategies
that become conscious after a handful of unconscious uses (Siegler &
Stern, 1998).

A study by Siegler and Jenkins (1989) illustrated the advantages of
microgenetic methods for studying cognitive change. First, 4- and 5-year-
olds were pretested to identify children who could solve some simple single-digit addition problems via the sum strategy or retrieval but who did
not yet use the min strategy. Children who met the criteria were then presented an 11-week practice period, which included roughly 25 sessions
(the exact number of sessions varied from child to child, due to absences
and family vacations). In each session, the child was asked to solve seven
addition problems, and was given a star each time his or her answer was
correct.

During Weeks 1–7 of the practice period, children were presented relatively easy problems, in which both addends were between 1 and 5 inclusive. A number of children discovered the min strategy during this period,
but none of them used it very much. Therefore, in Week 8, we presented
challenge problems, that is, problems such as 3 + 21, with one very large addend and one small one. On such challenge problems, the 4- and 5-year-
olds were very unlikely to generate a correct answer either via counting
from one or via retrieval. However, the min strategy could solve such
problems relatively easily, and its advantages would be especially evident
on them. Finally, in Weeks 9–11, we presented a mixed set of problems,
including small addend problems like those presented in the first 7 weeks,
challenge problems like those presented in Week 8, and in-between problems, such as 3 + 9, which had not been presented previously.

Overview of Findings

Children solved the addition problems quite accurately. They averaged
85% correct, with individuals ranging from 76% to 98% correct. Accuracy
improved over the 11 weeks of the practice period. On the small addend

TABLE 2.2
Eight Key Findings about Preschoolers'
Discovery of the Min Strategy

---

1. Almost all children discovered the min strategy.
2. Strategy use was highly variable.
3. The shortcut sum strategy was frequently transitional to the min strategy.
4. Discoveries occurred without impasses.
5. Generalization of the min strategy was slow.
6. Presenting challenge problems substantially increased use of the min strategy.
7. Strategy choices were adaptive.
8. Children discovered the min strategy without trial and error.

---

problems, which were the only problems presented throughout the study, percent correct increased from 80% in the first five sessions to 96% in the last five.

Children used a variety of strategies throughout the 11-week period. Across all sessions, the sum strategy was the most frequent approach (34% of trials) and retrieval the next most frequent (22%). The distribution of strategies changed over sessions, with the sum strategy becoming less prevalent and the min strategy more prevalent.

Findings About Discovery

The study yielded eight key findings, which are listed in Table 2.2.

Consistent Discovery of the Min Strategy. The most important finding, given the purpose of the study, was that almost all children (7 of 8) discovered the min strategy. Timing of the discoveries varied greatly. The first child to use the min strategy discovered it in the second session of the experiment, on the 8th problem that she encountered. The last child to discover the min strategy first used it in his 29th session, on the 209th of his 210 trials. The other discoveries were spaced fairly evenly in between.

Roughly half of the children indicated quite clearly not only what they had done on the trial on which they first used the min strategy but also why it was a good idea. Even for these children, however, the trials on which discoveries occurred also involved an unusual degree of inarticulateness and odd references. Consider the first trial on which one girl used the min strategy:

Experimenter: *How much is 6 + 3?*
Child: (long pause) *9*
E: *OK, how did you know that?*
C: *I think I said, . . . I think I said . . . oops, um . . . I think he said . . . 8 was 1 and . . . um . . . I meant 7 was 1, 8 was 2, and 9 was 3.*

E: *How did you know to do that? Why didn't you count "1,2,3,4,5,6,7,8,9"?*
C: *Cause then you have to count all of those numbers.*
(Siegler & Jenkins, p. 66).

This child's long pause before answering, and her "oopses" and "ums," suggest that the discovery process was quite cognitively demanding. Her protocol also raises the question of who the "he" was in her comment "I think *he* said . . . 8 was 1 . . ." Such long solution times and inarticulateness on the trial of discovery were typical.

Variability of Strategy Use. All children used at least four strategies, and the majority used six or more. The variability was present within a single session as well as across sessions. It was also present on a given problem encountered in different sessions. None of the children used a single strategy on more than 70% of the trials, and half did not use any one strategy on as many as 45% of trials.

Transition Strategies. One useful feature of microgenetic methods is that they can test hypotheses regarding short-lived transitional strategies that usher in more enduring approaches. Microgenetic methods can also lead to the discovery of unanticipated transition approaches. Both of these functions were evident in the Siegler and Jenkins data. Previous investigators had hypothesized that the strategy of counting from the first addend mediates the transition from the sum strategy to the min strategy (Resnick & Neches, 1984; Secada, Fuson, & Hall, 1983). This hypothesis was plausible, but the data provided no support for it; none of the children counted from the first addend before using the min strategy.

On the other hand, the results did reveal a different transition strategy, the *shortcut sum strategy.* As indicated in Table 2.1, the shortcut sum strategy shared some features with the sum strategy and others with the min strategy. Like the sum strategy, it involved counting from 1. Like the min strategy, it involved counting each number once rather than twice as in the sum strategy. Thus, it was a plausible transitional approach. There also were empirical grounds for viewing the shortcut sum strategy as transitional to the min strategy. Six of the seven children who discovered the min strategy used the shortcut sum strategy before it, and five of the six children who used the shortcut sum strategy started using it two sessions or less before they started using the min strategy.

Discovery Without Impasses. One common view of discovery is that new ideas typically are elicited by especially difficult problems. This view is prevalent in both the psychology and artificial intelligence (AI) literatures (Newell, 1990; van Lehn, 1988) and in everyday expressions such as "Necessity is the mother of invention." Children's discovery of the min strategy did not conform to this wisdom, however. The problems on which discoveries were made were quite representative of the overall set of problems, neither more difficult than nor less difficult than the overall set of problems. The same was true of the problems just before the discovery. Moreover, accuracy just before the discoveries was at the same high rate as in the experiment as a whole. Thus, in this case, necessity was not a necessity.

Slow Generalization. After discovering the min strategy, children only gradually increased its use. This was true even for children such as Lauren (the child quoted previously), who advanced compelling explanations for why the sum strategy was less good than the min strategy ("cause then you have to count all those numbers").

Effects of Challenge Problems. When children were presented the challenge problems, those who had previously used the min strategy greatly increased their use of it. In the five sessions before the challenge problems, those children who had discovered the min strategy used it on 14% of the trials on which they counted. In contrast, in the five sessions after the challenge problems, they used it on 45% of counting trials on the same types of problems (problems with both addends between 1 and 5).

Adaptive Strategy Choices. Children used each strategy most often on those problems on which it was most effective. Using the min strategy far more often on the challenge problems than they had used it on the small number problems was one example of this. Additionally, throughout the study, children used retrieval primarily on the easiest problems and counted from one primarily on the hardest.

Discovery Without Trial and Error. Perhaps the most surprising finding of the study was that the preschoolers did not require any trial and error to make the discoveries. Almost all views of discovery, going back at least to Thorndike (1898), assume that people initially try a mix of correct and flawed new approaches, and then they eliminate the flawed approaches. The trial-by-trial data from Siegler and Jenkins (1989), however, revealed that children discovered the min strategy, and also the shortcut sum strategy, without even once trying conceptually flawed approaches such as counting the first addend twice or adding the larger addend to the second addend.

It is important to note that most of these eight findings about discovery could only have emerged within a microgenetic design. Following children's changing competence on a trial-by-trial basis was essential for learning that their discoveries occurred without any trial and error; if we had not examined each trial before the discovery, how could we have known that children did not use an illegal strategy on even one of them. For the same reason, following performance on a trial-by-trial level was essential for determining that children discovered the new strategy without encountering failure or particularly difficult problems; that in the trials immediately after the discovery, they used the newly discovered strategy very little; that challenge problems were crucial for its wider use; and that the shortcut sum strategy, rather than counting from the first addend, was transitional to the min strategy. Microgenetic studies are time consuming and effortful, but they yield data about the process of change that simply cannot be obtained via other methods.

Conceptual Constraints on Strategy Discovery

Microgenetic studies often raise issues that are best addressed with more conventional experimental methods. One such issue raised by the microgenetic study of addition was how children could discover the min strategy without any trial and error. Not even on one trial did children use a strategy that violated the basic principles of addition, such as counting the first addend twice. The question was how they were able to discover legal new strategies without any direct instruction in them and without trying any illegal ones.

Siegler and Jenkins (1989) proposed that such errorless discovery was possible because the preschoolers' strategy generation process was constrained by a *goal sketch*, which specified the hierarchy of objectives that a satisfactory strategy must meet. In the case of addition, the goal sketch would consist of two goals: (1) represent each addend in the original problem once and only once; (2) quantitatively represent the objects in the combined symbolic representation. The prototypic strategy for making clear these requirements is the one that most parents teach first: Put up fingers to represent the first addend, put up fingers to represent the second addend, and then generate a single quantitative representation by counting the fingers on both hands and stating the last number as the answer. Such a goal sketch would direct searches of existing knowledge toward procedures that could meet goals in the domain. It also would lead to rejection of illegitimate procedures that were considered.

Goal sketches function much like the planning nets described by Greeno, Riley, and Gelman (1984), in that they are domain-specific mechanisms that identify critical goals and that are useful for generating new procedures to meet the goals. Thus, the goal sketch expresses conceptual knowledge of the domain, in particular knowledge of the goals that all satisfactory strategies must meet. The question was how we could test whether children who did not yet use the min strategy possessed the knowledge hypothesized by the goal sketch.

Siegler and Crowley (1994) tested the goal sketch hypothesis by first identifying a set of 5-year-olds who already knew the min strategy and a set of peers who did not. They then presented the children with the task of judging whether instances of three addition procedures were "very smart," "kind of smart," or "not smart." One of the procedures that they judged was the sum strategy, which all of the children already knew; another was the min strategy, which some children knew and others did not; the third was an illegal strategy, counting the first addend twice, which none of the children used. If children had a goal sketch for acceptable addition procedures even before they discovered the min strategy, they should rate more highly the min strategy than the illegal strategy, despite their not using either of them.

This hypothesis was supported. Both children who already used the min strategy and children who did not yet use it rated it as significantly smarter than the illegal strategy. In fact, both groups of children rated the min strategy as being at least as smart as the sum strategy, which both of them knew and used. Thus, even before children begin to use the min strategy, they already recognize its superiority to illegal alternatives. Such recognition is consistent with the hypothesis that children have a goal sketch for addition like that postulated by Siegler and Jenkins (1989) and provides a means by which children would attempt the min strategy without first attempting illegal procedures.

Goal sketches are not unique to addition. Siegler and Crowley (1994) demonstrated a similar pattern with 8-year-olds who did not know a relatively advanced strategy for playing tic-tac-toe (the forking strategy). They rated smarter a legal strategy that they did not use than the legal one that they did. It seems likely that conceptual constraints, such as those in goal sketches, help children minimize trial and error while discovering a variety of strategies.

A COMPUTER SIMULATION OF STRATEGY DISCOVERY

Background

The data from the microgenetic study, together with the data from the goal sketch study, provided a detailed empirical base for modeling the discovery process. The main goal was to formulate a model that would produce all eight of the key aspects of strategy discovery listed in Table 2.2, as well as the general characteristics of strategic development generated by the previous Siegler and Shipley (1995) model of development of addition skills. These general characteristics include increases in speed and accuracy, decreases in the use of the sum strategy, increases and then decreases in use of the min strategy, increases in use of retrieval, generalization of adaptive strategy choices to novel problems, and patterns of individual differences like those produced by children (Kerkman & Siegler, 1993; Siegler, 1988). To do this, we maintained the main features of the Siegler and Shipley model, but added components that, together with the previous model, would discover new addition strategies as well as perform, learn, and choose as children did.

To understand the new model, it is essential to have some understanding of its predecessor, ASCM (*A*daptive *S*trategy *C*hoice *M*odel). ASCM's basic organization is illustrated in Fig. 2.1. Strategies operate on problems to generate answers. This process yields information not only about the answer to the particular problem but also about the speed and accuracy of the strategy used to solve the problem. The information is incorporated into data bases regarding the strategy that was used and the problem that was solved; thus, the data bases are modified progressively as the system gains experience solving problems with different strategies.

The type of information in ASCM's data base is illustrated in Fig. 2.2. Children's knowledge of each strategy is hypothesized to include the data produced by past problem-solving experience, and also projections from that data concerning the strategy's likely effectiveness in solving problems. The database about each strategy includes information regarding its performance over all problems (global data); over problems with particular features, such as a large difference between the addends (featural data);

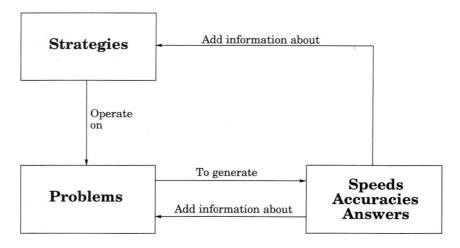

FIG. 2.1. Overview of ASCM.

Organization of Database

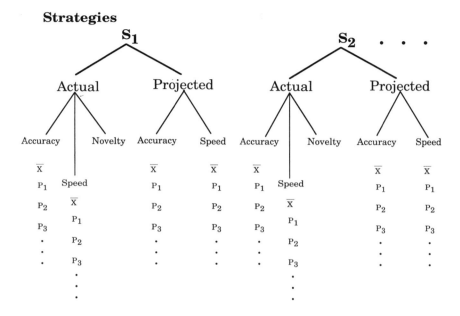

FIG. 2.2. Organization of ASCM's databases on strategies and problems.

and on individual problems that the system has encountered (problem-specific data).

Information about each strategy's novelty is also present in the data base. The motivation for including the novelty data was to answer the question: What leads children to use new strategies in situations where existing strategies work quite well? This question has become of increasing interest with documentation of numerous utilization deficiencies, that is, situations in which children use new approaches despite the approaches not immediately resulting in improved performance (Bjorklund & Coyle, 1995; Miller & Seier, 1994). In line with Piaget's construct of functional assimilation, Siegler and Shipley (1995) hypothesized that children are motivated to try out newly discovered approaches. Each use of the new strategy reduces its novelty by a certain amount. If the new approach yields relatively fast and accurate performance, the strength it gains through use more than outweighs the loss of novelty strength. If the new approach yields relatively slow or inaccurate performance, use results in a net loss of strength. The novelty points allow strategies to be used at a time when their effectiveness alone does not justify their use, and thus allows them to become more practiced and effective.

ASCM started with the two strategies, retrieval and the sum approach, that are most often used by 4-year-olds. At all points in the simulation's run, each strategy had a particular strength, which was a function of the speed, accuracy, and novelty of the strategy on problems in general, on problems with features like those being considered, and on the particular problem. Probability of choosing a given strategy was proportional to the relative strengths of the available strategies. Each time a strategy was used, the speed and accuracy it produced modified the data base relevant to it. After a preset number of problem presentations, the modeler added the min strategy to the mix, so that ASCM could choose among it, the sum strategy, and retrieval. Then, the run continued as previously.

ASCM produced addition performance much like that of children in many ways. Both speed and accuracy increased considerably over the course of its run. For example, early in one run, it solved 31% of problems accurately; toward the end, it solved 99%. Much of the improved accuracy was due to changes in strategy use. At first, it primarily used the sum strategy; then, it used the min strategy slightly more often than either the sum strategy or retrieval; later still, retrieval became the predominant approach. Throughout its run, ASCM used diverse strategies both within and between problems. It also chose adaptively among the strategies. It used retrieval most often on the easiest problems; used the min strategy most often on problems with large differences between the addends, where its advantage relative to counting from one was greatest; and used the sum strategy most often on hard problems with relatively small differences between the addends. ASCM's global and featural information allowed it to generalize these sensible strategy choices to novel problems. Finally, variations in two parameters—the rate of learning and the threshold for stating retrieved answers—allowed the model to generate the patterns of individual differences previously observed in children.

Thus, ASCM accurately reproduced a great many aspects of the development of arithmetic. It did not produce one crucial aspect, though: Discovery of new strategies. Generating such discoveries, while also generating the many aspects of the development of addition captured by ASCM, was the motivation for formulating a new model.

SCADS

A Description of the Model. SCADS (Strategy Choice And Discovery Simulation) maintains the main mechanisms within ASCM but changes the model in one important way and extends it in several others (Shrager & Siegler, 1998). The basic organization of the model is shown in Fig. 2.3. The boxes surrounded with dashed lines represent an initial strategy choice component similar to that in ASCM. Strategies operate on problems to produce answers, speeds, and accuracies; the data are fed back to modify the system's data base regarding the effectiveness of strategies and the difficulty of problems; these modification influence subsequent choices. The one dif-

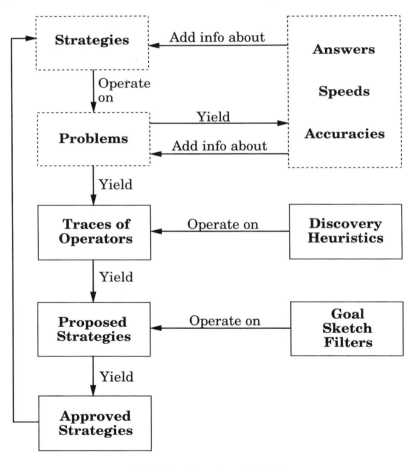

FIG. 2.3.   Overview of SCADS.

ference from ASCM is that within SCADS, each strategy is represented as a set of separable operators, rather than as a modular unit.

The solid boxes represent the new components of SCADS. These components make it possible for the simulation to discover new strategies as well as to learn and to choose adaptively among existing ones. SCADS forms a working memory trace of the sequence of operators used to execute a strategy. For example, the sum strategy is represented as the following sequence of operators: all fingers down, choose hand, choose addend, say addend, clear echoic buffer, count out fingers to represent addend, switch hands, switch addends, say addend, clear echoic buffer, count out fingers to represent addend, clear echoic buffer, count all fingers, end. As a strategy is being used to solve a problem and immediately afterward, the system has available the operations and partial results produced during the solution process.

SCADS also includes two discovery heuristics: (a) when a redundant sequence of behaviors is detected, delete one of the sets of operators that caused the redundancy; (b) if the data base indicates that a strategy is faster or more accurate when it is executed in a particular order, then create a specialized version of the strategy that always uses that order.

These discovery heuristics operate on the trace of the most recent strategy execution and on the data bases. When doing so, the discovery heuristics sometimes generate proposals for new strategies. These proposed strategies are then tested for consistency with the goal sketch filters. These goal sketch filters indicate criteria that legitimate strategies in the domain must meet. In the case of addition, one goal sketch filter requires that both addends be represented; the other goal sketch filter indicates that the reported result includes the representations corresponding to both addends. Only those proposed strategies that are consistent with the goal sketch filters are attempted. For example, if the discovery heuristics generate the strategy "count the second addend twice," the goal sketch filters will reject the strategy without it being tried, because it violates the criterion that both addends must be represented.

One other aspect of SCADS does not show up in Fig. 2.3 but is critical to the model's functioning. This is control of attentional resources. Attention is divided between the strategy being executed and the discovery process. To the degree that a strategy is not fully mastered, and the association between successive operators in it are weak, attention is devoted to insuring correct execution of the strategy, This leaves few or no resources available for discovering new strategies. Each time the strategy is executed, the operators in it become more strongly associated, which decreases the attentional demands of executing the strategy. This results in the model being able to devote more attentional resources to applying the discovery heuristics to the trace of the strategy's execution and thus for proposing new strategies. The mechanism is consistent with Case's (1992) emphasis on automatization of existing procedures as a crucial contributor to discovery of new approaches.

These new aspects of SCADS form an interlocking discovery mechanism. If the strategies were represented in the modular form used in ASCM, there would be no separate data on operators from which to form the working memory trace. Without the working memory trace, the discovery heuristics would lack data from which to detect redundancies in processing and to guide proposals for potential strategies. Without the discovery heuristics and traces of operators, there would be no need to divide attentional resources between strategy execution and strategy discovery. Without the proposals for potential strategies that are generated by the discovery heuristics, the goal sketch would lack possibilities from which to select promising new approaches. Thus, all of these components are integral parts of the strategy discovery process.

SCADS' Functioning. To understand how SCADS operates, it is useful to consider a representative run. At the outset, SCADS includes two strategies: retrieval and the sum strategy. Retrieval is a basic cognitive process that appears to be present from birth; most parents teach their children the sum strategy during the preschool period (Saxe, Guberman, & Gearhart, 1987).

When the model is presented a problem, the ASCM-like part chooses a strategy. Early in its run, SCADS usually chooses the sum strategy, because answers rarely are associated with problems sufficiently strongly to be stated. In this early phase, most of the system's attentional resources are devoted to supervising execution of the sum strategy, because its operators are not yet strongly associated.

On the run being described, after about 70 trials, the sum strategy was mastered sufficiently that attentional resources could begin to be devoted to discovering new strategies. The first three strategies that were generated all violated one or both of the goal sketch filters and therefore were not tried. Some trials later, SCADS generated a strategy that was legal, albeit inefficient: Counting both addends, starting over and doing the same thing, and then advancing the answer yielded on the second count. The strategy was then available for use, though its inefficiency resulted in it never being used much. Soon after, the simulation generated a specialized version of the sum strategy in which the larger addend was always represented before the smaller addend. This procedure was slightly more efficient than the usual form of the sum strategy, because errors become more frequent when the second number is large. A few problems later, the simulation noticed the redundant processing involved in the sum strategy and generated the shortcut sum approach. Finally, SCADS eliminated the redundant processing of starting from 1 within the shortcut sum strategy, and generated the min strategy.

To test the model, we ran it 30 times. On each run, we presented it 500 trials, 20 repetitions of each of the addition problems with addends 1 through 5 inclusive. These were the problems used in the first 7 weeks of Siegler and Jenkins (1989). Then, to test the effect of challenge problems, we ran the model an additional 30 times, substituting one block of 50 standard problems with a block of challenge problems (problems with one addend below 5 and the other above 20, such as 2 + 23). These were the types of problems presented in Week 8 of Siegler and Jenkins.

The model generated all eight of the major strategy choices and strategy discovery phenomena described earlier and in Table 2.2. Its strategy use was variable throughout its run. It chose adaptively among the strategies, especially as it gained experience solving problems. For example, the harder the problem, the more often it used backup strategies. It discovered the min strategy on all of its runs. It consistently discovered the shortcut sum strategy shortly before the min strategy, and used what it learned from discovering and executing the shortcut sum strategy to help generate the min strategy (i.e., it built on the knowledge that correct answers could be generated by counting objects only once, and that starting with the larger addend was more effective than starting from the smaller one). Discoveries followed incorrect as well as correct performance. Newly discovered strategies were generalized only slowly. It never executed illegal strategies. Finally, introduction of the challenge problems led to greater generalization of the min strategy than did the runs without challenge problems. The min strategy was used especially often during presentation of the challenge problems, and remained higher afterward than on the runs without such problems.

Particularly interesting were the interlocking roles of the different parts of the strategy discovery mechanism. On each run, SCADS proposed between 15 and 21 strategies. However, the model showed the same characteristic as the analogical reasoning model that Falkenheiner, Forbus, and Gentner (1989) labeled MACFAC: Many are called, few are chosen. In the present context, this meant that more than half of the proposed strategies were vetoed by the goal sketch filters and therefore were not used.

In addition to the strategies in Table 2.1 that are quite common in children's behavior, SCADS generated between one and four other strategies per run. These strategies were legal but inefficient. Some had previously been observed in children's addition. For example, SCADS, like preschoolers, occasionally uses the sum strategy, obtains an answer, and then goes through the same process again before advancing the answer. In future work, my colleagues and I plan to examine whether all these inefficient-but-legal strategies that SCADS generates are used by children.

More general aspects of SCADS' performance also fit those of children's arithmetic. Accuracy and frequency of use of retrieval increased substantially with experience, both reaching roughly 90% by the end of the runs. Mean number of trials before discovery of the min strategy closely resembled that of children, 107 problems in the simulation runs, 95 problems for the children in Siegler and Jenkins (1989). The close numerical similarity is almost certainly coincidental, but the fact that the time scale is on the right order of magnitude (as opposed to discoveries taking 10 trials or 1,000) is not.

SCADS thus illustrates how a combination of associative and metacognitive processes can give rise to a wide variety of phenomena concerning discovery of new strategies, as well as choices among existing ones. It generates performance and learning that resemble those of children in many ways. It also indicates how both metacognitive processes, which generate qualitative novelties, and associative processes, which produce gradual learning, contribute to strategic development.

More generally, the empirical experiments and simulation models illustrate how microgenetic studies build on other types of empirical work, yield predictions that can be tested via other types of empirical experimentation, and provide the detailed data needed to both guide and constrain modeling efforts. Normative descriptions of the development of addition strategies indicated an age range (4½ to 5½ years) at which most children would not have discovered the min strategy but would be close enough to discovering it that they could benefit from relevant experience. Previous normative studies also indicated the type of experience involved in discoveries of the min strategy: experiences solving problems. Thus, it was not coincidental that Siegler and Jenkins studied children close to their fifth birthday, nor that they presented the children problems to solve rather than didactic instruction. The data from the microgenetic study raised the issue of why children never tried illegal strategies and suggested the goal sketch hypothesis. This hypothesis, in turn, was the motivation for asking children to rate the smartness of strategies that they did not yet use, the procedure used by Siegler and Crowley (1994) to test the goal sketch hypothesis. The overall body of data from the microgenetic study, as well as from other studies of arithmetic, provided a large number of constraints that SCADS needed to meet. The model, in turn, suggested additional hypotheses, such as that children occasionally use inefficient but legal strategies of the type occasionally used by the model. In sum, microgenetic methods, used in conjunction with other empirical methods and formal modeling, can substantially advance understanding of cognitive change.

It seems likely that the advantages of microgenetic methods for understanding children's discoveries can also contribute to understanding of the rediscoveries that are made by brain damaged patients. The remainder of this chapter focuses on an illustration of this use of microgenetic methods.

A MICROGENETIC STUDY OF REDISCOVERY

Background

KBK was a successful businesswoman, who at age 42 moved to a city in upstate New York. Her goal was to expand her business. She had a B. A. from University of Massachusetts and had never encountered any special difficulty in mathematics. On arrival in her new city, she moved into an apartment in a former schoolhouse that had been renovated and divided into condos. Unfortunately, a heating contractor had improperly connected the pipes so that carbon monoxide was spewing into the bathroom. KBK was exposed to the carbon monoxide for 4 months, during which time she developed a number of symptoms, including headaches, nausea, vomiting, and feeling cold, weak, tired, and listless. She also noticed a number of cognitive deficiencies, including memory loss, poor comprehension while reading, and difficulties working with numbers. Finally, because her apartment was cold, she called the utility company. The repairman found that the bathroom had a level of carbon monoxide of 1500 parts per million; the standard for danger to one's health is 400 parts per million.

Repair of the heating reduced KBK's problems with headaches, nausea, and vomiting. Her right side remained weak, however. Despite her being right-handed, her grip with her right hand remained weaker than that with her left (right hand grip strength = 15.5 kg; left hand strength = 30 kg). She often walked with a cane and continued to suffer from superior altitudinal hemi-field loss, which prevented her from seeing material at the top of her visual field.

The carbon monoxide poisoning also had enduring cognitive effects. IQ tests given shortly after KBK was hospitalized and again 8 months later indicated WAIS-R scores considerably lower than typical for a graduate of a high-quality university. Her full scale score was 88 on the first testing, 102 on the second. On both occasions, her verbal score was considerably higher than her performance score (96 vs. 79 on the first testing; 113 vs. 93 on the second). Most striking, however, was the immense variability of her subtest scores. On the digit symbol test, she scored 1, a level ordinarily indicative of severe retardation. On the comprehension test, she scored 16, which is in the superior adult range. Achievement test scores also were highly variable. On the Wide Range Achievement Test, she scored at the 4th grade level (1st percentile) on the arithmetic subtest and at the 12th grade level (63rd percentile) on the reading subtest.

KBK read an article in her local newspaper about some of the research described earlier in this article, and wrote to ask if I would be interested in studying her strategies for solving math problems. This led Randi Engle and me to travel to her city of residence three times to examine her math proficiency and to examine microgenetically the effects of experiences intended to help her rediscover her sense of numbers (Siegler & Engle, 1994).

Pretest

Arithmetic. To assess KBK's level of mathematical functioning at the time of the study, we obtained several types of data. One involved presenting her the 100 basic addition facts with addends 0 to 9 and the 100 multiplication facts with multiplicands 0 to 9. On each trial, she was asked, "How much is N + M" or "How much is N × M?" For problems with the same operands in different orders (e.g., 3 + 6 and 6 + 3), one of the pair was in the first 50 problems and the other in the second 50.

KBK's performance on these simple problems was far inferior to that of a typical college-educated adult. She answered correctly only 53% of the single-digit addition problems and 81% of the multiplication problems. Although KBK answered more multiplication problems correctly, she indicated that her subjective understanding of multiplication was even less good than that of addition. She said that she understood addition to an extent, but had no understanding whatsoever of multiplication. Instead, she said that she remembered the answers to multiplication problems as if they were "part of a song," but had no idea what they meant.

Memory Span for Numbers and Letters. Another aspect of KBK's special difficulty with numbers was her poor memory for them. From early in childhood to adulthood, almost everyone has a longer digit span than letter span (Dempster, 1981). There are many reasons for there to be a difference in this direction: fewer numbers to choose among, greater experience remembering totally arbitrary sequences of numbers such as phone numbers, and chunking of numbers allowing their magnitudes to be used as a retrieval cue (Staszewski, 1988).

In contrast to the usual pattern, KBK remembered more numbers than letters on standard digit-span and letter-span tasks. We presented her lists of numbers or letters at a rate of one item per second and asked her to repeat them back to us in order as soon as we stopped reading the lists. We started with two elements in a list, and provided three opportunities at each list length. If she was correct on at least one of the three opportunities, we proceeded to the next longer list length. This continued until she failed all three lists at a given length.

The longest lists on which KBK succeeded in the forward direction were seven letters and five numbers. The longest lists on which she succeeded in the backward direction were five letters and three numbers. Using a threshold of 67% correct at a given length, KBK's forward letter span was five, as compared to a forward digit span of four; by the same criterion, her backward letter span was five, versus a backward digit span of three. The finding was consistent with her standardized test performance and also with the view of both KBK and her therapists that she had more difficulty processing numbers than would have been expected on the basis of her general level of cognitive functioning.

Transcoding of Numbers. KBK also had difficulty reading numbers and translating them into spoken form and in writing numbers after hearing spoken words. She read single-digit numbers quite fluently, requiring 3 seconds per number, but had more difficulty with multi-digit numbers. At first, she read them digit-by-digit (e.g., reading 437 as four-three-seven). When we asked her to use terms such as hundreds and thousands, she needed to use a place value chart to decide which terms to use, and even then proceeded digit-by-digit, as in "four hundreds, three tens, seven ones." She also stated numbers in nonstandard ways such as "and one left-over" and "the rest is eight." Writing down numbers that were spoken by the experimenter was a source of even greater difficulty. After a few requests to write single numbers, KBK became sufficiently frustrated that she refused to continue the task.

Learning

KBK's comments about not understanding multiplication, together with her difficulty in remembering and transcoding numbers, led us to hypothesize that one effect of her brain damage was loss of understanding of the number system. This hypothesis forced us to analyze what such understanding involves. Two characteristics seemed critical: a veridical representation of the structure of the decimal system, and procedures for fluent translation among alternative task-specific representations.

The number line task shown in Fig. 2.4 provided a means both for assessing KBK's understanding of the number system and for helping her gain greater understanding. Saying a number and asking someone to locate its place on a number line requires the person to translate from a verbal representation to a spatial one. Locating numbers on different number lines (0–10, 0–100, 0–1,000, etc.) demands a relatively complex understanding of the number system, in that it requires a grasp both of each number's cardinal value and its value relative to other numbers that could be located on the scale. Another advantage of the number line task

458
(Four Hundred Fifty-Eight)

FIG. 2.4.   Example of problem presented to KBK. On the particular problem, KBK was asked to place the number 458 on a 0–1,000 number line.

was that the quality of estimates could be assessed objectively by measuring the distance between each number's true position on the scale and the
slash that was drawn to represent it. To provide a comparative assessment,
we compared KBK's estimates to those of control participants (college students) who performed the same tasks.

The microgenetic study of KBK's learning involved presenting a series
of number lines that included larger and larger numbers (first 0–10, then
0–100, then 0–1,000, and finally 0–10,000) and providing her with feedback regarding the discrepancy between where she put her slash and
where it should go. The number lines were identical except for the number at the rightmost end of the line. Table 2.3 indicates the order in which
number lines were presented on each of the two days of the study (the
dashed lines indicate that the particular types of number line in that row
was not presented on that day). Within the notation for representing problem types, "X" represents an integer between 1 and 9, inclusive; "0" represents 0. Thus, X problems involved locating numbers between 1 and 9 on
a 0–10 number line; XO problems involved placing numbers such as 10,
20, and 30 on a 0–100 number line; XX problems involved placing numbers such as 17, 53, and 87 on a 0–100 number line; and so on. On Day 1,
KBK started with X problems and ended with XXX problems; on Day 2,
she started with XO problems and ended with XOOO problems. On both
days, she was given nine successive problems of a given type (e.g., nine XX
problems) before moving on to the next type of problem. The two sessions
each involved about 3 hours of work and took place on 2 successive days.

As Table 2.3 indicates, KBK's estimates of the position of numbers became considerably more accurate from Day 1 to Day 2. Reading the middle column of Table 2.3 from top to bottom indicates that the improvement started toward the end of Day 1. Despite the increasing difficulty of
the problems, the deviations from correct placements began to decrease
substantially. Examining the number of placements within 10% of the cor-

TABLE 2.3
Mean Absolute Deviations (cm) on Number Line Estimation Task

Problem Type	Day 1	Day 2
X	1	—
XO	3	1
XOO	8	1
XX	6	1
XXX	3	1
XXO	—	1
XOOO	—	1
MEAN	4	1

rect location, estimates that were sufficiently close that we considered them correct, substantiate this conclusion. The percentage of such placements increased from 45% on the relatively easy X00 problems presented in the middle of the first day to 89% on the considerably more difficult XXX problems that she encountered at the end of the first day (e.g., placing 789 on a 0–1,000 number line). The improvement continued on the second day. The same XOO problems that elicited 45% correct performance on Day 1 elicited 100% correct performance on Day 2.

What caused the change? Analysis of KBK's written representations suggested that they played an important role. KBK first used such a representation on the 40th problem that she encountered, which required her to estimated the position of 73 on a 0–100 number line. Her occupational therapy had emphasized use of coins and paying bills by giving slightly too much money and asking for change. From this, she knew that 73 was near 75, that a dollar had a value of 100 pennies and could be divided into 100 units, that a dollar also could be divided into 4 quarters, and that dividing a dollar into four quarters would place one division at 75. She used this knowledge to draw a dollar bill; to draw 3 vertical lines on the number line at places corresponding roughly to 25, 50, and 75; and to locate 73 between 50 and 75. At first, she placed the mark closer to the 50 than to the 75; then she crossed it out and placed a mark closer to the 75. Her verbal protocol on this trial is quite revealing of the cognitive processes that led to her placement.

E. Make a mark for the number 73.

K: No, oh, ok, ok. The first thing that's coming to my head is the concept of 75. OK (starts drawing a dollar bill) a quarter, a quarter, a quarter, that's 75 in a quarter, so I go (starts making marks on the number line) a quarter, a quarter, a quarter, and a quarter, so I go, a quarter, a quarter, that's 50, a quarter, and I have to back up a little, 10, 20, 30, 40, 50, 60, 70, and I have to back up a little, a quarter, a quarter, is 50, 75, and 100 (makes marks). Well, it's gonna be somewhere. I certainly know it's somewhere between here and here (50 and 75), uh, it's somewhere between here and here, and I don't know where.

E: OK, well make your best guess at where it would be 73.

K: Uh (makes an "X" between the 50 and the 75 but closer to the 50).

E: OK.

K: Wait a minute, wait 1 second, no Randi, that's wrong (crosses out the "X"). If seven *tee*- five (KBK's accenting) is seven-*tee*-three, seven*tee* four, seven*tee* five, seven*tee* three, seven*tee* four, seven*tee* five, uh, this is 75 so 73 should be here (makes another X closer to the 75). Sorry about that.

All of KBK's written representations immediately after this discovery involved dividing the number lines into quarters or halves. She did not make marks on the number line in this way on all trials, but when she used a written representation, she employed this format. She also continued to draw dollar bills on some trials to remind her of the conceptual framework.

At about the same time, KBK also began to make use of another prior source of knowledge: her occupational therapy experience with an abacus. The abacus that she had used contained 10 rows of 10 beads. She first noted a relation between the abacus and number lines on Problem 15, noted it again on Problem 37, and then referred to it on several other problems (38, 39, and 45). For example, on Problem 45, which involved a 0–100 number line, she noted that she was "imagining in my mind how many beads I would be moving."

On the next problem, which required her to locate 613 on a 0–1,000 number line, she extended the abacus analogy to the 0–1,000 context. Her protocol reflects increasing attention to the magnitude entailed by the number:

So, if that number (1,000) is 10 of the abacus, it (613) is gonna be half of the abacus plus a little more, like maybe here (pointing). OK, 1,000 is the whole, 1,000. So wait a minute now, just like I go 1, 2, 3, 4, 5, 6, 7, 8, 9, 10, and 10 becomes the whole, 10, 20, 30, 40, 50, 60, 70, 80, 90, 100. 100, 2, 100, 102, 103. I don't know, wait, I almost got that. This number (points to 6 in 613) is 6 of these, it's 6 of the abacus, no, I'm losing it, I almost had it, but I'm losing it.

As the last comments suggest, KBK did not succeed in applying the abacus to the problem. She divided the line into halves, and then placed her mark relatively far from 613. On the next problem, she again alluded to the abacus, but again was unable to apply it to her written representation. However, on the next problem, which involved locating 939, her analogy to the decimal-based abacus began to influence her written representations. She first drew two lines, one fairly close to the 0 end of the number line and one fairly close to the 1,000 end, and said the one near the origin stood for 100 and the other for 900. She used the mark near 900 as her estimate of 939. After three problems on which she either produced no written representation or just divided the line at the midpoint, she returned to partial application of the decimal notation. Asked to locate 772, she made marks for 800 and 900 and then made her "X" a little to the left of the mark that she said stood for 800. Her protocol reflects a deeper understanding of the relevance of the abacus to the number line task.

> So that makes sense . . . this would be 8 abacuses, this would be 9 abacuses, and this would be 10 abacuses (pointing to 1,000.) So this would be 8 abacuses, maybe and that (makes mark slightly to the left of the 800 mark) would be pretty close. So really what I have, Randi, is I have 7 abacuses, then I have 3 quarters of the eight abacus, almost but not quite.

After this trial, KBK used written representations on all remaining trials, relying on trials with the base 10 notation on 92% of them. Her performance also became far more accurate. More than 95% of her subsequent estimates were within 10% of the correct location, a number that compared favorably to that of control participants (undergraduates at Carnegie Mellon).

Part of the reason for her superior accuracy was that she continued to elaborate her written representations to encompass the base 10 system in increasingly detailed ways. She began to mark all 10 decimal intervals even when not all were needed to locate her target. She also began labeling each hundreds mark and drawing circles to represent the corresponding abacus. This generation of unncessarily elaborate representations is reminiscent of Karmiloff-Smith's (1992) descriptions of metarepresentational development.

Especially impressive, KBK began generating hierarchical written representations. The first of these occurred when she was asked to locate 828. After she marked 800 and 900 on the number line, she made marks that divided the space between them into four quarters, using a newly-drawn picture of a dollar bill as a guide. She then noted, albeit with a lot of hesitations and repetitions, that the mark must be between 8 and 9 on the larger scale and that it must be near the first of the quarter marks that she had made between 8 and 9. Just before she answered she said,

> So, I have my pencil on 8 and I have my finger on 9, and I'm breaking this (the space between 8 and 9) down into, to moving it just a little, like I would a quarter, if this was a whole of something, and I would say that this would be the answer.

In other words, with the help of the pencil and her finger in delimiting the smaller space of numbers, KBK extended the logic that she had used with the larger number line to a hierarchical representation of both hundreds and quarters of a hundred. She repeated this strategy on the next problem, which was the last problem of the first day.

On Day 2, the first problems that KBK encountered where the hierarchical representation would be useful were XX problems. On the first three such problems, she only used a hierarchical representation once, and it was the simplest such representation, dividing the space between 40 and 50 in half to locate the number 45. On the next problem, however, which involved locating 54 on a 0–100 number line, she made another breakthrough. For the first time, she applied the decimal concept to both 10s and 1s units, as the following protocol indicates:

> E: OK, make a mark for the number 54.
>
> K: OK, I establish 50 and 60, 70, 80, 90, 100, 50, 60, 70, 80, 90, 100. OK, it's gonna go somewhere between 5 and 6, it's gonna go somewhere between 5 and 6, somewhere between 5 and 6, somewhere between 5 and 6. The 4 is confusing me, um, I know we want to be somewhere in here. A 100 is the whole, halfway is 50, 1, 2, 3, 4, (pointing to 90). That can't be right, I can't go 1, 2, 3, 4, because they don't represent, they represent 10, it's not representing 4, this is the, this is the 5, this is the 6, this is the 7, this is the 8, the 9, this is the 10. OK, so it can't be 54. Each one of these equal 10. Oh, alright, wait a minute, OK, wait, I think I got this, wait. 10, 20, 30, 40, 50, OK. 54 is greater than 50, lesser than 6, but this is this, so this isn't this, this is, this is this column. There are 10 of these in this, ok, I got it. 50, 51, 52, 53, 54, 55, 56, 57, 58, 59, 60, 1, 2, 3, 4, 54 would be about here, got it. I wanna go back and do the other ones (*said with great excitement*).

Consistent with her comment, KBK's subsequent performance was consistently accurate, regardless of the type of number line. She used hierarchical written representations on almost every problem involving two or more nonzero numbers. The hierarchical embedding reached its peak on two XXX problems near the end of Day 2. On these, she divided a 1,000s line into three interlocking levels of decimals corresponding to hundreds, tens, and ones. At the end of the session, she also provided an eloquent summary of what she had learned:

> E: If there was any one thing that you were going to say that you understand after doing these problems that you didn't understand before, what would that be?
>
> K: It would be that I can, I can visualize having 10 of those (*abacuses*). I now have an idea that if you said to me that you wanted 1,000 of something, I would have an idea of how much it meant. Truly not just a number, not a

rhyme, not mnemonically, phonetically; it has a value, that we're using something up.

These comments suggest that KBK had at least partially regained her understanding of the decimal system and of what numbers meant.

CHILDREN'S DISCOVERIES AND KBK'S REDISCOVERIES

Considering together the data on children's discoveries and KBK's rediscoveries reveals some interesting commonalities. One of these is that discoveries occurred on all kinds of problems. KBK, like the children, made discoveries on problems not discernibly different than the problems on which she did not make them. She often had solved a number of problems of the same type before making a discovery. For example, nothing obvious distinguished the third XX problem, 73, on which she produced her first written representation, from the immediately preceding ones, 24 and 92, on which she did not. Similarly, there was nothing obvious that led to her first hierarchical representation appearing on 828 rather than on any of the seven XXX problems that preceded it. Nor did her discoveries typically follow failure. If anything, she typically seemed to be doing a little better than usual before most of her discoveries. Thus, in this brain-damaged patient's rediscoveries as in children's discoveries, generation of new approaches does not require an impasse.

Another commonality involved inarticulateness on the trials when discoveries were made. The frequent repetitions, comments to oneself, and odd references in KBK's protocols closely resembled those evident in many children's protocols on discovery trials. Graham and Perry (1993) noted similar phenomena in children performing other tasks. Both the children and KBK also produced unusually long solution times on these discovery trials.

A third commonality was the strong positive emotion that accompanied KBK's rediscoveries and the discoveries of some of the children. This is not a universal feature of children's discoveries—some seem quite indifferent to using a new strategy—but it certainly occurs on a higher percentage of discovery trials than trials on which they use their typical procedures.

KBK's rediscoveries also resembled children's discoveries in the relatively slow generalization of new approaches to similar problems. For example, after KBK began to divide the number line into four quarters, she sometimes did that, sometimes just divided it into two halves, and sometimes produced no written representation. As the evidence accumulated that more elaborate representations produced better results, she began producing them more frequently.

Another commonality was that KBK's discoveries, like those of the children studied previously, were constrained by conceptual understanding. This could be seen particularly clearly in her efforts to place the number 54 on the number line. She generated a flawed approach—counting up four 10-units from the mark that represented 50—but rejected it as incorrect without any feedback or external prompting. Her behavior was an audible version of the kind of strategy evaluation process envisioned in SCADS.

A sixth and final commonality between KBK's performance and that of children concerned the adaptiveness of strategy choices. Like the children, KBK consistently chose among the strategies available to her in adaptive ways. This was particularly evident in her use of reference points that were close to the number to be placed on the number line (e.g., placing 73 near the third of the four quarters and 45 near the half-mark).

This case study is, of course, only a beginning. It needs to be supplemented by analyses of rediscoveries in other brain-damaged patients; computational models of how rediscoveries are generated; derivation of predictions from those models; and experiments that test the predictions. It also would be illuminating to track neural level changes and to determine whether they, too, indicate parallels between children's discoveries and brain-damaged patients' rediscoveries. Nonetheless, the commonalities between children's discoveries and KBK's rediscoveries that are already apparent suggest that examining change in both children and brain-damaged patients will be mutually enriching. The microgenetic method clearly is applicable to studying change processes in brain-damaged patients as well as in children. Many of the basic phenomena regarding discoveries also appear to be comparable. Most important, use of the method in both areas promises to produce a more encompassing— and more interesting—depiction of cognitive functioning, than we have at present, one that illuminates how change occurs as well as how thinking proceeds at a given time.

REFERENCES

Bjorklund, D. F., & Coyle, T. R. (1995). Utilization deficiencies in the development of memory strategies. In F. E. Weinert & W. Schneider (Eds.), *Research on memory development: State of the art and future directions*. Hillsdale, NJ: Lawrence Erlbaum Associates.

Case, R. (1992). *The mind's staircase: Exploring the conceptual underpinnings of children's thought and knowledge*. Hillsdale, NJ: Lawrence Erlbaum Associates.

Dempster, F. N. (1981). Memory span: Sources of individual and developmental differences. *Psychological Bulletin, 89*, 63–100.

Falkenheiner, B., Forbus, K. D., & Gentner, D. (1989). The structure-mapping engine: Algorithm and examples. *Artificial Intelligence, 41*, 1–63.

Finger, S., & Stein, D. G. (1982). *Brain damage and recovery: Research and clinical perspectives.* New York: Academic Press.

Finger, S., LeVere, T. E., Almli, C. R., & Stein, D. G. (Eds.). (1988a). *Brain injury and recovery: Theoretical and controversial issues.* New York: Plenum.

Geary, D. C., & Burlingham-Dubree, M. (1989). External validation of the strategy choice nodel for addition. *Journal of Experimental Child Psychology, 47,* 175–192.

Geary, D. C., Fan, L., & Bow-Thomas, C. C. (1992). Numerical cognition: Loci of ability differences comparing children from China and the United States. *Psychological Science, 3,* 180–185.

Geary, D. C., & Wiley, J. G. (1991). Cognitive addition: Strategy choice and speed-of-processing differences in young and elderly adults. *Psychology and Aging, 6,* 474–483.

Graham, T., & Perry, M. (1993). Indexing transitional knowledge. *Developmental Psychology, 29,* 779–788.

Greeno, T. G., Riley, M. S., & Gelman, R. (1984). Conceptual competence and children's counting. *Cognitive Psychology, 16,* 94–134.

Groen, G. J., & Resnick, L. B. (1977). Can preschool children invent addition algorithms? *Journal of Educational Psychology, 69,* 645–652.

Karmiloff-Smith, A. (1992). *Beyond modularity: A developmental perspective of cognitive science.* Cambridge, MA: MIT Press.

Kerkman, D. D., & Siegler, R. S. (1993). Individual differences and adaptive flexibility in lower-income children's strategy choices. *Learning and Individual Differences, 5,* 113–136.

Kolb, B. (1992). Mechanisms underlying recovery from cortical injury: Reflections on profess and directions for the future. In F. D. Rose & D. A. Johnson (Eds.), *Recovery from brain damage: Reflections and directions* (pp. 169–186). New York: Plenum.

Ladd, S. F. (1987). *Mental addition in children and adults using chronometric and interview paradigms.* Unpublished doctoral dissertation, University of Northern Colorado, Greeley.

LeFevre, J. A., Bisanz, J., Daley, K. E., Buffone, L., Greenham, S. L., & Sadesky, G. S. (1996). Multiple routes to solution of single-digit multiplication problems. *Journal of Experimental Psychology: General, 125,* 284–306.

LeFevre, J., Sadesky, G. S., & Bisanz, J. (1996). Selection of procedures in mental addition: Reassessing the problem-size effect in adults. *Journal of Experimental Psychology: Learning, Memory, and Cognition, 22,* 216–230.

Lemaire, P., & Siegler, R. S. (1995). Four aspects of strategic change: Contributions to children's learning of multiplication. *Journal of Experimental Psychology: General, 124,* 83–97.

Miller, P. H., & Seier, W. L. (1994). Strategy utilization deficiencies in children: When, where, and why. In H. W. Reese (Ed.), *Advances in child development and behavior* (Vol. 25, pp. 107–156). New York: Academic Press.

Newell, A. (1990). *Unified theories of cognition.* Cambridge, MA: Harvard University Press.

Ostad, S. A. (1998). Developmental differences in solving simple arithmetic word problems and simple number-fact problems: A comparison of mathematically normal and mathematically disabled children. *Mathematical Cognition, 4,* 1–19.

Resnick, L. B., & Neches, R. (1984). Factors affecting individual differences in learning ability. In R. J. Sternberg (Ed.), *Advances in the psychology of human intelligence* (pp. 275–323). Hillsdale, NJ: Lawrence Erlbaum Associates.

Rose, F. D., & Johnson, D. A. (1992). Research on recovery: Ends and means. In F. D. Rose & D. A. Johnson (Eds.), *Recovery from brain damage: Reflections and directions* (pp. 187–198). New York: Plenum.

Saxe, G. B., Guberman, S. R., & Gearhart, M. (1987). Social processes in early number development. *Monographs of the Society for Research in Child Development, 52*(2, Whole No. 216).

Secada, W. G., Fuson, K. C., & Hall, J. W. (1983). The transition from counting-all to counting-on in addition. *Journal for Research in Mathematics Education, 14,* 47–57.

Shrager, J., & Siegler, R. S. (1998). SCADS: A model of children's strategy choices and strategy discoveries. *Psychological Science, 9,* 405–410.

Siegler, R. S. (1987). The perils of averaging data over strategies: An example from children's addition. *Journal of Experimental Psychology: General, 116,* 250–264.

Siegler, R. S. (1988). Individual differences in strategy choices: Good students, not-so-good students, and perfectionists. *Child Development, 59,* 833–851.

Siegler, R. S. (1989). Hazards of mental chronometry: An example from children's subtraction. *Journal of Educational Psychology, 81,* 497–506.

Siegler, R. S. (1995). How does change occur: A microgenetic study of number conservation. *Cognitive Psychology, 28,* 225–273.

Siegler, R. S. (1996). *Emerging minds: The process of change in children's thinking.* New York: Oxford University Press. Translated into French.

Siegler, R. S., & Crowley, K. (1991). The microgenetic method: A direct means for studying cognitive development. *American Psychologist, 46,* 606–620.

Siegler, R. S., & Crowley, K. (1994). Constraints on learning in non-privileged domains. *Cognitive Psychology, 27,* 194–227.

Siegler, R. S., & Engle, R. A. (1994). Studying change in developmental and neuropsychological contexts. *Current Psychology of Cognition, 13,* 321–350.

Siegler, R. S., & Jenkins, E. A. (1989). *How children discover new strategies.* Hillsdale, NJ: Lawrence Erlbaum Associates.

Siegler, R. S., & Lemaire, P. (1997). Older and younger adults' strategy choices in multiplication: Testing predictions of ASCM via the choice/no-choice method. *Journal of Experimental Psychology: General, 126,* 71–92.

Siegler, R. S., & Shipley, C. (1995). Variation, selection, and cognitive change. In T. Simon & G. Halford (Eds.), *Developing cognitive competence: New approaches to process modeling.* Hillsdale, NJ: Lawrence Erlbaum Associates.

Siegler, R. S., & Shrager, J. (1984). Strategy choices in addition and subtraction: How do children know what to do? In C. Sophian (Ed.), *The origins of cognitive skills* (pp. 229–293). Hillsdale, NJ: Lawrence Erlbaum Associates.

Siegler, R. S. & Stern, E. (1998.) A microgenetic analysis of conscious and unconscious strategy discoveries. *Journal of Experimental Psychology: General, 127,* 377–397.

Staszewski, J. J. (1988). Skilled memory and expert mental calculation. In M. T. H. Chi, R. Glaser, & M. J. Farr (Eds.), *The nature of expertise* (pp. 71–128). Hillsdale, NJ: Lawrence Erlbaum Associates.

Stein, D. G., & Glasier, M. M. (1992). An overview of developments in research on recovery from brain injury. In F. D. Rose & D. A. Johnson (Eds.), *Recovery from brain damage: Reflections and directions* (pp. 1–22). New York: Plenum.

Svenson, O. (1985). Memory retrieval of answers of simple additions as reflected in response latencies. *Acta Psychologica, 59,* 285–304.

Thorndike, E. L. (1898). Animal intelligence: An experimental study of the associative processes in animals. *Psychological Review, Monograph Supplement, 2,* No. 8.

Van Lehn, K. (1988). Towards a theory of impasse-driven learning. In H. Mandl & A. Lesgold (Eds.), *Learning issues for intelligent tutoring systems* (pp. 19–41). New York: Springer-Verlag.

Will, R. L., & Kelche, C. (1992). Towards a model of cognitive rehabilitation. In R. L. Wood & I. Fussey (Eds.), *Cognitive rehabilitation in perspective* (pp. 3–25). London: Taylor & Francis.

NEUROSCIENCE APPROACHES

Chapter 3

Cortical Plasticity Contributing to Child Development

Michael M. Merzenich
University of California, San Francisco

How does the brain change its representations of inputs and actions as we learn? What is the nature of the processes that control the progressive elaboration of performance abilities? What are the sources of variance for emergent performance abilities in different individuals? The answers to these fundamental questions are the subject of this review.

My colleagues and I have taken a direct approach for defining changes induced by learning, by documenting in detail the patterns of distributed neuronal response representations of specific inputs and actions prior to, during, and after the learning of a skill. Most studies have been conducted in the primary auditory, somatosensory, and motor cortices of young rats and monkeys. Many parallel electrophysiological and behavioral studies have been conducted in humans. The phenomenology recorded in such studies bears powerful inferences for the functional self-organizing processes that operate through the course of child development—indeed, throughout life.

Aspects of cortical plasticity processes contributing to learning demonstrated in these (and in many related) studies include:

The Distributed Cortical Representations of Inputs and Brain Actions "Specialize" in Their Representations of Behaviorally Important Inputs and Actions in Skill Learning

In all perceptual, cognitive, and motor skill learning tasks, the selective responses of cortical neurons specialize to meet the specific demands of the

67

task (see Merzenich, Allard, & Jenkins, 1991; Merzenich & deCharms, 1996; Merzenich & Jenkins, 1993; Merzenich, Tallal, Peterson, Miller, & Jenkins, 1998, for reviews). If an animal is trained to make progressively finer distinctions about specific sound stimuli, for example, then cortical neurons come to represent those stimuli in a progressively more specific and progressively amplified manner. In a learning phase of plasticity:

1. Cortical neuron populations that are directly excited by these behaviorally important stimuli grow progressively in number.
2. Growing neuronal populations respond with progressively greater specificity to the spectral (spatial) and temporal dimensions of the behaviorally important stimuli that are processed in the skill learning.
3. The growing numbers of selectively responding neurons discharge with progressively stronger temporal coordination (distributed synchronicity).

Thus, skill learning drives multidimensional changes in cortical responses that provide, through the course of progressive learning, a more refined basis for processing stimuli and for generating the actions that are critical to the task at hand. Not surprisingly, specific aspects of these distributed neuronal response changes are highly correlated with learning-based improvements in perceptual, motor, or cognitive capabilities (e.g., Nudo, Milliken, Jenkins & Merzenich, 1995; Recanzone, Merzenich, & Schreiner, 1992; Recanzone, Schreiner, & Merzenich, 1993; Xerri, Merzenich, Jenkins, & Santucci, 1999).

It should be emphasized that in these processes, the brain is not simply changing to record and store content. It is not merely a plastic machine that is filling its dictionaries and constructing its address systems to facilitate its complex associations and operations. By adjusting its spectral/spatial and temporal filters, *the cerebral cortex is actually selectively refining its processing capacities to fit each task at hand*—and *in toto*, establishes its own general processing capabilities. At the same time, this "learning how to learn" determines the fidelity and facility with which specific classes of information *can* be recorded, associated, and manipulated.

Across the period of child development, these powerful self-shaping processes of the forebrain machinery are in operation on a massive scale, as the brain is refining and developing, through progressive learning, its great array of hierarchically organized perceptual, cognitive, motor, and executive skills, and as it is receiving, sorting, and storing massive quantities of externally and internally generated information.

Selection of Behaviorally Important Inputs Is a Product of an Input–Coincidence-Based Connection- (Synapse-) Strengthening Process

Experimental results from many lines of study have collectively demonstrated that cortical representational changes recorded in plasticity experiments involve processes of coincident input coselection. *Inputs that fire together are strengthened together.* In skill learning, this simple principle of *concurrent input coselection* results inexorably with repetitive practice in: (1) a progressive amplification of cell numbers engaged by repetitive inputs (e.g., Jenkins, Merzenich, Ochs, Allard, & Guic-Robles, 1990; Recanzone, Merzenich, & Schreiner, 1992; Recanzone, Schreiner, & Merzenich, 1993; Xerri, Merzenich, Jenkins, & Santuccci, 1999); (2) an increase in the temporal coordination (near-simultaneity) of distributed neuronal discharges evoked by successive events marking features of behaviorally important inputs, which is a consequence of a progressive increase in positive coupling between nearly simultaneously engaged neurons within cortical networks, (e.g., Recanzone, Merzenich, & Schreiner, 1992; Wang, Merzenich, Sameshima, & Jenkins, 1995); and in (3) a progressively more specific "selection" of all of those input features that collectively represent behavioral important inputs, expressed moment by moment in time (Jenkins, Merzenich, Ochs, Allard, & Guic-Robles, 1990; Wang, Merzenich, Sameshima, & Jenkins, 1995). Note that these mechanisms also underlie the linking of representations of consistently immediately successive inputs in time, because temporal neighbors are mapped in networks to adjacent spatial network locations (see Allard, Clark, Jenkins, & Merzenich, 1991; Clark, Allard, Jenkins, & Merzenich, 1988; Merzenich, Recanzone, & Jenkins, 1991). A Hebbian change principal (see Hebb, 1949) achieving this temporal-to-spatial representational transformation is the basis of the functional creation of the many detailed cortical maps that occupy large sectors of the cortical mantle.

Note that the majority of excitatory and inhibitory inputs impinging on the principal excitatory neuronal population in the cortex, the cortical pyramidal neurons, come from neighboring local-network neurons. Hebbian plasticity applies to these interconnections within cortical network neuronal elements just as well as to their extrinsic inputs. Hebb understood that the operation of coincidence-based synaptic plasticity in cortical networks would result in the formation, strengthening and continuous recruitment of neurons within neuronal assemblies that would together (cooperatively) represent behaviorally important stimuli. Indeed, in skill learning, cortical plasticity results in both an increased strengthening and selectivity of behaviorally important extrinsic inputs *and* in the progressive strengthening of intrinsic interconnections of the neuronal populations of cortical cells assemblies that cooperatively represent those inputs.

Plasticity Is Constrained by Anatomical Sources and Convergent-Divergent Spreads of Inputs

Every cortical field has specific extrinsic and intrinsic input sources, and specific dimensions of anatomical divergence and convergence of its inputs, that limit dynamic combinatorial Hebbian input coselection capacities (Merzenich, 1990; Merzenich & Sameshima, 1993). Those anatomical input sources and limited projection overlaps both: enable change by establishing input-selection repertoires; and determine the limits for change. At lower system levels, because of relatively strict anatomical constraints, only spatially (spectrally) limited input–coincidence-based combinatorial outcomes are possible. By contrast, at the top of system hierarchies, anatomical projection topographies are nearly all-to-all, and correspondingly more powerful combinatorial capacities are revealed by neurons and neuronal assemblies that respond to complex combinations of features of real-world objects, events, and actions.

Plasticity Is Also Constrained by the Time Constants Governing Coincident Input Coselection, and by the Time Structures and Potentially Achievable Coherences of Extrinsic and Intrinsic Cortical Input Sources

Temporally coordinated inputs are prerequisite for effectively driving representational changes by the operations of coincident input-dependent Hebbian mechanisms, given the short durations (milliseconds to tens of milliseconds) of the time constants that govern synaptic plasticity in the adaptive cortical machinery (see Buonomano & Merzenich, 1998, for review). In fact, consistently noncorrelated or low-discharge-rate inputs induce *negative* changes in synaptic effectivenesses. These negative change effects also contribute importantly to the learning-driven selection of behaviorally important inputs.

Different Representational Structures Are Created Within Different Cortical Areas Because of Cortical Field-Specific Differences in Input Sources and Distributions, and in the Time Structures of Inputs

Consider the differences in the statistics of activity from afferent inputs from the retina, skin, or cochlea generated in a relatively strictly topographically wired VI (area 17), SI proper (area 3b), or A1 (area 43); in contradistinction to the statistics of afferent barrages from the highly diffused inputs feeding "higher" inferotemporal visual or insular somatosensory or dorsotemporal auditory or prefrontal cortical fields. In the former cases, very heavy schedules of repetitive, temporally coherent inputs are delivered from powerful, redundant projections from relatively strictly topographically organized thalamic nuclei and lower-level, associated cortical areas. In inferotemporal, insular, dorsotemporal, and prefrontal cortices, afferent input projections from any given source are greatly dispersed; highly repetitive inputs are uncommon; inputs from multiple, diffuse cortical sources are the rule; and, far more varied and complex input combinations are in play. These differences in input schedules, spreads, and combinations presumably largely account for the dramatic differences in the patterns of representation of behaviorally important stimuli at "lower" versus "higher"' levels (Merzenich & Sameshima, 1993). At lower levels, cortical areas are occupied by topographically ordered representations of the retina, cochlea, or skin, and in learning, large, continuous, strongly interconnected neuronal ensembles emerge to represent behaviorally important stimuli. At higher system levels, while neighboring neurons can share some response properties and can be more strongly coupled than are more distant neurons, representations of behaviorally important stimuli are sparse. That is, neurons or clusters of neurons that respond selectively to a learned input are distributed widely across cortical areas, and generally share much less information with neighboring neurons or cooperative neuron clusters.

One of the most remarkable aspects of the cortical processing machinery is this striking difference in system-level functional organization, given the basic uniformity in the numbers and distributions of neuron types all across the neocortex. Originating from a common developmental germinal source, the remarkably uniform cortex in the newborn is progressively differentiated to accomplish very different specific operational tasks *by its inputs*. Field by field, in a remarkable serial progression that occupies the many years of human childhood, cortical fields progressively mature as they organize functionally to progressively master more—and more elaborated and differentiated—aspects of perceptual, cognitive, motoric, and executive skills.

Representational Specialization Also Applies for the Temporal Dimensions of Behaviorally Important Inputs

In learning, the cortex refines its representations of the temporal aspects of behaviorally important inputs in at least four ways.

First, as noted earlier, the cortex generates more synchronous representations of successive input perturbations or events, thereby more strongly recording their identities and marking their occurrences (for examples, see Kilgard & Merzenich, 1999; Recanzone, Merzenich, & Schreiner, 1992; Wang, Merzenich, Beitel, & Schreiner, 1995; Wang, Merzenich, Sameshima, & Jenkins, 1995). *This learning-induced increase in distributed response coordination is a primary effect of learning.* It appears to be primarily achieved through increases in positive coupling strengths between interconnected neurons participating in stimulus- or action-specific neuronal cell assemblies. This change in distributed response coordination has four important consequences (Merzenich, Tallal, Peterson, Miller, & Jenkins, 1998; Nagarajan et al., 1999; Wang, Merzenich, Beitel, & Schreiner, 1995).

1. *It increases representational salience.* That is because downstream neurons are excited as a direct function of the degree of temporal synchronization of *their* inputs.

2. *It increases the power of the outputs of a cortical area to drive downstream plasticity.* Hebbian plasticity mechanisms operating within downstream cortical (or other) targets also have relatively short time constants. The greater the syncronicity of inputs, the more powerfully those change mechanisms are engaged.

3. *It confers immunity to noise.* By this simple information abstraction/ coding strategy, the distributed neuronal representation of the signal (a temporally coordinated, distributed neuronal response pattern representing the input or action) is converted at the entry levels in the cortex into a form that is not as easily degraded or altered by noise.

4. *It confers robustness of complex signal representation for spatially or spectrally incomplete or degraded inputs.* In the auditory domain, for example, this information abstraction quantitatively explains how the distributed neuronal response representations of speech substantially survive spectral degradation or spectral simplification or reduction up to the distortion levels at which signal intelligibility actually breaks down in human perception. It further explains how complex information delivered in strikingly different spectral forms (e.g., whispered vs. phonated speech) can be equally and interchangeably intelligible. In such examples, the distributed abstracted patterns of temporally coordinated neuronal discharge are very similar for normal, spectrally incomplete, and substantially noisy inputs.

Second, the cortex can select specific inputs through learning to exaggerate the representation of specific input time-structures. Conditioning a monkey or a rat with stimuli that have a consistent, specific temporal modulation rate or interstimulus time, for example, results in a selective exaggeration of the responses of neurons at that rate or time separation. In effect, the cortex specializes for expected relatively higher-speed or relatively lower-speed signal event reception.

Both electrophysiological recording studies and theoretical studies present a compelling argument that cortical networks richly encode temporal interval as a simple consequence of cortical network dynamics (e.g., Buonomano & Merzenich, 1995; Buonomano, Hickmott, & Merzenich, 1997). It is hypothesized that the cortex accomplishes time-interval and duration selectivity in learning by positively changing synaptic connection strengths for input circuits that can respond with recovery times and circuit delays that match behavioral important modulation frequency periods, intervals, or durations.

Third, the cortex links representations of immediately successive inputs that are presented in a learning context. As a result of Hebbian plasticity, it establishes overlapping and neighborhood relationships between immediately successive parts of rapidly changing inputs, irresistibly "mapping" input continuum dimensions (Merzenich, Grajski, Jenkins, Recanzone, & Peterson, 1991; Merzenich, Recanzone, & Jenkins, 1991).

Fourth, the cortex generates stimulus sequence-specific (combination-sensitive) responses, by which neuronal responses are selectively modulated by the prior application of stimuli in the learned sequence of temporally separated events. We and others have recorded the emergence of such associative or combination-sensitive responses, and have recorded evidence of strengthened interconnections between cortical cell assemblies representing successive event elements separated by hundreds of milliseconds to seconds in time (Naya, Sakai, & Miyashita, 1996; Sakai & Miyashita, 1991). The mechanisms of origin of these effects have not yet been established.

The Integration Time (Processing Time) in the Cortex Is Itself Subject to Powerful Learning-Based Plasticity

When cortical networks are engaged by strong input perturbations, both excitatory and inhibitory neurons are excited. Within a given processing channel, cortical pyramidal cells cannot be effectively reexcited by a following perturbation for tens to hundreds of milliseconds. These *integration times* are primarily dictated by the time for recovery from inhibition, which ordinarily dominates poststimulus excitability. This integration time, "processing time," or "recovery time" is commonly measured by deriving a modulation transfer function, which defines the ability of cortical neurons to respond to identical successive stimuli within cortical processing channels.

In primary auditory receiving areas, for example, these integration times normally range from about 15 ms to about 200 ms (Bieser, Müller-Preuss, 1996; Eggermont, 1994; Schreiner & Urbas, 1988). Progressively longer processing times are recorded at higher system levels. In the auditory cortex, they are approximately syllable-length (200–500 ms in duration) for the belt cortex surrounding the primary auditory cortex (Kaas, Hackett, & Tramo, 1999), and roughly 1 s in duration for dorsolateral temporal cortex (Wernicke's area or its presumed homologs).

These time constants govern—and limit—the cortex's ability to chunk (i.e., to separately represent by distributed, coordinated discharge) successive events within its processing channels. Both neurophysiological studies in animals and behavioral training studies in human adults and children have shown that *the time constants governing event-by-event complex signal representation are highly plastic.* With intensive training in the right form, cortical processing times reflected by the ability to accurately and separately process events occurring at different input rates can be dramatically shortened, or lengthened (e.g., Ahissar & Hochstein, 1993; Karni & Sagi, 1991; Kilgard & Merzenich, 1999; Merzenich, Jenkins, Johnston, Schreiner, Miller, & Tallal, 1996).

Plasticity Processes Are Competitive

If two spatially or spectrally different inputs are consistently delivered non-simultaneously to the cortex, cortical networks generate input-selective cell assemblies for each input, and actively segregate them from one another (Allard, Clark, Jenkins, & Merzenich, 1991; Grajski & Merzenich, 1990a, 1990b; Kilgard & Merzenich, 1998; Somers, Todorov, Siapas, Toth, Kim, & Sur, 1998; Wang, Merzenich, Sameshima, & Jenkins, 1995). Boundaries between such inputs grow to be sharp and are substantially intensity-independent. Computational models of Hebbian network behaviors indicate that this sharp segregation of nonidentical, temporally separated inputs is accomplished as a result of a wider distribution of inhibitory versus excitatory responses in the emerging, competing cortical cell assemblies that represent them.

This Hebbian network cell assembly formation and competition appears to account for how the cortex creates sharply sorted representations of the fingers in primary somatosensory cortex (e.g., see Allard, Clark, Jenkins, & Merzenich, 1991; Merzenich & Jenkins, 1993b), and almost certainly accounts for how the cortex creates sharply sorted representations of native aural-language-specific phonemes in lower-level auditory cortical areas in the auditory/speech processing system humans (see McClelland, this volume).

Consider this competitive process from the perspective of a competitive input source. If inputs are delivered in a constant and stereotyped way from a limited region of the skin or cochlea in a learning context, that skin surface or cochlear sector is an evident competitive winner (Recanzone, Merzenich, & Dinse, 1992; Recanzone, Merzenich, & Schreiner, 1992). By

Hebbian plasticy, the cortical networks will select that specific combination of inputs, and represent it within a competitively growing Hebbian cell assembly. The competitive strength of that cooperative cell assembly will grow progressively because more and more neurons are excited by behaviorally important stimuli with increasingly coordinated discharges. That means that neurons outside of this cooperative group have greater numbers of more coordinated outputs contributing to their later competitive recruitment.

Now, consider the normal case of the competitive landscape in a cortex engaged by larger numbers of competitive inputs (see McClelland, this volume). Through progressive functional remodelling, the cortex clusters and competitively sorts information across sharp boundaries dictated by the spectrotemporal statistics of its inputs. It is obviously highly flexible in what can be achieved by that sorting. On the one hand, if it receives information on a very heavy schedule that sets up competition for a limited input set—for example, the vowels of one's native language—at the appropriate system level, it will sort competitive inputs into a correspondingly small number of largely discontinuous response regions (see McClelland, this volume; Kuhl, 1991, 1994). If the stimulus set is larger—for example, if the native language contains a larger number of vowels—then the cortex segregates inputs into this correspondingly larger number of neuronal assemblies.

In humans, speech inputs are the predominant source of aural inputs that shape the auditory cortical processing machinery. By the time a child is 9 months old, it has been in the presence of more than 4 million words in its native language, more than a half-million of which have been directed to it. This massive experiential exposure results in progressive competitive input selection mechanisms that set the spectral and temporal filters at multiple system levels to specialize with high specificity in the representation of that native language.

Competitive outcomes are, again, cortical-level dependent. Moreover, the cortex links events that occur in different competitive groups if they are consistently excited synchronously in time. To cite a simple example, neurons within the primary auditory cortex respond to inputs that are usually spectrally limited to less than a critical band in hearing. In speech representation terms, they are largely limited in the spectral domain to the representations of individual speech formants, and in the temporal domain, to the responses of inputs separated by about 20 ms to 100 ms in interevent times.

When the primary auditory cortex of a monkey is excited by a speech-like stimulus (e.g., a species-specific vocalization), each competitive column responds selectively to only a relatively limited part of the complex stimulus (Wang, Merzenich, Beitel, & Schreiner, 1995). At the same time, competitively formed groups of neurons come to be synchronously linked in their representations of different parts of the complex stimulus, and *collectively* represent successive complex features of the vocalization through the coordinated activities of *many* groups.

Neurons within the two levels of the cortex surrounding A1 have greater spectral input convergences and longer integration times that enable their facile combination of information representing different spectrotemporal details, that is, have the combinational capacity to represent different vowels and consonants. Their information extraction is greatly facilitated by the learning-based linkages of cooperative groups that deliver behaviorally important inputs in a highly salient, *temporally coordinated* form to these fields. With their progressively greater space and time constants, still higher-level areas organize competitive cell assemblies that represent still more complex spectral and serial-event combinations.

Note that these organizational changes apply over a large cortical scale. In skill learning over a limited period of training, participant neuronal members of such assemblies can easily be increased by many hundredfold, even within a primary sensory area like S1, area 3b, or A1 (e.g., Byl, Merzenich, & Jenkins, 1996; Kilgard & Merzenich, 1998; Recanzone, Merzenich, Jenkins, Grajski, & Dinse, 1992; Recanzone, Merzenich, & Schreiner, 1992; Wang, Merzenich, Sameshima, & Jenkins, 1995). In extensive training in complex signal recognition, more than 10% of neurons within temporal cortical areas can come to respond highly selectively to a specific, normally rare, complex training stimulus (Kobatake, Wang, & Tanaka, 1998). The distributed cell assemblies representing those specific complex inputs involve tens or hundreds of millions of neurons, and are achieved by enduring effectiveness changes in many billions of synapses.

Learning Is Modulated as a Function of Behavioral State

At lower levels of cortex, changes are generated only in attended behaviors (Ahissar, Vaddia, Ahissar, Bergman, Arieli, & Abeles, 1992; Jenkins, Merzenich, Ochs, Allard, & Guic-Robles, 1991; Recanzone, Merzenich, & Jenkins, 1992; Recanzone, Merzenich, Jenkins, Grajski, & Dinse, 1992; Recanzone, Schreiner, & Merzenich, 1993; Weinberger, 1993). Trial-by-trial change magnitudes are a function of the importance of the input to the animal as signaled by (a) the level of attention; (b) the cognitive values of behavioral rewards or punishments; and (c) internal judgments of practice trial precision or error, based on the relative success or failure of achieving a target goal or expectation. Note that little or no enduring change is induced when a well-learned automatic behavior is performed from memory without attention.

It is also interesting to note that at some higher levels of cortex, activity changes can be induced even in nonattending subjects under conditions in which priming effects of nonattended reception of information can be demonstrated.

The modulation of progressive learning is achieved by the activation of powerful reward systems releasing the neurotransmitters noradrenaline (NA), norepinephrine (NE), and dopamine (DA) (among others) through widespread projections to the cerebral cortex. Acetylcholine plays a particularly important role in modulating learning-induced change in the cortex (see Kilgard & Merzenich, 1998, 1999; Weinberger, 1993). In studies conducted in nonbehaving animals, chronic trial-by-trial pairing of acoustic stimuli with stimulation of the nucleus basalis (the nucleus containing the ACh neurons that enable plasticity in the cortex) results in plasticity-induced changes that closely parallel those induced by intense, closely attended, highly rewarded skill learning (Kilgard & Merzenich, 1998, 1999).

Note that the cortex is a *learning machine*, in the sense that during the learning of a new skill, NA, NE, DA, et al. are all released trial-by-trial with application of the behaviorally important stimulus and/or behavioral rewards. If the skill can be mastered and thereafter replayed from memory, its performance can be generated without attention. That results in a profound attenuation of the modulation signals from these neurotransmitter sources; plasticity is no longer positively enabled in cortical networks. As first understood by William James (1890) more than a century ago, through the course of child development, the cortex is acquiring a progressively broader and deeper repertoire of automatically performable skills that can be brought into play from memory. In this way, the cortical learning machinery can be continuously dedicated to the development of cutting edge learning, that is, to a continuous progression of behavioral differentiation and elaboration.

Top–Down Influences Constrain Cortical Representational Plasticity

Attentional control flexibly defines an enabling window for change in learning (e.g., Ahissar & Hochstein, 1993). Moreover, progressive learning generates progressively more strongly represented goals and expectations (e.g., see Chelazzi, Duncan, Miller, & Desimone, 1998; Kaster, Pinsk, deWeerd, Desimone, & Ungerleider, 1999), which feed back (1) all across representational systems that are undergoing change, and (2) to modulatory control systems weighing performance success and error. Strong intermodal behavioral and representational effects have also been recorded in experiments that might be interpreted as shaping expectations in monkeys (e.g., Haenny, Maunsell, & Schiller, 1988; Hsiao, O'Shaughnessy, & Johnson, 1993)—and almost certainly, by analogy, in a human subject that employs (for example) auditory, visual, and somesthetic information to create integrated phonological representations, or to create the movement trajectory patterns that underlie precise hand control or vocal production.

The Scale of Plasticity in Progressive Skill Learning Is Massive

Cortical representational plasticity must be viewed as arising from multiple-level systems that are broadly engaged in learning, perceiving, remembering, thinking, and acting. Any behaviorally important input (or consistent internally generated activity) engages many cortical areas, and with repetitive training, *drives all of them to change* (see Merzenich & deCharms, 1996; Merzenich & Jenkins, 1993a; Merzenich & Sameshima, 1993, for review). Different aspects of any acquired skill are contributed from field-specific changes in the multiple cortical areas that are remodeled in its learning.

It should be noted that in this kind of continuously evolving representational machine, perceptual constancy cannot be accounted for by locationally constant brain representations: *Relational* representational principals must be invoked to account for it (Merzenich & deCharms, 1996; also see Phillips & Singer, 1997). Moreover, representational changes must obviously be coordinated level-to-level.

It should also be understood that plastic changes are also induced extracortically. Although we believe that learning at the cortical level is usually predominant, plasticity induced by learning within many extracortical structures significantly contributes to learning-induced changes that are expressed within the cortex.

Enduring Cortical Plasticity Changes Appear to Be Accounted for by Local Changes in Neuropil Anatomy

Dramatic changes in synapse turnover, synapse number, synaptic active zones, dendritic spines, and the elaboration of terminal dendrites have all been demonstrated to occur in a behaviorally engaged cortical zone, and in several models, in direct parallel with changes in synaptic effectivenesses (e.g., Engert & Bonhoeffer, 1999; Geinisman, deToledo-Morrell, Morrell, Heller, Rossi, & Parshall, 1993; Greenough & Chang, 1988; Keller, Arissian, & Asanuma, 1992; Kleim, Swain, Czerlanis, Kelly, Pipitone, & Greenough, 1997). Through a myriad of changes in local structural detail, the learning brain of a child is continuously *physically* remodeling its processing machinery across the course of child development.

CORTICAL PLASTICITY PROCESSES IN CHILD DEVELOPMENT

Progressive, Multiple-Stage Skill Learning

There are two remarkable achievements of brain plasticity in child development. The first is the progressive shaping of the processing machinery to handle the accurate, high-speed reception of the rapidly changing streams of information that flow into the brain. In the cerebral cortex, that shaping almost certainly begins most powerfully within the primary receiving areas of the cortex, because with their early myelination, these main gateways for information into the cortex are receiving strongly coherent inputs from subcortical nuclei, and they can quickly organize their local networks on the basis of coincident input coselection (Hebbian) plasticity mechanisms. The self-organization of the cortical processing machinery spreads outward from these primary receiving areas over time, to ultimately refine the basic processing machinery of all of the cortex.

The second great achievement, which is strongly dependent on the first, is the efficient storage of massive content compendia, in richly associated forms.

The brain accomplishes its functional self-organization through a long parallel series of small steps that cumulatively comprise child development. At each step, the brain masters a series of elementary processing skills, and establishes reliable information repertoires that enable the accomplishment of subsequent skills. Those second- and higher-order skills can be viewed as both elaborations of more basic, mastered skills *and* the creation of new skills dependent on second- and higher-order combinatorial processing. That hierarchical processing is enabled by greater cortical anatomical spreads, by more complex convergent anatomical sources of inputs, and by longer integration (processing, recovery) times at progressively higher cortical system levels, which allow for progressively more complex combinations of information integrated over progressively longer time epochs as one ascends across cortical processing hierarchies.

As this machinery functionally evolves and consequently physically matures through childhood developmental stages, information repertoires are represented in progressively more salient forms, that is, with more powerful distributed response coordination. That growing salience directly controls the power of emerging information repertoires for driving the next level of elaborative and combinatorial changes. We hypothesize that it also directly enables the maturation of the myelination of projection tracts that deliver outputs from functionally refined cortical areas. More mature myelination of output projections also greatly contributes to the power of this newly organized activity to drive strong, downstream plastic changes through the operation of Hebbian plasticity processes.

As each elaboration of skill is practiced, in a learning phase, neuro-modulatory transmitters enable change in the cortical machinery and the cortex functionally and physically adapts to generate the neurological representations of the skill in progressively more selective, predictable, and statistically reliable forms. Ultimately, the performance of the skill concurs with the brain's own accumulated, learning-derived expectations. The skill can now be performed from memory, without attention. With this consolidation of the remembered skill and information repertoire, the modulatory nuclei enable no further change in the cortical machinery. *The learning machine—the cerebral cortex—moves on to the next elaboration.* In this way, the cortex constructs highly specialized processing machinery that can progressively produce: (a) great towers of automatically performable behaviors; (b) great progressively maturing hierarchies of information processing machinery that can achieve progressively more powerful complex signal representations, retrievals, and associations; and (c) with this machinery in a mature and thereby efficiently operating form, a remarkable capacity for reception, storage, and analysis of diverse and complex associated information.

The flexible, self-adjusting capacity for refinement of the processing capabilities of this remarkable machine confers the ability in this later epoch in the history of our species to represent complex language structures, to develop high-speed reading abilities, to develop a remarkably wide variety of complex modern-era motor abilities, to develop the abstract logic structures of the mathematician or software engineer or philosopher—to create elaborate, ideosyncratic, experience-based behavioral abilities in all of us.

HOW ARE LEARNING SEQUENCES CONTROLLED? WHAT CONSTRAINS LEARNING PROGRESSIONS?

Perhaps the most important basis of control of learning progressions is *representational consolidation.* By directly specializing its processing machinery, the trained cortex creates progressively more specific and more salient distributed representations of behaviorally important inputs. Growing representational salience increases the power of a cortical area to effectively drive change wherever outputs from this evolving cortical processing machinery are distributed—for example, to higher system levels—because distributed and coordinated (synchronized) responses more powerfully drive downstream, Hebbian-based plasticity changes.

A second very powerful basis of sequencing learning is *progressive myelination.* At the time of birth, only the core primary extrinsic information entry zones (Al, SI, VI) in the cortex are heavily myelinated (Fleschig, 1920; Yakolev & Lecours, 1967). Across childhood, connections to and interconnections between cortical areas are progressively myelinated, proceeding from these core areas out to progressively higher system levels. Myelination in the posterior parietal, anterior, and inferior temporal and prefrontal cortical areas is not mature in the human forebrain until 8 to 20 years of age. Even in the mature state, it is far less developed at the highest processing levels.

Myelination controls the conduction times and therefore the temporal dispersions of input sources to and within cortical areas. With initially poor myelination at higher levels in the young brain, inputs are temporally diffuse. They cannot generate reliable representational constructs of an adult quality because they do not as effectively engage input–coincidence-based Hebbian plasticity mechanisms. That assures, in effect, that plasticity is not enabled for complex combinatorial processing until lower level input repertoires are consolidated, that is, until they evolve into stable, statistically reliable forms.

Myelination has usually been thought to be genetically programmed. However, we hypothesize that myelination in the central (but not the peripheral) nervous system is more likely controlled by emerging temporal response coherence, achieved through temporally coordinated signalling from the multiple branches of oligodendrocytes that terminate on different projection axons in central tracts and networks. It has been argued that central myelination is positively and negatively activity-dependent, and that distributed syncronization may contribute to positive change (see Demerens, Stankoff, Logak, Anglade, Alliquant, Courand, Zalc, & Lubetski, 1996). If the hypothesis that coherent activity controls myelination proves to be true, then emerging temporal correlation of distributed representations of behaviorally important stimuli generated level-by-level by changes in coupling in local cortical networks in the developing cortex of the child would directly drive changes in myelination for the outputs of that cortical area. That would, in turn, enable the generation of reliable and salient representational constructs at that higher level—just when, and only when, the combinatorial inputs fed to that higher level are in appropriately reliable and powerful plasticity-transforming forms. By this kind of progression, skill learning is hypothesized to directly control progressive functional and physical brain development through the course of child development, both by refining ("maturing") local interconnnections and response dynamics of information processing machinery at successive cortical levels, and by the coordinated refinement ("maturing") of the critical information transmission pathways that interconnect different processing levels.

Progressive Changes in Sleep Patterns.   Sleep patterns almost certainly also constrain plasticity, especially within the first year of life (Hopson, 1990). There is a large body of evidence that argues that sleep both enables the strengthening of learning-based plastic changes, and resets the learning machinery by erasing temporary nonreinforced and nonrewarded in-put-generated changes generated over the preceding waking period (e.g., see Buzsáki, 1998; Karni, 1995; Qin, McNaughton, Skaggs, & Barnes, 1997). The dramatic shift in the percentage of time spent in rapid eye movement sleep is consistent with a strong early bias toward noise removal in a very immature and poorly functionally unorganized brain in which most extrinsic and intrinsic inputs are effectively contributing to the din. Sleep patterns change dramatically in the older child, in parallel with a strong increase in its daily schedule of closely attended, rewarded, and goal-oriented behaviors.

Top–Down Modulation Controlling Attentional Windows and Learned Predictions (Expectations and Behavioral Goals) Must All Be Constructed by Learning. Their necessarily delayed development also contributes an im-portant constraint for the progression of early learning. In the very young brain, prediction and error-estimation processes must necessarily be weaker because stored higher-level information repertoires are ill-formed and statisticallly unreliable. As the brain matures, stored information pro-gressively more strongly and reliably enables top–down attentional and predictive controls, progressively provides a stronger basis for success- and error-signalling for modulatory control nuclei, and progressively enables top–down syntactic feedback to increase representational reliability for accu-rate reception of high-speed complex-signal input streams such as speech.

The Modulatory Control Systems That Enable Learning Are Also Plas-tic. Their maturation provides another constraint on the progression of learning. Modulatory control nuclei gate learning as a function of atten-tion, reward and punishment, accuracy of achievement of goals, and error feedback. Signalling in these subcortical nuclei is largely based on complex information fed back from the cortex itself. The salience and specificity of that feedback information grows over time, and its ability to provide accu-rate error judging or goal-achievement signalling must grow progressively. The nucleus basalis, nucleus accumbens, ventral tegmentum, and locus coerulus must undergo their own functional self-organization based on Hebbian plasticity principals to achieve their mature modulatory selectivity and power. Their progressive maturation provides another important con-straint on skill development progressions in child development.

IN WHAT SENSE ARE LEARNING PROGRESSIONS SUBJECT TO CRITICAL PERIODS OF DEVELOPMENT?

In its self-organization, the brain changes from a machine with incredible capacities to specialize in its representation of the specific inputs that en-gage it, into a highly specialized machine that is powerfully adapted to meet the specific experiential inputs that have historically challenged it. To cite a simple example, the spectral and temporal filters of the native English speaker are specialized at multiple processing levels by the statistics of English language inputs. By the time that a child first puts meaning to words, its brain has been exposed to many millions of English phonemes. By the time it enters school, its plastic brain has been modified by hundreds of millions of phonemes received in attended listening in tens of millions of words. The consequence: massive, language-specific machine specialization.

It should be noted that as this self-organizing machine stores its increasingly massive content, it progressively more powerfully employs top–down prediction to facilitate high-speed operations. Prediction strength (syntax) also grows continuously.

Obviously, the earlier this machine is exposed to the phonology, syntax, and grammar of a second language, the easier it is for these competitive plastic processes to represent them with acceptable levels of competitive interference. Growing specificity and salience for speech inputs and progressively more powerful syntactic predictions must necessitate proportionally greater learning, to develop equally complete and facile mastery of a second language in older children.

At the same time, it is important to recognize that the learning machine is still completely functionally operational throughout life. We believe that the critical period for language primarily reflects the progressively more powerful specialization of the cerebral cortex for the first language, and NOT any shutdown of the cortical plasticity machinery. This distinction is important, for example, for understanding the operational capacities for driving improvements in children whose abnormal language learning progressions have left them severely impaired in language and reading skills.

It is also important to understand that learning progressions for complex abilities are necessarily staged (see Neville, this volume). The development and performance of complex processing abilities are dependent on the genesis of reliable and salient inputs, which are most effectively created by staged functional and physical brain maturation.

ALTERNATIVE LEARNING PROGRESSIONS CONTRIBUTING TO A CLASS OF DEVELOPMENT IMPAIRMENTS: ORIGINS OF LANGUAGE LEARNING IMPAIRMENTS AND DYSLEXIA

To further understand how the progressive, hierarchical, functionally self-organizing processes in the cortex create complex neurobehavioral abilities, it is useful to consider how predictable alternative functional and physical brain development progressions in early childhood can result in variances in emergent behavioral capabilities. The neurological origins of specific language learning impairments and dyslexia represent a relatively well-understood case in point.

First, we outline the changes that underlie normal language development. Acoustic inputs strongly statistically dominated by aural speech inputs in humans rapidly drive selective refinements in infants in spectral and temporal processing of information as aural information streams into the primary auditory cortical field. At this principal gateway, cortical level, spectral, and temporal filters are refined, and separate representations of rapidly successive perturbations in the speech stream are generated through learning in increasingly more highly salient (distributed-synchronous-response) forms. That is achieved in large part because the cortical networks refine their cortical integration (chunking, processing, integration) times so that they can generate accurate representations of rapidly successive input perturbations within individual system-entry-level processing channels. The development of progressively more coherent response activity enables progressive myelination of the output tracts that convey information fed forward into the two levels of the auditory belt cortex flanking A1 on the superior temporal plane. That myelination, and the increasingly temporal coherent activities evolving through learning-based refinement of the A1 processing machinery, coupled with greater anatomical spreads of inputs and longer integration time constants in the belt cortex, altogether enable the emergence of sharply separated native-language-specific, combination-selective neuronal cell assemblies that represent the reliably separable sounds of the native language, its *phonemes*. The maturation of responses in the belt cortex enables the maturation of myelination of its forward projections to the crown of the superior temporal gyrus (Wernicke's area) where still more complex spectral and temporal-sequence combinations specific to the native language are generated. The progressive maturation (growing temporal coherence) of responses in Wernicke's area results in the delivery of progressively more coherent activity via the arcuate faciculus to frontal cortex speech areas, which enables the maturation of myelination of this great central projection tract.

In parallel with those changes, similar change progressions are occuring in the other great forebrain information processing systems. Information converges between these systems to a limited extent in Wernicke's area, but to a substantially greater extent in the frontal lobes by processes that employ these plastic correlational mechanisms to bring complex multimodal information sources into register to achieve an integrated representation of speech. This richly combinatorial information is processed within higher temporal and frontal cortical areas to create reliable longer-speech-string, semantic, syntactic and grammatical representational constructs. It enables speech production learning. It also generates an extensive array of remembered and associated constructs that enable powerful top–down biasing, for increasing the efficiencies of learning, for providing better error-feedback information to subcortical modulatory control nuclei, and for facilitating high-speed reception by implementing ongoing prediction (syntax).

A THOUGHT QUESTION: WHAT HAPPENS TO THE PROGRESSIVE DEVELOPMENT OF THIS COMPLEX, SERIALLY SELF-ORGANIZING SYSTEM IF IT DEVELOPS WITH CONSISTENTLY DEGRADED (MUFFLED, NOISY) AURAL SPEECH INPUTS?

The short answer is that it would develop limitations in processing that would result in language and reading impairments. The processing systems that emerge under these conditions shall be effectively delayed and degraded in their experience-driven functional and physical maturation. As a consequence, multiple deficits in language processing and usage and in higher-level, language-dependent operations (e.g., reading) must necessarily develop in a delayed and degraded manner.

What *Specific* Predictions Could Be Made About the Consequences of the Muffled or Noisy Inputs for the Functional Self-Organization of Speech and Language Representations?

1. Plasticity studies indicate that muffled or noisy inputs should result in the establishment of relatively long integration (processing, recovery) times at entry-levels of auditory signal processing. Processing times cannot be progressively shortened postnatally as in the normal case because the brain has to effectively integrate information for longer time epochs to make reliably rewardable decisions. Plasticity only occurs if the machine gets the answers right. With degraded inputs, it can only accomplish that with minimum requisite reliability with a relatively long epoch-chunking of the speech stream. Syllable-rate modulations in the speech stream are relatively powerful, and remain robust even in muffled and noisy speech. We have hypothesized that in the face of muffled or noisy inputs the primary auditory cortex would be expected to adopt syllable-length integration time constants because only they can provide a basis for minimally reliable, integrated, rewardable, event-by-event distinctions (Merzenich & Jenkins, 1995).

2. Temporal- and spectral-processing filter bandwidths established by the stimulus-driven maturation of the primary auditory cortex would be expected to be wider and less sharply bounded than normal.

3. Abnormal processing of slowly modulated stimuli that (unlike higher-rate modulation) derive from the normal, experience-driven shortening of cortical integration (recovery) times should be recorded.

4. Because the postnatal refinement of processing/recovery time constants is a basis of left hemisphere dominance for language, weaker-than-normal dominance is expected.

5. Given the weaker correlation strengths and immature, imprecise spectral and temporal filters for representations at the entry levels of cortex, distributed cortical responses representing signals in a successive-signal (masking) context should be much less strongly temporally coordinated than normal.

6. Given the less coordinated (less coherent) forms of neurological representations of these complex signals, they should also be more easily degraded by noise.

7. Processing of rapidly successive inputs should show strong, abnormal within-channel interferences, but if the spectral components of successive elements are very different, little interference and more normal speech element recognition should be recorded because in those circumstances, successive events would be treated by the cortex as separate perturbations, even when they occur in rapid succession.

8. The distributed neuronal representations of *brief* speech sounds occurring in a successive signal (masking) context will be most vulnerable to interference because they cannot be reliably separately represented in the cortex because of its long integration time constants.

9. Complex signal combinations resulting in phonological parsing in the belt cortical zones shall occur more slowly and in less reliable, less sharply bounded and more strongly intensity-dependent forms. That is a predicted consequence of more poorly resolved temporal and spectral filtering, and of the far less salient forms of representation of the intrasyllabic spectrotemporal fine-structure from which they are derived. An emergent fuzzy aural phonological representational sorting is the predicted result.

10. Myelination of the projection tracks from A1 to the belt cortex and Wernicke's area and myelination on the arcuate fasciculus should remain in a relatively immature state.

11. A fuzzy aural phonological representation would be integrated with visual, tactile, and proprioceptive information in the temporal and frontal (premotor, motor) cortex to generate much less sharply defined and much more poorly segregated neuronal cell assemblies representing native-

language phonemes. The creation of reliable frontal lobe representations would be further impeded by the greater-than-normal temporal dispersion of fed-forward information contributed by an enduringly immature arcuate fasciculus.

12. Because the development of useful aural/visual speech representations delivered in a minimally coherent form to the frontal lobe would be delayed, and because it shall be delivered in a weak and unreliable representational form in any event: (a) speech production development shall be delayed, and (b) the development of normal syntactic and grammatical abilities shall be delayed. As these higher-level neurological constructs evolve, because they shall necessarily be constructed from less salient distributed neuronal response representations, they shall emerge in weaker-than-normal forms.

13. Attentional control, short-term memory, and information storage processes as they apply for the language domain shall all be weaker than normal, because these cortically engendered processes will also be poorly enabled, refined, and elaborated because of the relatively nonsalient inputs that feed them.

14. These combined problems, but especially unreliable phonological representations and a related weakness in parsing words into their sound parts (phonological awareness deficits) will result in delayed and impaired reading initiation.

Note that this kind of functional self-organizing fault is especially destructive because a signal-to-noise problem would be strongly expressed at the bottom of the cortical processing hierarchy. All higher-level processing is ultimately dependent on high-fidelity processing at this level. Very widespread multiple-dimensional deficits would be the result.

Nearly All of These Predictions Have Been Confirmed to Apply for at Least the Great Majority of Language Impaired and Poor Reading Individuals

Babies destined to be language impaired have abnormally long cortical integration (processing) times, manifested by a striking deficit in their ability to distinguish between similar, rapidly successive stimuli (Benasich & Tallal, 1996; Spitz, Tallal, Flax, & Benasich, 1997). Specific language-impaired children and poor readers have longer-than-normal integration times (e.g., Eden, Stein, Wood, & Wood, 1995; Farmer & Klein, 1995; Helenius, Uutela, & Hari, 1999; Tallal & Piercy, 1973, 1974). Response dynamics recorded electrophysiologically from A1 and surrounding entry-level cortical areas on the superior temporal plane in language-impaired and reading-impaired subjects reveal longer-than-normal integration time constants (Nagarajan, Mahncke, Salz, Tallal, Roberts, & Merzenich, 1999). Spectral discrimination is degraded (spectral filters and critical bandwidths appear to be wider than normal) for language-impaired and dyslexic individuals (Ahissar, Protopapas, Reid, & Merzenich, 2000; de Wierdt, 1989; McAnally & Stein, 1996; Wright, Lombardino, King, Puranik, Leonard, & Merzenich, 1997). That degradation is equivalent for both nonspeech (e.g., tonal stimuli) and speech (e.g., formant) stimuli. Poorer temporal interval and duration discrimination abilities for within-channel tasks, and errors in speech discrimination for contrasts that are dependent on temporal interval or duration judgments are recorded (see Ahissar, Protopapas, Reid, & Merzenich, 2000; Hari & Kiesila, 1996; Helenius, Uutela, & Hari, 1999).

Abnormal behavioral discrimination abilities for lower-frequency (primary auditory cortex-dependent) modulation differences, but not for higher-frequency modulation (subcortically processed) rates are recorded in several related behavioral tasks (Witton, Talcott, Hansen, Richardson, Griffiths, Rees, Stein, & Green, 1998). Weaker-than-normal hemispheric dominance is recorded in these impaired populations (e.g., Harel & Nachson, 1997). Distributed cortical responses representing brief signals in a masking or successive-signal context are much less temporally coordinated than in a normal individual (Nagarajan, Mahncke, Salz, Tallal, Roberts, & Merzenich, 1999). Within a given processing channel, *almost no* additional distributed response coherence is added with the presentation of successive, within-channel acoustic features unless they are separated by more than 200 ms from an initial stimulus perturbation, that is, by roughly a syllable-length duration. Distributed responses representing rapidly successive inputs and brief sounds in the speech stream are much more strongly degraded by noise than in normals.

In general, as predicted, speech reception is more strongly affected by noise. Processing of rapidly successive inputs show grossly abnormal within-channel interferences, but not when the spectral components of successive elements substantially differ spectrally, that is, if successive sounds engage different, parallel entry-level processing channels (see Ahissar, Protopapas, Reid, & Merzenich, 2000; Merzenich & Jenkins, 1995; Tallal & Piercey, 1973, 1974; Wright, Lombardino, King, Puranik, Leonard, & Merzenich, 1997). Relatively brief speech sounds in a successive signal (masking) context are especially vulnerable to destructive interferences. Their poor representations largely account for differences in grammatical impairments for language and reading impaired children with different native languages (e.g., Leonard & Bortolini, 1998; see Leonard, 1998, for review).

Phonological sorting (and "phonological awareness") emerges more slowly and in a less reliable form in these populations (see Bishop, 1992;

Leonard, 1998; Lundberg, 1998, for reviews). Myelination of the arcuate fasciculus remains immature, and the state of its myelination is highly correlated with reading and language abilities (Klingberg et al., 2000). Speech production development, and the development of syntactic and grammatical abilities are delayed and impoverished (see Bishop, 1992; Leonard, 1998, for reviews). The development of refined attentional, short term memory and a wide variety of other higher-order cognitive abilities is delayed, and enduringly represented in weaker-than-normal forms (see Baddeley, 1996; Bishop, 1992; Gillam, Cowan, & Day, 1995; Leonard, 1998; Lundberg, 1998, for reviews). Phonological fidelity and parsing are impaired, and reading initiation is delayed. Impairments in accurate, high-speed signal reception account for the majority of the variance in reading ability (Ahissar, Protopapas, Reid, & Merzenich, 2000).

We conclude from this analysis (also see Plaut, McClelland, Seidenberg, & Patterson, 1996) that there is a high probability that the most common basis of origin of language and reading impairments is a poorer-than-normal signal-to-noise representation of speech in the functionally self-organizing auditory/speech processing system in early childhood. Interestingly, if a source of muffled inputs is corrected before the end of the first year of life—for example, by correcting a cleft palate that generates muffled hearing through chronically and continuously blocking both eustachian tubes—then relatively normal speech, language, and reading abilities are developed. However, after about the end of the first year, such corrections have no impact on speech and language development; such children are almost invariably language-learning-impaired and reading-impaired. We hypothesize that this occurs because learning with abnormally wide and long filters and time constants over the first year of life results in widespread, multiple-level progressive changes that become *dependent* on this processing machine configuration. They now provide the best way, at least in the present moment, for these brains reared with muffled inputs to extract information from the speech stream.

A PROOF OF PRINCIPLE. LANGUAGE LEARNING IMPAIRMENTS AND DYSLEXIA CAN BE REMEDIATED IN CHILDREN OF ALL AGES BY APPROPRIATE INTENSIVE RETRAINING

Animal and human electrophysiological and behavioral studies of plasticity indicate that spectral and temporal filtering and shortening of cortical integration (chunking, processing, recovery) times can be achieved at any age. Can a more salient representation of speech and language information be generated in a child with a language and/or reading impairment? Will the developing ability of such a child to make more accurate distinctions about the fine-structure of rapidly successive nonspeech and speech inputs manifest more normal filtering and chunking of information within the critical entry-level processing areas on the superior temporal plane? Can the ability of such a child to make accurate phonological distinctions and to accurately parse the sound parts of words be advanced to a normal performance level with such training? Will such improvements have the positive, predicted impacts on speech production and on syntactic, grammatical, short-term memory and attentional abilities?

The answer to all of these questions appears to be "yes" (Merzenich, Jenkins, Johnston, Schreiner, Miller, & Tallal, 1996; Merzenich, Miller, Jenkins, Saunders, Protopapas, Peterson, & Tallal, 1998; Merzenich, Tallal, Peterson, Miller, & Jenkins, 1998; Tallal, Miller, Bedi, Byma, Wang, Nagarajan, Schreiner, Jenkins, & Merzenich, 1996). Impaired children (> 1 SD below the normal mean in their overall language abilities) trained for an average of about 50 hours at tasks designed to provide them with more salient representations of spectrotemporal details of the speech input stream show: (a) dramatic improvements in their abilities to distinguish rapidly successive acoustic and speech inputs (an average effect size of about (+1.6 SD); (b) dramatic improvements in the immunity of speech reception in noise (+2.5 SD); (c) strong improvement in all aspects of speech reception, speech production, and language usage abilities manifested by major advances in language battery subtest and overall language quotient scores (for Testing of Language Development [TOLD] and Clinical Evaluation of Language Function [CELF] language assessment batteries, a +1.4 SD effect size); (d) substantial improvements in phonological parsing, and in reading; and (e) highly significant improvements in measures of specific, language-related, short-term memory, attentional, comprehension (and many other) abilities.

SUMMARY AND CONCLUSIONS

Studies from normal child development, studies documenting the specific deficits of language and reading impaired individuals, and studies documenting neurological and behavioral changes induced by training designed to reverse the early-childhood learning progressions that hypothetically cause them, collectively provide a powerful confirmation of this new neurologically based brain plasticity view of the origins of complex human behaviors from powerfully, functionally self-organizing cortical machinery. This and related research holds the promise of developing more complete neurological descriptions of the complex perceptual, cognitive, and executive skills processes that emerge from adaptive human forebrain machinery. They also provide us with an increasingly deeper understanding of the true sources of variation in human ability. Finally, as is illustrated by this final example, they provide us with new methodologies for remediating neurological impairments by specifically correcting abnormally operating and immature cortical processing machinery.

ACKNOWLEDGMENTS

Research summarized in this review was supported by NIH Grant NS-10414, the Coleman Fund, HRI, and by Scientific Learning Corporation, Berkeley, CA.

REFERENCES

Ahissar, E., Vaadia, E., Ahissar, M., Bergman, H., Arieli, A., & Abeles, M. (1992). Dependence of cortical plasticity on corelated activity of single neurons and on behavioral context. *Science, 257*, 1412–1415.

Ahissar, M., & Hochstein, S. (1993). Attentional control of early perceptual learning. *Proc. Natl. Acad. Sci. USA, 90*, 5718–5422.

Ahissar, M., Protopapas, A., Reid, M., & Merzenich, M. M. (2000). *Proc. Natl. Acad. Sci. (USA), 97*, 6832–6837.

Allard, T. A., Clark S. A., Jenkins W. M., & Merzenich M. M. (1991). Reorganization of somatosensory area 3b representation in adult owl monkeys following digital syndactyly. *J. Neurophysiol., 66*, 1048–1058.

Baddeley, A., & Wilson, B. A. (1993). A developmental deficit in short-term phonological memory: Implications for language and reading. *Memory, 1*, 65–78.

Benasich, A. A., & Tallal, P. (1996). Auditory temporal processing thresholds, habituation, and recognition memory over the first year. *Infant Behav. Develop., 19*, 339–356.

Bieser, A., & Müller-Preuss, P. (1996). Auditory responsive cortex in the squirrel monkey: Neural responses to amplitude-modulated sounds. *Exptl. Brain Res., 108*, 273–284.

Bishop, D. V. (1992). The underlying nature of specific language impairment. *J. Child Psychol. Psychiat. Allied Discip., 33*, 3–66.

Buonomano, T. V., Hickmott, P. W., & Merzenich M. M. (1997). Context-sensitive synaptic plasticity and temporal-to-spatial transformations in hippocampal slices. *Proc. Natl. Acad. Sci., USA, 94*, 10403–10408.

Buonomano, D. V., & Merzenich, M. M. (1995). Temporal information transformed into a spatial code by a network with realistic properties. *Science, 267*, 1028–1030.

Buonomano, D. V., & Merzenich, M. M. (1998). Cortical plasticity: From synapses to maps. *Ann. Rev. Neurosci., 21*, 149–186.

Buonomano, D. V., & Merzenich, M. M. (1998). Net interaction between different forms of short-term synaptic plasticity and slow-IPSPs in the hippocampus and auditory cortex. *J. Neurophysiol., 80*, 1765–1774.

Buzsáki, G. (1998). Memory consolidation during sleep a neurophysiological perspective. *J Sleep Res., 1*, 17–23.

Byl, N. N., Merzenich, M. M., & Jenkins, W. M. (1996). A primate genesis model of focal dystonia and repetitive strain injury. *Neurology, 47*, 508–520.

Chelazzi, L., Duncan, J., Miller, E. K., & Desimone, R. (1998). Responses of neurons in inferior temporal cortex during memory-guided visual search. *J. Neurophysiol., 80*, 2918–2940.

Clark, S. A., Allard, T., Jenkins, W. M., & Merzenich, M. M. (1988). Receptive fields in the body-surface map in adult cortex defined by temporally correlated inputs. *Nature, 332*, 444–445.

Demerens, C., Stankoff, B., Logak, M., Anglade, P., Alliquant, B., Courand, F., Zalc, B., & Lubetzki, C. (1996). Induction of myelination in the central nervous system by electrical activity. *Proc. Natl. Acad. Sci., USA, 93*, 9887–9992.

DeWierdt, J. (1989). Spectral processing deficit in dyslexic children. *Appl. Psychol., 9*, 163–174.

Doupe, A. J., & Kuhl, P. K. (1999). Birdsong and human speech: common themes and mechanisms. *Ann. Rev. Neurosci., 22*, 567–631.

Eden, G. G., Stein, J. F., Wood, H. M., & Wood, F. B. (1995). Temporal and spatial processing in reading disabled and normal children. *Cortex, 31*, 451–468.

Eggermont, J. J. (1994). Temporal modulation transfer functions for AM and FM stimuli in cat auditory cortex. Effects of carrier type, modulating waveform and intensity. *Hear. Res., 74*, 51–66.

Engert, F., & Bonhoeffer, T. (1999). Dendritic spine changes associated with hippocampal long-term synaptic plasticity. *Nature, 399*, 66–70.

Farmer, M. E., & Klein, R. (1995). The evidence for a temporal processing deficit linked to dyslexia: A review. *Psychonom. Bull. Rev., 2*, 460–493.

Fleschig, P. (1920). *Anatomie des menschlichen Gehirns und Ruckenmarks auf myelogenetischen Grundlage.* Leipzig: Georg Thieme:

Geinisman, Y., deToledo-Morrell, L., Morrell, F., Heller, R. E., Rossi, M., & Parshall, R. F. (1993). Structural synaptic correlate of long-term potentiation: Formation of axospinous synapses with multiple, completely partitioned transmission zones. *Hippocampus, 3*, 435–445.

Gillam, R. B., Cowan, N., & Day, L. S. (1995). Sequential memory in children with and without language impairment. *J. Speech Hear. Res., 38*, 393–402.

Grajski, K. A., & Merzenich M. M. (1990a). Hebb-type dynamics is sufficient to account for the inverse magnification rule in cortical somatotopy. *Neural Computation, 2*, 74–81.

Grajski, K. S., & Merzenich, M. M. (1990b). Neuronal network simulation of somatosensory representational plasticity. In D. L. Touretzky (Ed.), *Neural Information Processing Systems*, vol. 2. San Mateo: Morgan Kaufman.

Greenough, W. T., & Chang, F. F. (1988). In A. Peters & E. Jones (Eds.), *Cerebral cortex, vol. 7* (pp. 335–392). New York: Plenum.

Haenny, P. E., Maunsell, J. H., & Schiller, P. H. (1988). State dependent activity in monkey cortex. II. Retinal and extraretinal factors in V4. *Exptl. Brain Res., 69*, 245–259.

Harel, S., & Nachson, I. (1997). Dichotic listening to temporal tonal stimuli by good and poor readers. *Percept. Motor Skills, 84*, 467–473.

Hari, R., & Kiesila, P. (1996). Deficit of temporal auditory processing in dyslexic adults. *Neurosci. Letters, 205*, 138–140.

Hebb, D. O. (1949). *The organization of behavior.* New York: Wiley.

Helenius, P., Uutela, K., & Hari, R. (1999). Auditory stream segregation in dyslexic adults. *Brain, 233*, 907–913.

Hopson, J. A. (1989). *The dreaming brain.* New York: Basic.

Hsiao, S. S., O'Shaugnessy, D. M., & Johnson, K. O. (1993). Effects of selective attention on spatial form processing in monkey primary and secondary somatosensory cortex. *J. Neurophysiol, 70*, 444–457.

James, W. (1890). *The principles of psychology, Vol. 1.* New York: Dover.

Jenkins, W. M., Merzenich, M. M., Ochs, M., Allard, T. T., & Guic-Robles, E. (1990). Functional reorganization of primary somatosensory cortex in adult owl monkeys after behaviorally controlled tactile stimulation. *J. Neurophysiol., 63,* 82–104.

Kaas, J. H., Hacket, T. A., & Tramo, M. J. (1999). Auditory processing in primate cerebral cortex. *Current Opin. Neurobiol., 9,* 164–170.

Karni, A. (1995). When practice makes perfect. *Lancet, 345,* 395.

Karni, A., & Sagi, D. (1991). Where practice makes perfect in texture discrimination: Evidence for primary visual cortex plasticity. *Proc. Natl. Acad. Sci., USA, 88,* 4966–4970.

Kaster, S., Pinsk, M. A., deWeerd, P., Desimone, R., & Ungerleider, L. G. (1999). Increased activity in human visual cortex during directed attention in the absence of visual stimulation. *Neuron, 22,* 751–761.

Keller, A., Arissian, K., & Asanuma, H. (1992). Synaptic proliferation in the motor cortex of adult cats after long-term thalamic stimulation *J. Neurophysiol., 68,* 295–308.

Kilgard, M. P., & Merzenich, M. M. (1998). Cortical map reorganization enabled by nucleus basalis activity. *Science, 279,* 1714–1718.

Kilgard, M. P., & Merzenich, M. M. (1999). Plasticity of temporal information processing in the primary auditory cortex. *Nature Neurosci., 1,* 727–731.

Kleim, J. A., Swain, R. A., Czerlanis, C. M., Kelly, J. L., Pipitone, M. A., & Greenough, W. T. (1997). Learning-dependent dendritic hypertrophy of cerebellar stellate cells: plasticity of local circuit neurons. *Neurobiol. Learn. Memory, 67,* 29–33.

Klingberg, T., Hedekus, M., Temple, E., Salz, T., Gabrieli, J. D. E., Moseley, M. E., Poldrade, R. A. (2000). Microstructure of temporo-parietal white matter as a basis for reading ability: Evidence from diffusion tensor magnetic resonance imaging. *Neuron, 25,* 493–500.

Kobatake, E., Wang, G., Tanaka, K. (1998). Effects of shape-discrimination training on the selectivity of inferotemporal cells in adult monkeys. *Journal of Neurophysiology, 80,* 324–330.

Kuhl, P. K. (1991). Human adults and human infants show a "perceptual magnet effect" for the prototypes of speech categories, monkeys do not. *Percept. Psychophy., 50,* 93–107.

Kuhl, P. K. (1994). Learning and representation in speech and language. *Current Opinion in Neurobiology, 4,* 812–822.

Leonard, L. B. (1998). *Children with language impairment.* Cambridge, MA: MIT Press.

Leonard, L. B., & Bartolini, U. (1998). Grammatical morphologiy and the role of weak syllables in the speech of Italian-speaking children with specific language impairment. *J. Speech Hear. Res., 41,* 1363–1374.

Lundberg, I. (1998). Why is learning to read a hard task for some children. *Scand. J. Psychol., 39,* 155–167.

McAnally, K. E., & Stein, J. F. (1996). Auditory temporal coding in dyslexia. *Proc. Royal Soc. London, 263,* 961–965.

Merzenich, M. M. (1990). Development and maintenance of cortical somatosensory representations: Functional 'maps' and neuroanatomical repertoires. In K. E. Barnard & T. B. Brazelton (Eds.), *Touch: The foundation of experience* (pp. 47–71). Madison, WI: Intl. Univ. Press.

Merzenich, M. M., Allard, T., & Jenkins, W. M. (1991). Neural ontogeny of higher brain function; Implications of some recent neurophysiological findings. In O Franzén & P. Westman (Eds.), *Information processing in the somatosensory system* (pp. 293–311). London: Macmillan.

Merzenich, M. M., & deCharms, R. C. (1996). Neural representations, experience and change. In R. Llinas & P. Churchland (Eds.), *The mind–brain continuum* (pp. 61–81). Cambridge, MA: MIT Press.

Merzenich, M. M., Grajski, K. A., Jenkins, W. M., Recanzone, G. H., & Peterson, B. (1991). Functional cortical plasticity. Cortical network origins of representational changes. *Cold Spring Harbor Symp. Quant. Biol., 55,* 873–887.

Merzenich, M. M., & Jenkins, W. M. (1993a). Cortical representation of learned behaviors. In P. Andersen (Ed.), *Memory concepts* (pp. 437–453). Amsterdam: Elsevier.

Merzenich, M. M., & Jenkins, W. M. (1993b). Reorganization of cortical representations of the hand following alterations of skin inputs induced by nerve injury, skin island transfers, and experience. *J. Hand Ther., 6,* 89–104.

Merzenich, M. M., & Jenkins, W. M. (1995). Cortical plasticity, learning and learning dysfunction. In B Julesz & I. Kovacs (Eds.), *Maturational windows and adult cortical plasticity* (pp. 247–272). New York: Addison-Wesley.

Merzenich, M. M., Jenkins, W. M., Johnston, P., Schreiner, C., Miller, S. L., & Tallal, P. (1996). Temporal Processing deficits of Language-Learning Impaired Children Ameliorated by Training. *Science, 271,* 77–81.

Merzenich, M. M., Miller, S., Jenkins, W. M., Saunders, G., Protopapas, A., Peterson, B., & Tallal, P. (1998). Amelioration of the acoustic reception and speech reception deficits underlying language-based learning impairments. In C. V. Euler (Ed.), *Basic neural mechanisms in cognition and language* (pp. 143–172). Amsterdam: Elsevier.

Merzenich, M. M., Recanzone, G. H., & Jenkins, W. M. (1991). How the brain functionally rewires itself. In M. Arbib & J. A. Robinson (Eds.), *Natural and Artificial Parallel Computations* (pp. 201–234). New York: MIT Press.

Merzenich, M. M., & Sameshima, K. (1993). Cortical plasticity and memory. *Current Opin. Neurobiol., 3,* 187–196.

Merzenich, M. M., Tallal, P., Peterson, B., Miller, S. L., & Jenkins, W. M. (1998). Some neurological principles relevant to the origins of—and the cortical plasticity based remediation of—language learning impairments. In J. Grafman & Y. Cristen (Eds.), *Neuroplasticity: Building a bridge from the laboratory to the clinic* (pp. 169–187). New York: Verlag.

Nagarajan, S., Mahncke, H., Salz, T., Tallal, P., Roberts, T., & Merzenich, M. M. (1999). Cortical auditory signal processing in poor readers. *Proc. Natl. Acad. Sci. USA, 96,* 6483–6488.

Naya, Y., Sakai, K., & Miyashita, Y. (1996). Activity of primate inferotemporal neurons related to a sought target in pair-association task. *Proc. Natl. Acad. Sci. USA, 93,* 2664–2669.

Nudo, R. J., Milliken, G. W., Jenkins, W. M., & Merzenich, M. M. (1995). Use-dependent alterations of movement representations in primary motor cortex of adult squirrel monkeys. *J. Neurosci., 16,* 785–807.

Phillips, W. A., & Singer, W. (1997). In search of common foundation for cortical computation. *Behav. Brain Sci., 20,* 657–722.

Plaut, D. C., McClelland, J. L., Seidenberg, M. S., & Patterson, K. (1996). Understanding normal and impaired word reading: Computational principles in quasi-regular domains. *Psychol. Rev., 103,* 56–115.

Qin, Y. L., McNaughton, B. L., Skaggs, W. E., & Barnes, C. A. (1997). *Phil. Trans. Roy. Soc. London, 352,* 1525–1533.

Recanzone, G. H., Merzenich, M. M., & Dinse, H. R. (1992). Expansion of the cortical representation of a specific skin field in primary somatosensory cortex by intracortical microstimulation. *Cerebral Cortex, 2,* 181–196.

Recanzone, G. H., Merzenich, M. M., & Jenkins, W. M. (1992). Frequency discrimination training engaging a restricted skin surface results in an emergence of a cutaneous response zone in cortical area 3a. *J. Neurophysiol., 67,* 1057–1070.

Recanzone, G. H., Merzenich, M. M., Jenkins, W. M., Grajski, K. A., & Dinse, H. A. (1992). Topographic reorganization of the hand representational zone in cortical area 3b paralleling improvements in frequency discrimination performance. *J. Neurophysiol., 67,* 1031–1056.

Recanzone, G. H., Merzenich, M. M., & Schreiner, C. S. (1992). Changes in the distributed temporal response properties of SI cortical neurons reflect improvements in performance on a temporally-based tactile discrimination task. *J. Neurophysiol., 67,* 1071–1091.

Recanzone, G. H., Schreiner, C. E., & Merzenich, M. M. (1993). Plasticity in the frequency representation of primary auditory cortes following discrimination training in adult owl monkeys. *J. Neurosci., 13*, 87–103.

Sakai, K., & Miyashita, Y. (1991). Neural organization for the long-term memory of paired associates. *Nature, 354*, 152–155.

Schreiner, C. E., & Urbas, J. V. (1988). Representation of amplitude modulation in the auditory cortex of the cat. II. Comparison between cortical fields. *Hear. Res., 32*, 49–63.

Somers, D. C., Todorov, E. V., Siapas, A. G., Toth, L. J., Kim, D. S., & Sur, M. (1998). A local circuit approach to understanding integration of long-range inputs in primary visual cortex. *Cerebral Cortex, 8*, 204–217.

Spitz, R. V., Tallal, P., Flax, J., & Benasich, A. A. (1997). Look who's talking: A prospective study of familial transmission of language impairments. *J. Speech Lang. Hear. Res., 40*, 990–1001.

Tallal, P., Miller, S. L., Bedi, G., Byma, G., Wang, X., Nagarajan, S. S., Schreiner, C., Jenkins, W. M., & Merzenich, M. M. (1996). Acoustically modified speech improves language comprehension in language-learning impaired children. *Science, 271*, 81–84.

Tallal, P., & Piercy, M. (1973). Defects of non-verbal auditory perception in children with developmental aphasia. *Nature, 241*, 468–469.

Tallal, P., & Piercy, M. (1974). Developmental aphasia: Rate of auditory processing and selective impairment of consonant perception. *Neuropsychologia, 13*, 69–74.

Wang, X., Merzenich, M. M., Beitel, R., & Schreiner, C. (1995). Representation of specifies-specific vocalizations in the primary auditory cortex of the marmoset monkey. Spectral and temporal features. *J. Neurophysiol., 74*, 1685–1706.

Wang, X., Merzenich, M. M., Sameshima, K., & Jenkins, W. M. (1995). Remodelling of Hand Representation in Adult Cortex Determined by Timing of Tactile Stimulation. *Nature, 378*, 71–75.

Weinberger, N. M. (1993). Learning-induced changes of auditory receptive fields. *Current Opin Neurobiol, 3*, 570–577.

Witton, C., Talcott, J. B., Hansen, P. C., Richardson, A. J., Griffiths, T. D., Rees, A., Stein, J. F., & Green, G. G. (1998). Sensitivity to dynamic auditory and visual stimuli predicts nonword reading ability in both dyslexic and normal readers. *Current Biol, 8*, 791–797.

Wright, B., Lombardino, L. J., King, W. M., Puranik, C. S., Leonard, C. M., & Merzenich, M. M. (1997). Deficits in auditory temporal and spectral resolution in language-impaired children. *Nature, 387*, 176–177.

Xerri, C., Merzenich, M. M., Jenkins, W., Santucci, S. (1999). Representational plasticity in cortical area 36 paralleling tactual-motor skill acquisition in adult monkeys. *Cerebral Cortex, 9*, 264–276.

Yakolev, P. I., & Lecours, A. R. (1967). The myelogenetic cycles of regional maturation of the brain. In A. Minkowski (Ed.), *Regional development of the brain in early life* (pp. 3–70). Oxford: Blackwell.

Chapter 4

Failures to Learn and Their Remediation: A Hebbian Account

James L. McClelland
Carnegie Mellon University

This Carnegie Symposium celebrates a growing convergence of behavioral and neural approaches to the mechanisms of cognitive change and reflects an overall convergence of behavioral and neural approaches to all aspects of cognition and cognitive development. This chapter is a part of a corresponding convergence in my own research. In my own formative years as a psychologist, facts about underlying neural mechanisms were considered to be nearly irrelevant to understanding cognition and its development. Where I went to graduate school, we studied cognition and cognitive development on the one hand and physiological psychology on the other, and they seemed almost completely disconnected subjects. My own early experimental work stayed strictly on the cognitive side of this huge divide.

Things began to change for me in the late 1970s, with the emergence of connectionist models. They seemed to many of us who were involved in their development to represent a clear step toward bridging the gap between mind and brain (Hinton & Anderson, 1981). We saw information processing as arising from the interactions of vast numbers of neurons. We began to think of propagation of activation among neurons via their synaptic connections as producing cognitive outcomes ranging from perception to comprehension to problem solving, and we thought of changes in the synaptic connections among the neurons as the basis for changes in these processes. Memory, learning, and development were all taken to be based on changes in the strength and distribution of synaptic connections.

This way of thinking could be traced back to 19th century physiologists and psychologists and was quite fully articulated 50 years ago by Donald

Hebb (1949). It died out for a time in the 1960s but emerged again around 1980, stimulated in part by the advent of computers that were fast enough and cheap enough to allow the exploration of computer simulations of neural processes. In my own case, I found this approach appealing as it gave answers to many questions about cognition that had previously puzzled me, questions that I had not previously found answers to when considering them within other information processing frameworks (McClelland, 1979; McClelland & Rumelhart, 1981). The connectionist (Feldman & Ballard, 1982) or parallel-distributed processing (Rumelhart, McClelland, & the PDP Research Group, 1986) framework for modeling cognitive processes arose around this approach, and several interesting and successful models of cognitive and developmental processes have been constructed within it.

However, the promise of connectionism for aiding the rapprochement between psychology and neuroscience was not immediately kept. One reason for this, in my view, was that many clever connectionists focused more on computational than on neurobiological considerations. They wanted procedures for adjusting connection weights that would solve the hard learning problems, and they did not constrain themselves to consider only mechanisms that seemed consistent with the details of the underlying neurobiological processes. The canonical example is the back-propagation algorithm (Rumelhart, Hinton, & Williams, 1986). Back propagation provided a powerful solution to a problem previously viewed as intractable (Minsky & Papert, 1969) and has been at the heart of many successful cognitive and developmental models, but it did not seem to neurobiologists to be at all plausible in terms of the underlying neurobiological mechanisms (see Mazzoni, Andersen, & Jordan, 1991 for discussion). In neurobiological circles, the use of back propagation has sometimes been tolerated as a vehicle for constructing networks that solve biologically interesting problems, such as mapping from eye position to position in a head-centered coordinate system, taking position of the eyes in the head into account (Zipser & Andersen, 1988). Yet, back propagation is generally not thought to reflect the way in which adjustments to connection weights are actually carried out in the brain, and so is generally considered irrelevant to the study of the neural basis of learning from a biological point of view. Some effort has been made to find computationally powerful algorithms that are more biologically plausible (Hinton & McClelland, 1988; Mazzoni et al., 1991; O'Reilly, 1996), but to my knowledge the psychological or functional implications of these proposals have not been very fully explored.

The key point is that in spite of all that was gained by adopting the computationally powerful back-propagation algorithm, something important was lost. It was not just the chance to sit down and have a good conversation with a neurobiologist. Rather, it was the opportunity to exploit what we do know about the neurobiology of learning in our efforts to understand learning at the behavioral level: Particularly, when improvements in functions such as perception or skilled performance will occur due to experience, and when they will not occur.

The work described here began with the following thought: Hebb's original proposal for how learning may occur in the brain is strongly supported by current research in neurophysiology. Perhaps this proposal can help us understand some cases in which experience can fail to lead to cognitive change, and perhaps it can also suggest procedures we can use to induce change in these cases. I hope this chapter will indicate that this thought has been a useful guide for some new research, leading to some interesting new findings, some of which support the Hebbian proposal. However, I suggest below that Hebb's proposal, even if it is partially correct, may need further elaboration before it can stand as an adequate guide to the conditions under which experience will lead to improvements in functional outcomes.

I begin with two puzzling failures of cognitive change that began to intrigue me a few years ago. Next, I suggest how they might be explained as arising from a paradoxical property of Hebb's proposed mechanism of learning. I then consider supporting data from the literature and new experimental and modeling studies that we have carried out to explore the issues further, particularly in the domain of inducing change in adults' perception of phonological distinctions not present in their native language.

I would like to acknowledge that this work has been deeply influenced by two other participants in this symposium. First, over many years, Michael Merzenich's work has inspired me to think about the mechanisms of cognitive change in the brain. Merzenich has had many things to say about the neurobiology of learning (see his chapter in this volume for a presentation of some of his ideas), and many of the things I have to say appear to me to be explicit or at least implicit in his work. Second, one of the two puzzles I introduce was put to me by Helen Neville at a previous symposium. At the time, I was already thinking about the first of the two puzzles, and it was the juxtaposition of the two that led me to think about the implications of Hebbian learning. So Merzenich and Neville deserve the credit (or, alternatively, the blame) for setting me off on the trajectory of this work. Many other colleagues, to whom I refer where relevant, have also played key roles in shaping the ideas.

TWO PUZZLES

The first puzzle lies in the pattern of spared and impaired learning seen in individuals with amnesia arising from damage to the medial temporal lobes. Such patients have profound deficits in the ability to acquire new explicit memories for the contents of particular episodes and experiences, but show apparently normal improvements in performance in many different kinds of tasks where explicit memory of prior performance appears not to be required (See Squire, 1992 for a review). According to a theory that McNaughton, O'Reilly, and I developed, amnesia results from the loss of a system in the medial temporal lobes that is specialized for the rapid acquisition of the arbitrary conjunctions of elements that together make up an episode or specific experience. In this theory (McClelland, McNaughton, & O'Reilly, 1995), spared learning in amnesia reflects changes in the strengths of connections among neurons outside the medial temporal areas that occur in the course of carrying out information processing operations. These changes are thought to be relatively small in magnitude, so that they cannot lead to the formation of an arbitrary association in one or even a few repetitions, but they are thought to be sufficient to produce item-specific priming effects that can arise from a single processing event, and their accumulation over many processing events is thought to lead to the acquisition of cognitive skill. Furthermore, gradual accumulation of such changes through repeated activation of the arbitrary association is thought to result in the establishment of these associations outside the hippocampal system, accounting for the amnesic's ability to retain arbitrary knowledge acquired well before the onset of amnesia.

If the theory of McClelland et al. (1995) is correct, it becomes puzzling that it has sometimes proven impossible to teach amnesics new arbitrary associations even with extensive repetition. A clear case in point is provided by an experiment by Gabrieli, Cohen, and Corkin (1988). They tried to teach patient HM the meaning of eight rare words that are not known by most people. Each word was presented once with a defining phrase as its definition. After this first presentation, a series of 60 blocks followed; in each block, HM was required to choose the appropriate definition, synonym, or sentence context for each word. Whereas normal controls mastered the task within a very few blocks, HM never mastered the task. He appeared to know the meaning of one of the words from the start, but he never mastered the meanings of any of the other words. A similar dismal picture emerges from reports of HM's performance on paired-associate learning tasks, though the testing has not been as extensive. On the McClelland et al. (1995) theory, we should have expected that with enough repetitions, brain regions outside the medial temporal areas would have acquired the new word meanings through the gradual accumulation of small changes, just as they acquire new cognitive skills. The first puzzle, therefore, is:

Puzzle #1: Given their impressive gains from repeated practice in some tasks, why has it proven impossible in some other tasks to teach amnesic subjects new information, even with massive repetition?

The second puzzle lies in the apparent dramatic reduction of the ability to acquire new phonological distinctions in adulthood. As before, the puzzle arises within the framework of a connectionist approach to processing and learning, in which cognitive change occurs through the adjustment of connection weights during the course of processing. Such connection adjustments are thought to underlie children's and adults' acquisition of many sorts of knowledge and different kinds of skills. These include their ability to differentiate contrasting phonemes in their native language, their semantic knowledge of the world, their acquisition of cognitive skills such as reading, and many other things. The puzzle arises from the fact that many kinds of learning that occur in childhood are still quite possible in adulthood, suggesting that connection strengths are still subject to change in many domains, but the ability to learn to distinguish some phonetic contrasts not used in a person's native language appears to be lost or at least much diminished in adulthood. One famous case in point is the distinction between the English phonemes /r/ and /l/, which is extremely difficult for many Japanese adults, whose native language lacks such a distinction. Such people experience great difficulty in both the production and the perceptual identification of these sounds, and attempts at remediation have generally shown very gradual progress (details are presented below). Thus, the second puzzle is:

> *Puzzle #2:* Why is it that adults can still learn and adapt many skills, yet the ability to adapt the perception and production of speech appears to diminish drastically in adulthood?

Others have offered answers to both of the puzzles described here, but each puzzle has generally been considered independently of the other and the answers have not been at all related. I will suggest that both of these puzzles can be resolved by considering the functional consequences of the mechanism of learning that was proposed by Hebb. I now turn to an examination of Hebb's proposal.

HEBB'S PROPOSAL FOR LEARNING

Hebb outlined his proposal for learning in his book, *The Organization of Behavior*:

> When an axon of cell A is near enough to excite a cell B and repeatedly or persistently takes part in firing it, some growth process or metabolic change takes place in one or both cells such that A's efficacy, as one of the cells firing B, is increased. (Hebb, 1949, p. 62).

This proposal has been the subject of intense study by neurobiologists and has received a great deal of experimental support through research on the neurobiological phenomenon of long-term potentiation (see McNaughton, 1993 for a review). Briefly, in a slice of brain that has been placed in a dish, one can study the effect of electrically supplied pulses that emulate the firing of neuron A (and probably several others) on the activation (excitatory postsynaptic potential; ESP) and subsequent firing of neuron B. The magnitude of the EPSP is taken to reflect the synaptic efficacy. One can repeatedly excite A but, if strong depolarization of B does not occur, there is little or no change in the EPSP. But if conditions are arranged to pair the EPSP produced by A with a strong-enough depolarization of B, a long-lasting change in synaptic efficacy results (Barrionuevo & Brown, 1983). Interestingly, some evidence suggests that the EPSP from A must just barely precede the depolarization of B in order to have this effect (Markram & Sakmann, 1995); thus synaptic input from A that could be causal in firing B at the right time is strengthened, whereas synaptic input that comes too late to be causal is not.

Let us think for a minute about Hebb's rule, in the context of a situation in which some input (say the sound of a word) is presented. This input elicits activation of neurons throughout the auditory system and may elicit subsequent semantic or other associations. What Hebb's rule suggests is that the mechanisms of synaptic modification will tend to stamp in whatever pattern it was that the input turned out to elicit. Experiments on potentiation suggest that the stronger the elicited activation, the stronger the effect will be and the longer it will last. The result will be an increase in the probability and the efficiency of a subsequent, very similar input to produce the same activation. To the extent that the activation is appropriate and useful, effective acquisition and maintenance of desirable cognitive abilities will occur. Yet, to the extent that the activation is inappropriate, Hebbian synaptic adjustment will tend to stamp in existing tendencies, and progress in acquiring the desired or appropriate response will not occur. I argue next that failures of learning in amnesia and failures to acquire non-native speech contrasts in adult second-language learners might both arise from undesirable strengthening of inappropriate preexisting activations.

IMPLICATIONS OF HEBB'S PROPOSAL I: LEARNING IN AMNESIA

We can now consider why it may have proven difficult to teach amnesics arbitrary paired associates or meanings of unknown words. The answer may lie in part with the arbitrariness of the pairings to be learned, in part with the procedure used to teach them, and in part with the nature of the amnesic syndrome. In paired-associate learning, the standard procedure is to select a set of arbitrary word-pairs, including, say locomotive–dishtowel, and then to present the pairs, one at a time, for a single study trial. On subsequent learning trials, only the first or stimulus item is shown, and the subject is required to try to produce the second or response item. Generally, even normal subjects get few if any items correct on the first such trial, so after the subject guesses, a correction is provided. Now, the nature of the amnesic syndrome becomes crucial.

As previously noted, the syndrome is thought to prevent the rapid formation of arbitrary associations that bind the specific stimulus and response items in a pair together. In normals, we assume, such associations are formed during learning trials, and serve as the basis for the correct response, but the amnesic subject is impaired in the formation of such associations, so no retrieval can occur. Instead, the subject must rely on the small adjustments that were made to the strengths of connections outside the medial temporal region and on whatever preexisting associations he or she may have to the stimulus. By assumption (motivated in McClelland et al., 1995), the small adjustments in the neocortex that occur during any given processing event are too small to create a sufficient association between arbitrarily paired items, so the subject is very unlikely to be able to produce the correct response. In paired-associate learning, the subject is strongly enjoined to produce some response, which is very unlikely to be correct. Yet, according to the Hebbian hypothesis, synaptic modification processes strengthen whatever neural activity occurs in response to the stimulus. The effect will be to strengthen the incorrect response that is elicited, rather than the desired one.

If there is any validity to this explanation, one would expect that amnesics would learn better if conditions are arranged to prevent them from making errors. In fact, such procedures have been used to induce learning of arbitrary associations in amnesics. One of these is called the *method of vanishing cues*, and essentially it involves providing enough of the correct answer to ensure that the patient never makes a mistake, then gradually diminishing the cues (Glisky, Schacter, & Tulving, 1986). Unfortunately, such training is usually carried out in practical contexts, and so it is difficult to run control groups to demonstrate unequivocally the importance of the prevention of error responses.

There are, however, two experiments that provide more specific tests of the effects of allowing patients to make nontarget responses on their learning of target responses. In both cases, we see that if the training context is set up so that amnesics (or normal subjects, for that matter) will make nontarget responses, subsequent performance on a test for desired or target responses will be impaired.

Baddeley and Wilson (1994) selected lists of single words, and studied amnesic's learning, in a test where they were required to generate the whole word from the first two letters. Two conditions were run. In the *errorful* condition, subjects were required to guess responses for each of the target words. For example, the experimenter might say, "I am thinking of a five-letter word beginning with 'qu.' Can you guess what the word might be?" If the subject did not guess the predesignated word (e.g., quote) within four guesses or 25 seconds, the experimenter told the subject the word. If the correct response was the subject's first response, an alternate word was substituted as the correct response (e.g., quill), so that at least one incorrect guess was elicited for each cue. After either guessing the word correctly (on the second or a later try) or after being told the correct word, the subject was asked to write down the correct word. Three trials of this type were run for each item, although the target remained fixed for the second and third trials, and correct responses at the beginning of these trials were allowed to stand. In the contrasting, *errorless* condition, no opportunity to guess was provided. Instead, the subjects were told (for example) "I am thinking of a word beginning with 'qu.' The word is 'quote.' Please write it down." Again, three trials of this type with each item were presented. It is important to note that in both conditions, the subject always wrote down the correct target word on each trial.

After this initial phase, the subjects, who were amnesics, elderly controls, or young controls, were given standard memory instructions. The experimenter said (for example), "Earlier I told you I was thinking of a word beginning with 'qu.' What word was it?" The striking finding was that amnesics in the errorful condition only obtained about 30% correct, whereas in the errorless condition they obtained 70% correct. Both control groups showed similar effects, though of lesser magnitude.

Another experiment by Hayman, MacDonald, and Tulving (1993) produces a similar finding. In this case a single amnesic subject known as KC was pretested on a set of silly definitions for words. Two examples were "a talkative featherbrain" and "Marlon Brando's wife." The targets for these definitions are "parakeet" and "the godmother." The set was then divided into two sets, one containing all those for which KC was able to provide a preexisting, but incorrect answer (high-interference items) and one containing only items for which he had no answer (low-interference items). These two sets were then further subdivided into two halves, each of which was used in one of the two following conditions. In the *evaluate only* condition, KC was simply given a silly definition and the target and asked to evaluate how easily another person like himself would be able to grasp the connection between the definition and the word. In the other, *generate and evaluate* condition, KC was first asked to generate his own response to the silly definition, and only then was he presented with the target and the definition to perform the same evaluation. In a final test, KC was asked to generate a response for all items. The findings were particularly striking for the high interference items. In the evaluate only condition, KC gave 67% correct responses, and the probability of making a nontarget response declined from a pretest value of 79% to 29% in the posttest. In the generate and evaluate condition, however, KC gave only 12% correct target responses after training, and during the posttest he made incorrect, nontarget responses at the same high 79% rate as in the pretest.

Together, these experiments seem to show that amnesics are particularly prone to fail to learn target responses when training trials give them opportunities to generate preexisting, nontarget responses to the stimulus item. Preventing such responses leads to far better and more rapid learning. This finding is certainly consistent with the idea that the elicitation of a response strengthens it, whether that response is desirable or not, in accordance with Hebbian learning.

It may be worth coming back to the learning of arbitrary associations, as it can be noted that the material considered in Hayman et al. (1993) and Baddeley and Wilson (1994) is hardly arbitrary. Indeed, both experiments rely on materials where there is either a clear preexisting connection between cue and target (as in the first two letters of a familiar word and the rest of that word) or a possible way of making some meaningful connection, however silly, between the target and the cue, which makes the association far from completely arbitrary. We would expect amnesics to make more rapid progress when there is some preexisting basis for association than they would when the associative relationship between cue and target is completely arbitrary. Our theory claims that the mechanisms of synaptic modification for synapses outside the medial temporal area are deliberately set up so that connection weights change only a little on each episode of processing. This gradual learning facilitates discovery of useful internal representations and avoids catastrophic interference (McClelland et al., 1995), but the downside is that the acquisition of truly arbitrary associations in these connections is necessarily very gradual. Thus, one would not expect errorless learning conditions to lead to rapid learning of arbitrary associates; these would still require many repetitions, even if inappropriate associative activations are prevented.

This discussion of amnesia brings out one important aspect of the role of the hippocampus in individuals with intact hippocampal systems. In the McClelland et al. (1995) model, the hippocampal system plays a very important role in allowing the contents of recent episodic memories to override whatever preexisting associations may be present in connections outside the medial temporal area. This role seems particularly apparent in the control groups of the Baddeley and Wilson study, where there is only a very small decrement in performance in the errorful condition. This hints at a general point, which is that for Hebb's proposal to be at all feasible as a mechanism for learning, there must be a great deal of support for desired responses, including both supporting mechanisms within the brain and contextual support from the environment. Without this support, Hebbian systems would be far too susceptible to the undesireable effects of the self-reinforcing characteristic of Hebbian learning.

IMPLICATIONS OF HEBB'S PROPOSAL II: CRITICAL PERIODS IN LEARNING PHONOLOGY?

We have seen how the mechanism of Hebbian learning might reinforce incorrect responses elicited in memory experiments, thus leading to failures to benefit from practice in experiments where such responses are routinely elicited. We now consider how the same idea may help us understand why adults sometimes fail to learn to distinguish speech sounds that are not contrasted in their native language. To begin to see how this might arise, imagine an individual who has had years of experience carving up phonological space according to the phonological structure of her native language. Research on speech perception in infants suggests that when this person first came into the world, she would have been capable of distinguishing the phonemes characteristic of any natural human language. We also know that within a relatively short period of life, the baby would lose the ability to make most of the distinctions not made in her language environment. There are different theories of how this loss may arise. In our view, a useful framework is provided by Kuhl's (1991; Kuhl & Iverson, 1995) perceptual magnet theory, in which the phonetic prototypes of one's native language act like magnets or (in neural network terms) attractors, distorting perception of items in their vicinity to make them more similar to the prototype (see also Samuel, 1982).

Suppose the individual we are considering had been born into a Japanese language environment, where there is only a single alveolar liquid phoneme instead of the two distinct phonemes /r/ and /l/. For this individual, a single perceptual magnet would arise in the perceptual space spanned by English /r/ and /l/. As the child grew older, this magnet would grow stronger and stronger, so that different items within the vicinity of this magnet would come to be more and more strongly attracted to it. Once this process had become quite strong, presentation of either an /r/ or an /l/ would, through the operation of the perceptual magnet effect, result in a pattern of neural activation corresponding to the single Japanese alveolar liquid. Let us suppose this person then moved to the United States, where there would be separate /r/ and /l/ phonemes. What would happen when this Japanese adult hears examples of these phonemes? Based on past experience, tokens of either phoneme will give rise to the alveolar liquid representation. Now, the mechanisms of Hebbian synaptic adjustment will have their paradoxical, undesirable effect: Every time either an /r/ or an /l/ is presented, Hebbian synaptic modification will simply stamp in the tendency for the sound presented to activate this Japanese perceptual representation. The result will be that the phonological categories the person brought with her from Japan will simply be reinforced rather than eliminated by experience in the new environment.

A Computational Model.   For these suggestions to be sufficiently explicit and concrete, and to show that they actually lead to the envisioned effects, it seemed clear that a computational model was needed. Adam Thomas and I have developed such a simulation model (McClelland & Thomas, 1998), and I describe it briefly here. As in other modeling work I have been involved in, the approach has been to develop a model with the minimal structure necessary to capture the intuitive account with which we began. In my experience, even the simplest models are complex enough, and one can learn a great deal from them. In addition to simplicity, we wanted a model that would begin to make some contact with domain-general aspects of the physiology of perceptual representations. These characteristics were already present in a model of Kohonen (1982, 1990) called the Self-Organizing Map, so we adopted the Kohonen model, adapting it slightly for our purposes. The architecture of the model is shown in Fig. 4.1. It consists of two layers of 49 units, with the units in each layer arranged in a two-dimensional sheet. These layers are called the *input* layer and the *representation* layer. There are modifiable connections to each representation layer unit from each input layer unit; the connections to one of the representation units are illustrated in the figure. Others have previously used a very similar model to capture perceptual magnet effects (Guenther & Gjaja, 1996), although we will not be considering these further here.

The model is thought of as experiencing patterns generated from a simple environment. There are two environments, one analogous to Japanese and one analogous to English, and these are shown on the bottom panel of the figure. In the English environment, inputs are generated from each of six prototypes. Four are called *corner prototypes* (labeled A in Fig. 4.1), and they are thought of as analogous to background phonemes in English. The other two are called *overlapping prototypes* (B), and they are thought of as analogous to English /r/ and /l/. In the Japanese environment, inputs are generated from the same four corner prototypes, but in place of the two overlapping prototypes there is a single *central prototype* (C) analogous to the Japanese alveolar liquid.

The model works as follows. Inputs are generated by selecting one of the prototypes, perturbing it with a bit of noise, and then setting the acti-

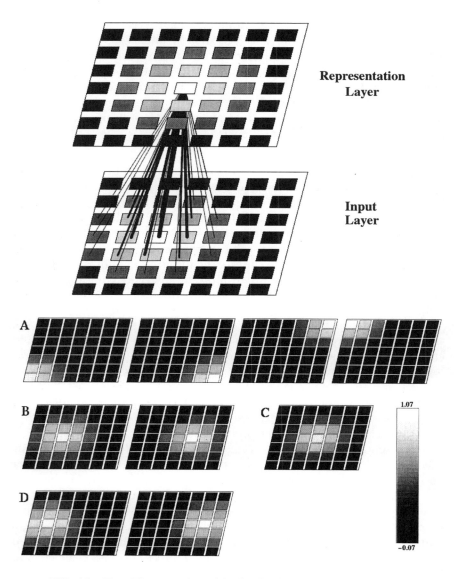

FIG. 4.1. Top: The network used in the simulations, based on the self-organizing map model of Kohonen (1982, 1990). Bottom: The prototypes of the input patterns used in training and testing the network. (A) The four corner prototypes. (B) The two overlapping prototypes. (C) The single central prototype. (D) Exaggerated versions of the central prototypes used in remediation of the network.

vations of the input units to the resulting perturbed values. An example case from the left overlapping prototype is shown imposed on the input units in the diagram of the model. Based on the existing connection weights, this input pattern generates excitatory influences on all the units in the representation layer (note that activations and weights are both constrained to range from 0 to 1 in this model). The net input to each representation unit is calculated: This is the sum, over all of the unit's incoming connections, of the strength of the weight on the connection times the activation of the input unit at the other end of the connection ($net_r = \sum_i w_{ri} a_i$, where net_r is the net input to representation unit r, a_i is the activation of input unit i, and w_{ri} is the weight to representation unit r from input unit i). The unit with the largest net input is chosen as the winner, and its activation is set to 1.

The activations of other units around the winner are set according to a Gaussian function of distance from the winner, so that near neighbors of the winner have reasonably strong activations and far neighbors are inactive. These aspects of the model are taken to reflect the influence of short-range excitatory and longer-range inhibitory interconnections among neurons in sensory cortex. Together with the connection weight adjustment rule about to be described (and variants of it), they have been used successfully to account for a number of aspects of neural organization, including the emergence and maintenance of topographic maps in sensory cortical areas as well as for the tendency for neurons representing similar higher-order features to be located near each other in inferotemporal cortex (Wang, Tanaka, & Tanifuji, 1996).

In the model, once a representation has been assigned to an input, the connection adjustment rule is applied. The form of the rule that we use is often called the *competitive learning* rule, and is also known as the *Oja rule*, based on its uses by Oja (1982) as in a principal-component analyzer. The rule is essentially Hebbian in nature, but with a built-in normalization factor. The rule is

$$\Delta w_{ri} = \epsilon a_r (a_i - w_{ri}).$$

In this rule, ϵ is the learning rate constant, and $a_r a_i$ is the Hebbian element, increasing strengths of connections to active representation units from active input units. The remaining subtractive element of the rule ($-a_r w_{ri}$) tends to cause weights from inactive input units to be reduced, and normalizes the weights to a representation unit so that $\sum_i w_{ri} = \sum_i a_i$. In our model, $\sum_i a_i$ is approximately constant (except for effects arising at the edges of the network), so that the sum of the weights coming to any representation unit tend to be fairly constant as well.

We can consider the learning rule at a slightly more abstract level and note that what it tends to do is to align the weights of the winning unit and its neighbors to match the current blob of activity on the input units. The rule can be seen as pulling the existing blob of weights toward the current input blob, and thus as implementing Hebbian learning at a system level by reinforcing the tendency of the current input to activate the same winning unit next time.

We simulated an analog of critical period effects in learning the phonological structure of English using this model, as follows. Each of 1,000 different learners was simulated with a different copy of the same network. Initial connection weights to each representation layer were set to random values with a loose topographic structure so that the initial weights to a particular representation unit tended to be strongest from corresponding points on the input layer and to fall off with distance. (This appears to characterize at least roughly the initial topographic biases in layer-to-layer projections in the brain.)

Our first baseline simulation illustrates what happened when networks were trained from the beginning with the "English" environment consisting of patterns generated from the four corner prototypes and patterns generated from the two overlapping prototypes. Then, 1,000 different networks, analogous to 1,000 different children learning English from birth, were randomly initialized and trained according to the following protocol. Training occurred in epochs. In each epoch, one training example was generated from each of the six prototypes in random order. After the presentation of each pattern, the winner was chosen and the weights were then adjusted as described before. Periodically we tested the network, with learning turned off, and looked to see whether the network differentiated the two overlapping patterns, as English children differentiate the phonemes /r/ and /l/. Differentiation was indexed by whether the two patterns activated different representation units.

The results of the first simulation are shown in the top panel of Fig. 4.2. Right from the start, about 92% of the networks assigned distinct representations to the two overlapping prototypes, and this tendency continued throughout 300 epochs of training. The 8% of the networks that failed could have been manipulated up or down; we chose to leave the parameters as is, to invite consideration of possible bases of perceptual deficits among native speakers of the English language. Not apparent in the figure is the considerable tuning and structuring of the representations of the inputs that actually results from the synaptic modification process as this occurs during learning. The effects of this will be apparent, however, in the next simulation.

The second simulation considered the effects of initial exposure to the "Japanese" environment. Again, 1,000 networks were initialized as before, but this time initial training over 300 epochs was undertaken using the four corner prototypes and the single central prototype. Periodic testing

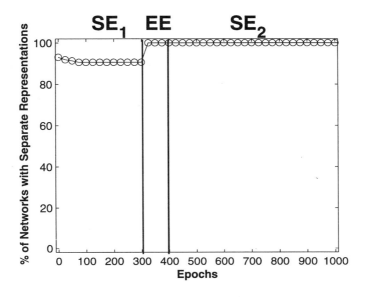

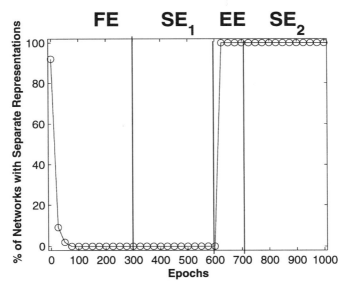

FIG. 4.2.   Simulation results from the model of McClelland and Thomas (1998). Vertical axis indicates the number of networks that assigned distinct representations to the two overlapping prototypes at each time point. (A) Results of training on the six-prototype "English" environment. Remediation training with exaggerated versions of the overlapping prototypes occurred between epochs 300 and 400. (B) Results of training on the six-prototype "English" environment (starting at Epoch 300) after initial training on the five-prototype "Japanese" environment. Remediation training with exaggerated versions of the overlapping prototypes occurred between Epochs 600 and 700.

using the overlapping prototypes with learning turned off revealed that over the 300 epochs, the networks learned to treat the two overlapping prototypes from the "English" environment (analogous to English /r/ and /l/) as the same: All the networks came to assign both overlapping prototypes to the same representation as shown in the bottom panel of the figure. A case of a network with this tendency is actually shown in Fig. 4.1: An example of the left overlapping prototype is shown activating the representation unit in the middle of the representation layer. This same unit is also activated in this network by presentations of the right overlapping prototype.

At this point, we are finally ready to ask what happens to such a network when its language environment changes, as would happen in the case of a Japanese adult who moves to the United States. In the model, the analog of this is a shift in the training environment from the "Japanese" (five-pattern) environment to the "English" (six-pattern) case. The result is shown in the next section of the bottom panel of Fig. 4.2. At Epoch 300, the training environment was switched from the five-prototype, Japanese-like environment to the six-prototype, English-like environment. Yet, it is apparent in the figure that the exposure to the six-prototype environment does not result in a change in the number of categories represented by the network. Every one of the 1,000 networks persists in maintaining a single representation, spanning both the left and right overlapping prototypes. This behavior continues indefinitely, and reflects the fact that whenever an example of either the left or the right prototype is presented, the same unit in the representation layer wins. The examples do jockey the weights back and forth a bit, but the jockeying averages out, and the network remains stuck with a single representation that spans the two overlapping prototypes.

In short, the network shows how a Hebbian learning mechanism can lead to a situation in which the network learns to treat two classes of inputs as the same and then diabolically maintains this tendency, even when faced with input that would at first have caused it to represent the classes separately. We consider later how this aspect of the network's behavior actually depends on additional aspects of the Kohonen network beyond the use of a Hebbian learning rule, but for now, we may consider whether the Hebbian perspective can suggest any steps we might take to overcome the network's tendency to fuse the two overlapping categories.

Note that our approach suggests that the mechanisms of synaptic modification are still at work in the network; the only problem is that whenever an example from the right or the left protoype is presented, the same representation is activated, and the mechanisms of synaptic modification are actively maintaining the network's tendency to map examples of both prototypes to the same representation. To fix the problem, we need to find some way of having the inputs activate different representations. In this context, consider what might happen if we used inputs that exaggerate the difference between the left and the right prototypes. If they are exaggerated enough, they will activate distinct representations. Once this happens, the mechanisms of synaptic modification will strengthen separate representations of the exaggerated items. We can then gradually reduce the difference between the inputs, making sure that we maintain the network's tendency to assign them distinct representations. Under these circumstances, the Hebbian synaptic modification process will tend to reinforce the tendency to treat the stimuli as distinct. Following this procedure we may eventually be able to move the inputs back to their original positions. We could even make them more similar to each other than they ordinarily would be in the natural situation, and still maintain separate representations.

Because of the relatively coarse grain of the model, the procedure we used in the simulations was even simpler than just described. We simply added examples taken from exaggerated versions of the overlapping prototypes, shown in Fig. 4.1D, to the six-prototype training environment, so that now there were eight stimuli in each epoch, one from each of the corner prototypes, one from each of the original overlapping prototypes, and one from each of the two exaggerated versions of the overlapping prototypes. This approach was very effective.

Within 25 epochs, all networks had split the central representation into two distinct representations. When after 100 epochs, the training environment reverted to the standard, English-like, six-prototype environment, all networks maintained separate representations of each of the two overlapping prototypes. The same intervention was also effective in remediating all of the networks that had been trained from the start with the English-like environment but had not separated the overlapping prototypes. In fact, in other simulations we have shown that various conditions analogous to possible causes of language inpairment can lead the network to form a fused representation of the two overlapping prototypes; unless these conditions are extreme, the remediation training causes the representations to separate (McClelland, Thomas, McCandliss, & Fiez, 1999).

The separation will be maintained even after the exaggerated examples are removed, except when the initial deficit is extremely severe. As one example of the possibilities we considered, one might suppose that in some cases of language impairment, crucial distinctions between certain phonemes are less salient than in most children, due perhaps to differences in processing rapid transitions (this might affect discrimination of stop consonants such as /b/ and /d/ more than liquids). A simple analog of this in our simulations involves increasing the overlap of the left and right overlapping prototypes, to the point where they almost always result in a single, fused representation. In this situation, remediation with exaggerated examples results in differentiation of the representations of examples of the overlapping prototypes, and unless the increase in overlap of these prototypes is extreme, the separation will be maintained after the exaggerated training examples are removed.

The remediation training we used in our network is similar to the procedure used by Merzenich, Tallal, and their collaborators in their intervention studies with language-impaired children (Merzenich, Jenkins, Johnson, Schreiner, Miller, & Tallal, 1996; Tallal, Miller, Bedi, Byma, Wang, Nagaraja, & others, 1996). Children received extensive exposure to speech with exaggerated (amplified and temporally extended) transitions, and also received ongoing exposure to normal speech from their parents and peers. The children also received extensive training with computer games in which they received initially exaggerated contrasts, which were then gradually reduced as their ability to discriminate improved. Another, less extensive study performed sometime earlier (Alexander & Frost, 1982) found that only a few sessions of exposure to exaggerated contrasts could lead to improvement in discrimination of stop consonants. In both studies, controls who received no exposure to exaggerated speech showed much less improvement on average.

Experimental Test of the Implications of Hebb's Proposal. The thoughts about the implications of Hebbian learning that gave rise to our model also give rise to hypotheses about the key processes at work in the remediation studies just discussed. With several colleagues, I have sought to explore this matter further by running additional remediation studies (McCandliss, Fiez, Protopapas, Conway, & McClelland, 1998). In these studies, we had two goals.

First, we wanted to demonstrate remediability in adults, to test predictions of loss-of-plasticity approaches against those of our plasticity-maintains-stability account. It has been shown in some domains (e.g., plasticity of the coordination of auditory and visual representation of external space in the barn owl; Brainard & Knudsen, 1998) that animals are less susceptible to remapping after puberty than they are earlier in life. These studies, coupled with age effects on degree of mastery of a non-native language (Johnson & Newport, 1989), have led some to suppose that synaptic modification may be shut off or at least greatly reduced after puberty in task-relevant brain regions. Evidence that the right intervention could effectively reopen the critical period might lead to some rethinking of those views. Of course, it is quite possible that mechanisms like those we have described play an important role in maintenance of prior language habits, and that at the same time there is some more maturationally based reduction of the degree of plasticity in adults. We do not intend to suggest that such maturationally based effects do not occur, but only that, in addition to any such effects, the Hebbian mechanism we have described may be a significant contributing factor. If indeed the mechanisms of synaptic modification are working strongly against learning under naturalistic circumstances, that it may be possible to alter circumstances so that these mechanisms can be turned to advantage instead of disadvantage. The question becomes, can we find an intervention that would exploit a Hebbian learning process and lead to rapid progress learning to distinguish /r/ and /l/ in Japanese adults, indicating that indeed considerable plasticity still exists if only we can harness it?

There have been several studies that have demonstrated training effects on discrimination of /r/ and /l/ in Japanese adults. Strange and Dittmann (1984) produced noticeable changes after several days of training with synthetic speech stimuli from a continuum of spoken inputs ranging from "rock" to "lock." In an extensive series of studies, Pisoni and colleagues have shown that they can improve discrimination of /r/ and /l/ through the use of natural speech tokens (see Akahane-Yamada, Tohkura, Lively, Bradlow, & Pisoni, 1997 for a summary of this work). Their approach involves presenting subjects with a naturally spoken word (such as "rake" or "eagle") and then giving the subject a two-alternative forced choice between the item and a minimal contrast foil ("rake" vs. "lake"; "eagle" vs. "eager") with feedback. Although there is progress in the Pisoni studies, it is fairly slow; for example in the most recent study, an improvement of about 20% required 45 hours of training (Akahane-Yamada et al., 1997). Thus, the existing evidence does not provide very dramatic support for the idea that there is still a lot of underlying plasticity for learning phonology in adults. On the other hand, the Pisoni group's training procedure does not seem particularly conducive to revealing the existence of such plasticity, as it is quite possible that the conditions of the experiment are promoting self-maintenance of existing tendencies to hear /r/ and /l/ as the same, at least for many of their stimuli. Their data show that for a subset of the stimuli they used, initial discrimination ability was very close to chance. Our goal was to see whether, with the use of exaggerated stimuli that subjects can initially discriminate, we could more rapidly induce learning of the /r/–/l/ contrast in Japanese adults.

Second, we wanted to test two aspects of the Hebbian approach to critical period effects in phoneme learning. One of these is that progress in learning can occur without feedback about correct performance. According to the Hebbian approach, as instantiated in our model, connections that map given inputs to the perceptual representations they produce are reinforced by Hebbian learning, even in the absence of feedback. The other aspect of the Hebbian approach is that success in learning to discriminate stimuli should depend critically on whether the stimuli used in training elicit different representations. As illustrated before in our simulations, the use of stimuli that are discriminable by native speakers of a language may not lead to differentiation of perceptual representations if they are not discriminable to the second-language learner. Yet, the use of exaggerated stimuli that are discriminable by the second-language learner should lead to much more rapid progress in learning.

We therefore contrasted two training regimes (See Fig. 4.3). One regime, called *fixed* training, used difficult minimal pair stimuli, specifically chosen to be very hard for our Japanese adults, while yet being reliably identifiable as /r/ and /l/ by native English speakers. The other, called *adaptive* training, used an entire continuum of stimuli ranging from exaggerated versions of one member of the difficult minimal pair at one end of the continuum for exaggerated versions of the other member of the difficult pair at the other end. Each subject in this condition was started with exaggerated items, and whenever the subject made a mistake, the stimuli were made even more exaggerated, until the outer limits of our continuum were reached. Whenever the subject correctly identified eight stimuli in succession, the discrimination was then made more difficult. According to the Hebbian approach and the model outlined previously, we would expect the adaptive training regime to lead to learning. On the other hand, we would expect that subjects in the fixed training condition would have great difficulty, given that they were not initially able to distinguish the stimuli used in that condition.

Two different continua were used in the study, one based on "rock" and "lock" (as shown in the figure) and one based on "road" and "load." For

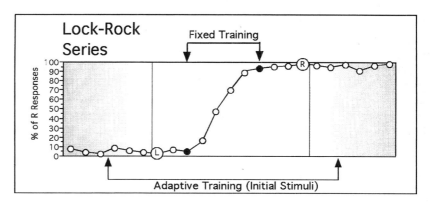

FIG. 4.3.   One of the two /r/-/l/ continua used in the training experiment with Japanese adults. Data indicating the phoneme identity assigned to these stimuli come from a group of University of Pittsburgh undergraduates. The stimuli used in the fixed training condition and the starting stimuli used for the adaptive training condition are marked. The two anchor stimuli which are resynthesized from natural tokens or "lock" and "rock" are circled and labeled with the letters L and R, respectively.

screening and to provide a pretraining baseline, each subject was pre-tested on both continua. During pretesting and posttesting, we assessed each subject's identification of the anchor stimuli, which were resyn-thesized from natural tokens of "rock" and "lock," and of all of the stimuli intermediate between these two items. We also measured discrimination of several pairs of stimuli, including the fixed training stimuli, several pairs that were less separated than the fixed training stimuli, and several pairs that were more separated. Subjects were allowed to participate in the experiment if their discrimination of the fixed training stimuli was below 70% correct on one of the two continua, which was then used for training in the experiment. After 3 days of 480 training trials, the same two tests that had been used during pretraining were readministered. Eight sub-jects were run in the adaptive training condition, and eight in the fixed condition. In addition, we ran eight control subjects who received the pre-tests and posttests with no training, only a 3-day gap between the two admininistration of these tests.

As predicted, all of the subjects in the adaptive condition made consid-erable gains in their identification of /r/ and /l/ stimuli as a result of their participation in the experiment, as indexed by the results of a logistic re-gression analysis of their performance in identifying the anchor stimuli and the stimuli interpolated between them (McCandliss et al., 1998). Prior to training, these subjects showed relatively flat identification functions, but after training all subjects showed an increase in slope of the identifica-tion function, indicating the acquisition of the ability to assign the correct identity to the anchor stimuli and neighboring interpolated stimuli with a rapid transition in their labeling functions at some point between the two anchors. Results in the fixed condition were very different. Some subjects did appear to show improvements in their identification performance over training, and there was a slight overall improvement in the slope of identification functions, but it appeared that even this slight gain could be attributed to the effects of the pretest, because a gain of the same size was also obtained for the control subjects who received only the pretests and posttests, with no intervening training. Thus, as expected on the Hebbian account, the use of exaggerated stimuli with subsequent adaptive modifi-cation led to rapid progress, but the use of difficult stimuli with no adap-tive modification produced little or no benefit.

The results of the experiment are promising in suggesting that there may indeed be a great deal of residual plasticity in the adult brain that may allow the acquisition of non-native speech discriminations. However, it is important to note that in our study, we have not completely reorganized the perceptual categories of our Japanese subjects. On the contrary, it appears that what they have learned is quite specific, in that the benefits of training on one of our two continua did not transfer to the other continuum. As others have found (Akahane-Yamada et al., in press), training across a wide range of tokens appears to be necessary to produce a generalizable improvement that transfers to other speakers and other items.

Discussion. Our experiment has succeeded in demonstrating that rapid learning of the /r/-/l/ distinction is possible in Japanese adults, even without feedback, if stimuli they can discriminate are used in training. We also showed that learning will not proceed if stimuli they cannot reliably discriminate are used, even when these stimuli are reliably identifiable by native speakers of English. These findings are consistent with the Hebbian proposal and the account we have offered with it of the basis for critical period effects in learning phonological contrasts. However, there are several additional points that have emerged from our research that indicate to me and my collaborators that the Hebbian proposal in its simplest form is at best only a partial guide to the conditions under which there may be successes or failures in learning.

First, Thomas and I have observed in simulations that our Kohonen network will eventually learn to separate two overlapping inputs that it has previously come to map to the same representation, unless additional inputs that compete for space in the representation layer are interleaved with presentations of the two overlapping inputs. That is, we train networks that initially learned on the Japanese environment in a training situation where they are exposed only to exemplars of the overlapping prototypes, the network will eventually come to assign distinct representations to the two inputs. Initially we considered this to be a characteristic of the Kohonen network that might be unrealistic, but our experience in pilot studies has begun to suggest to us that this property of the networks may not be so unrealistic after all. Based on this feature of our networks we have continued some of the subjects in the fixed training condition for several additional days beyond the three training sessions used in the experiment reported above, and we have found that several of these subjects eventually did learn the /r/-/l/ discrimination. Thus, it appears that an important factor in the maintenance of perceptual representations is competition from other perceptual representations that are maintained through exposure to language inputs. As a corollary of this observation, the fact that in our experiments we focus training on just a single contrast may be part of the reason for the rapid success of our adaptive training regime. We intend to test this point explicitly in future investigations.

A second observation that has emerged from our studies to date may appear to contradict a fundamental aspect of the Hebbian proposal. This is the finding that learning to discriminate /r/ and /l/ appears to be facilitated by the use of feedback. We have found in a small study with eight subjects (four receiving adaptive training and four receiving fixed training) that with feedback, progress is possible both for the adaptive subjects and for subjects receiving fixed training. Clearly, this contradicts a narrow version of the Hebbian proposal, in which no role is given for outcome information in learning. However, it may not be inconsistent with other versions, in which Hebbian synaptic modifications are moderated by feedback signals. One algorithm called the *reinforcement learning algorithm* (Barto, 1992) modulates the degree of Hebbian learning with outcome information. This algorithm has been shown to be quite effective in training networks to solve problems thought to require a more explicit error-correcting algorithm, and there is circuitry in the brain that may well allow the broadcast of reinforcement information to many of the synapses that may participate in perceptual learning. Thus, it is quite possible that Hebb's proposal was partially correct, but that it should be expanded to allow for modulation by outcome information, and possibly by other factors such as emotional state (McGaugh, 1989).

Conclusion

In this chapter, I have considered Hebb's proposal for learning in the brain, and I have suggested that his proposal may provide a partial guide to understanding some of the circumstances under which experience fails to lead to improvements in performance. The proposal provides a common account of failures of learning in amnesics and in normal second language learners, and has been incorporated in a simple simulation model that captures some aspects of critical period effects in the acquisition of contrasts between phonemes not distinguished in one's own native language. It also leads to testable predictions about what sorts of training regimes might lead to success or failure in teaching second language learners speech contrasts that are otherwise quite difficult for them to learn. That said, some additional observations that have arisen both from our modeling work and our empirical investigations suggest that Hebb's proposal for learning may turn out only to be a partial guide to the conditions under which humans and other organisms can improve their performance as a result of experience. There are many reasons to suppose that a complete account of how learning occurs in the brain will go beyond Hebb's initial proposal, and many reasons to suppose that additional insights into the successes and failures of learning at a behavioral or functional level will emerge as this more complete account is developed.

REFERENCES

Akahane-Yamada, R. A., Tohkura, Y., Lively, S. E., Bradlow, A. R., & Pisoni, D. B. (1997). Effects of extended training on English /r/ and /l/ identification by native speakers of Japanese. ATR Technical Report TR-H-212. Human Information Processing Research Lab, Japan.

Alexander, D. W., & Frost, B. P. (1982). Decelerated synthesized speech as a means of shaping speed of auditory processing of children with delayed language. *Perceptual and Motor Skills, 55,* 783–792.

Baddeley, A., & Wilson, B. A. (1994). When implicit learning fails: Amnesia and the problem of error elimination. *Neuropsychologia, 32,* 53–68.

Barrionuevo, G., & Brown, T. H. (1983). Associative long-term synaptic potentiation in hippocampal slices. *Proceedings of the National Academy of Science, USA, 80,* 7347–7351.

Barto, A. G. (1992). Reinforcement learning and adaptive critic methods. In D. A. White, & D. A. Sofge (Eds.), *Handbook of intelligent control: Neural, fuzzy, and adaptive approaches* (pp. 469–491). New York: Van Nostrand Reinhold.

Brainard, M. S., & Knudsen, E. I. (1998). Sensitive periods for visual callibration of the auditory space map in the barn owl optic tectum. *Journal of Neuroscience, 18,* 3929–3942.

Feldman, J. A., & Ballard, D. H. (1982). Connectionist models and their properties. *Cognitive Science, 6,* 205–254.

Gabrieli, J. D. E., Cohen, N. J., & Corkin, S. (1988). The impaired learning of semantic knowledge following bilateral medial temporal-lobe resection. *Brain and Cognition, 7,* 157–177.

Glisky, E. L., Schacter, D. L., & Tulving, E. (1986). Learning and retention of computer-related vocabulary in memory-impaired patients: Method of vanishing cues. *Journal of Clinical and Experimental Neuropsychology, 8,* 292–312.

Guenther, F. H., & Gjaja, M. N. (1996). The perceptual magnet effect as an emergent property of neural map formation. *The Journal of the Acoustical Society of America, 100,* 1111–1121.

Hayman, C. A. G., MacDonald, C. A., & Tulving, E. (1993). The role of repetition and associative interference in new semantic learning in amnesia: A case experiment. *Journal of Cognitive Neuroscience, 5,* 375–389.

Hebb, D. O. (1949). *The organization of behavior.* New York: Wiley.

Hinton, G. E., & Anderson, J. A. (Eds.). (1981). *Parallel models of associative memory.* Hillsdale, NJ: Lawrence Erlbaum Associates.

Hinton, G. E., & McClelland, J. L. (1988). Learning representations by recirculation. In D. Z. Anderson (Ed.), *Neural information processing systems* (pp. 358–366). New York: American Institute of Physics.

Johnson, J., & Newport, E. (1989). Critical period effects in second-language learning: The influence of maturational state on the acquisition of english as a second language. *Cognitive Psychology, 21,* 60–99.

Kohonen, T. (1982). Self-organized formation of topologically correct feature maps. *Biological Cybernetics, 43,* 59–69.

Kohonen, T. (1990). The self-organizing map. *Proceedings of the IEEE, 78,* 1464–1480.

Kuhl, P. K. (1991). Human adults and human infants show a 'perceptual magnet effect' for the prototypes of speech categories, monkeys do not. *Perception & Psychophysics, 50,* 93–103.

Kuhl, P. K., & Iverson, P. (1995). Linguistic experience and the "perceptual magnet effect". In W. Strange (Ed.), *Speech perception and linguistic experience: Issues in cross-language research* (pp. 121–154). Baltimore: York Press.

Markram, H., & Sakmann, B. (1995). Action potentials propagating back into dendrites triggers changes in efficacy of single-axon synapses between layer V pyramidal neurons. *Society for Neuroscience Abstracts, 21,* 2007.

Mazzoni, P., Andersen, R. A., & Jordan, M. I. (1991). A more biologically plausible learning rule for neural networks. *Proceedings of the National Academy of Sciences USA, 88,* 4433–4437.

McCandliss, B. D., Conway, M., Fiez, J. A., Protopapas, A., & McClelland, J. L. (1998). Eliciting adult plasticity: Both adaptive and non-adaptive training improves Japanese adults' identification of English /R/ and /L/. *Society for Neuroscience Abstracts*, Vol. 24, pg. 1898.

McClelland, J. L. (1979). On the time relations of mental processes: An examination of systems of processes in cascade. *Psychological Review, 86*, 287–330.

McClelland, J. L., McNaughton, B. L., & O'Reilly, R. C. (1995). Why there are complementary learning systems in the hippocampus and neocortex: Insights from the successes and failures of connectionist models of learning and memory. *Psychological Review, 102*, 419–457.

McClelland, J. L., & Rumelhart, D. E. (1981). An interactive activation model of context effects in letter perception: Part 1. An account of basic findings. *Psychological Review, 88*, 375–407.

McClelland, J. L., & Thomas, A. (1998). *Dynamic stability and adaptive intervention: Consequences of Hebbian learning?* Unpublished manuscript, Center for The Neural Basis of Cognition, 115 Mellon Institute, Pittsburgh, PA.

McClelland, J. L., Thomas, A., McCandliss, B. D., & Fiez, J. A. (1999). Understanding failures of learning: Hebbian learning, competition for representational space, and some preliminary experimental data. In J. Reggia, E. Ruppin, & D. Glanzman (Eds.), *Brain, behavioral, and cognitive disorders: The neurocomputational perspective* (pp. 75–80). Oxford, England: Elsevier.

McGaugh, J. L. (1989). Involvement of hormonal and neuromodulatory systems in the regulation of memory storage. *Annual Review of Neuroscience, 12*, 255–287.

McNaughton, B. L. (1993). The mechanism of expression of long-term enhancement of hippocampal synapses: Current issues and theoretical implications. *Annual Review of Physiology, 55*, 375–396.

Merzenich, M. M., Jenkins, W. M., Johnson, P., Schreiner, C., Miller, S. L., & Tallal, P. (1996). Temporal processing deficits of language-learning impaired children ameliorated by training. *Science, 271*, 77–81.

Minsky, M., & Papert, S. (1969). *Perceptrons: An introduction to computational geometry*. Cambridge, MA: MIT Press.

Oja, E. (1982). A simplified neuron model as a principal component analyzer. *Journal of Mathematical Biology, 15*, 267–273.

O'Reilly, R. (1996). *The LEABRA model of neural interactions and learning in the neocortex*. Ph.D. thesis, Department of Psychology, Carnegie Mellon University, Pittsburgh, PA.

Rumelhart, D. E., Hinton, G. E., & Williams, R. J. (1986). Learning internal representations by error propagation. In D. E. Rumelhart, J. L. McClelland, & the PDP Research Group (Eds.), *Parallel distributed processing: Explorations in the microstructure of cognition*, Vol. 1 (pp. 318–362). Cambridge, MA: MIT Press.

Rumelhart, D. E., McClelland, J. L., & the PDP Research Group (1986). *Parallel distributed processing: Explorations in the microstructure of cognition. Volume I: Foundations & Volume II: Psychological and biological models*. Cambridge, MA: MIT Press.

Samuel, A. G. (1982). Phonetic prototypes. *Perception & Psychophysics*, 307–314.

Squire, L. R. (1992). Memory and the hippocampus: A synthesis from findings with rats, monkeys, and humans. *Psychological Review, 99*, 195–231.

Strange, W., & Dittmann, S. (1984). Effects of discrimination training on the perception of /r–l/ by Japanese adults learning English. *Perception & Psychophysics, 36*, 131–145.

Tallal, P., Miller, S. L., Bedi, G., Byma, G., Wang, X., Nagaraja, S. S., Schreiner, C., Jenkins, W. M., & Merzenich, M. M. (1996). Language comprehension in language-learning impaired children improved with acoustically modified speech. *Science, 271*, 81–84.

Wang, G., Tanaka, K., & Tanifuji, M. (1996). Optical imaging of functional organization in the monkey inferotemporal cortex. *Science, 272*, 1665–1668.

Zipser, D., & Andersen, R. A. (1988). A back propagation programmed network that simulates response properties of a subset of posterior parietal neurons. *Nature, 331*, 679–684.

Chapter 5

PET Studies of Learning and Individual Differences

Richard J. Haier
University of California, Irvine

One of the surprising consequences following the explosion of functional brain imaging research over the last 10 years, is the rediscovery of individual differences. Early Positron Emission Tomography (PET) studies focused on cerebral blood flow or glucose metabolic rate comparisons between groups, say schizophrenics and normal controls. Most early studies measured the brain during a resting state, often eyes closed and lying quietly. As many psychologists discovered the PET technique, studies began to focus on comparisons between periods of different task activations in normal controls. Only a few early studies correlated actual task performance with regional cerebral function. These studies deliberately selected tasks that showed individual differences in performance rather than tasks where performance was uniform from subject to subject. Moreover, finding negative as well as positive correlations highlighted that regional brain deactivation may be as important for cognition as activation. Currently, many functional brain imaging studies examine correlations between task performance and regional brain function, even in cognitive psychology where individual differences historically have not been emphasized.

Most researchers using functional brain imaging techniques also realized, usually after testing their first two subjects, that the human brain varies in size and shape considerably among individuals. The anatomical location of functional activation or deactivation during a specific task also varies. The extent of these individual differences continues to cause worry about techniques that stretch or morph brain images from many individu-

123

als to a single standard size and shape. As a result, imaging data from individual subjects are often presented instead of or in addition to group images created by averaging images from individuals forced into a standard shape.

The focus of this chapter is on PET studies of learning and individual differences. We present an overview of our work, a brief review of other relevant studies, and present issues for future research on learning using functional imaging. We are guided by the view that all functional imaging techniques are inherently psychological tools and are best used in collaborations where psychology input is central (Haier, 1990). There are three main points to this review: (a) Individual differences in performance on all cognitive measures, especially measures of learning, exist and can have profound effects on brain imaging experiments, (b) Whereas many PET studies of learning primarily address the normative "where" in the brain learning occurs, questions can also be addressed with respect to whether the amount of activation (or deactivation) in a brain area is related to the amount of learning and, how the functional relationships among brain areas can be studied to help understand the neurocircuitry of learning, and (c) for maximum interpretability, brain imaging experiments need to use tasks that encourage individual differences in performance and incorporate variables of difficulty or effort (i.e., easy/hard), subject ability (i.e., good/poor performers), cognitive strategy, age/gender comparisons, and drug challenges.

EARLY UNIVERSITY OF CALIFORNIA, IRVINE (UCI) STUDIES WITH PET

Over the last 12 years, we have published a series of studies using PET and the metabolic tracer ^{18}flurodeoxyglucose (FDG) to examine individual differences. The PET FDG technique is particularly well suited for studying complex tasks because the uptake period for labeling the metabolic rate in the brain is about 32 minutes and a large number of trials can be tested during this time; spatial resolution can also be maximized. As the half-life of F^{18} is about 2 hours, however, only a single experiment or condition can be tested on one day. By contrast, the PET O^{15} technique requires a 40 to 60 second uptake period. The half-life of O^{15} is only 2 minutes so many short experiments can be performed on a single subject in one session albeit with poorer spatial resolution than obtained with FDG. Moreover, the FDG technique allows quantification of glucose metabolic rate so brain activation studies can provide information about neuronal work. Blood flow determinations provide only relative increases or decreases and fMRI signal changes are not closely coupled to neuron activity (see Haier, Alkire, et al., 1997 and Haier, 1998 for a more detailed discussion of PET techniques).

Abstract Reasoning. Our series of individual difference experiments began in 1988 with the first PET study of cortical functioning during abstract reasoning. A number of twin studies, especially of identical twins reared apart, had demonstrated a strong genetic component to psychometric measures of intelligence. The fact of a genetic component justifies questions about the biological basis of intelligence because genes act through biological processes. We used PET to examine whether there were specific areas in the brain that were activated during the performance of an intelligence test (Haier, Siegel, et al., 1988).

Eight right-handed normal males were recruited at random from a list of volunteers for brain imaging studies. Each completed the Raven's Advanced Progressive Matrices (RAPM), a standard test of 36 nonverbal abstract reasoning problems highly correlated with psychometric measures of intelligence, while PET was done with FDG as the metabolic uptake marker. Regional glucose metabolic rate (GMR) was compared between this group and two other groups of normal controls who had completed either a degraded Continuous Performance Test (CPT; subjects viewed single digits flashing on a screen out-of-focus for 40 msec each and pressed a button whenever the digit zero appeared, $n = 13$) or a stimuli-only, no-task version of the CPT ($n = 9$). Each task was performed for a 32-minute period following injection of the FDG.

ANOVA and follow-up t-tests revealed several cortical regions were uniquely activated during the RAPM (i.e., higher GMR) compared to the two attention conditions. Because there was a range in scores for the RAPM in the eight subjects (11 to 33 out of a perfect score of 36), we correlated each person's score with their GMR in each cortical region that was significantly different from the comparison tasks. Our expectation was that the higher the score, the higher the GMR in the salient brain areas. Several correlations were statistically significant but, surprisingly, they were all negative (−.72 to −.84). That is, high RAPM scores were correlated to low GMR. We interpreted this as evidence consistent with a brain efficiency hypothesis for complex problem solving. We also correlated attention scores (d′) with GMR in the CPT group but, because every subject performed nearly perfectly, the performance variance was small and only one area showed a significant correlation. In this study, the usual group analyses had identified cortical areas activated by the RAPM task but the individual difference analysis provided an additional and unexpected richness of data. It should be noted that a recent fMRI study in 7 normals performing the Raven's showed activation in similar cortical areas but individual differences in performance were not studied (Prabhakaran et al., 1997).

Shortly thereafter, two other PET studies with larger sample sizes reported inverse correlations between complex task performance on measures related to intelligence and regional brain function (Berent et al., 1988; Parks et al. 1988; see also Parks et al. 1989). We also reanalyzed our data with a more accurate method for defining anatomical localization of cortical areas. Although primitive by today's standard (Haier, 1993), we found even stronger inverse correlations in some areas, especially in the temporal lobes bilaterally. We decided to pursue the efficiency results with a second PET study to determine if learning decreased global and/or regional GMR).

Learning. We chose eight new right-handed male volunteers at random (Haier, Siegel, MacLachlan, et al., 1992a). Each one completed PET with FDG on two separate occasions. On the first day, the subject was instructed on how to perform a complex visuospatial motor task, the computer game Tetris (at this time, Tetris had just been introduced in the United States and none of the subjects had ever seen or played it previously). This task requires the subject to manipulate and place moving shapes together in a particular way by pressing keys on the keyboard. The better the performance, the faster the shapes move and the harder the task. After 1 minute of practice, FDG was injected and the next 32 minutes of performance determined regional GMR. Each subject than practiced Tetris for 4 to 8 weeks; performance increased on average by more than seven-fold. A second PET scan was obtained following this practice period. Again, performance over 32 minutes determined regional GMR.

The group comparisons of GMR in the naïve minus the practiced conditions showed overall decreases, as predicted by an efficiency hypothesis. The parietal lobe showed the most significant decreases. There were also significant inverse correlations in a number of areas between change in GMR (practice minus naïve) and change in Tetris score showing that the most improvement was associated with the largest GMR decreases. Some of the areas implicated in learning changes were areas identified in rat/lesion studies of problem solving (Thompson, Crinella, & Yu, 1990) but given small sample sizes and the difficulties in matching rat and human brain areas, these comparisons are tentative, as detailed in Haier et al., (1993).

Moreover, each subject in the Tetris experiment completed the RAPM and the WAIS-R on separate occasions. Those subjects with the highest scores on the RAPM showed the largest GMR decreases with practice, especially in frontal cortical and in cingulate areas (Haier, Siegel, Tang, et al., 1992b). This was one of the first PET studies to examine the effects of learning in humans and to relate these effects to intelligence. The addition of individual differences in task performance added an important dimension to the standard before-and-after group analysis.

Brain Efficiency

These studies led us to think more about a concept of brain efficiency. The Tetris study suggested to us that practice taught the brain what areas, circuits, and strategies *not* to use for better performance. Hence, a more efficient use of GMR in the practiced condition than in the naïve condition during which many areas and strategies may have been tried.

Concepts of brain efficiency have been proposed previously. Researchers working with EEG/evoked potentials (Chalke & Ertl, 1965; Ertl & Schafer, 1969), reaction time (Vernon et al., 1985, 1993), eye-pupil dilation (a sign of increased cognitive capacity; Ahern & Beatty, 1979), and even psychometrics (Maxwell et al., 1974, 1976) have discussed brain efficiency ideas (see Haier, 1993). Ertl (1969, 1971), for example, used EEG techniques to define an index of neural efficiency. The EEG efficiency index was thought to reflect elemental information processing in the brain. Low ability subjects had poor brain efficiency indexes whereas high ability subjects showed more efficiency. Ertl hoped such a measure would have practical screening uses to identify poor learners for early remedial attention. He even marketed the Brain Wave Analyzer for this purpose. Subsequent attempts to replicate the brain efficiency index were inconsistent and Ertl's work was critiqued on a number of technical grounds (Callaway, 1975). Others have pursued using EEG and evoked potential measures to assess IQ or learning potential but findings continue to be inconsistent and controversial (see Barrett & Eysenck, 1994).

An early attempt to define brain efficiency is found in a theory first proposed by Thompson (1939). This work was reviewed by Maxwell et al. (1974) in the context of a psychometric study of good and poor readers. The basic idea was that general cognitive ability would require many neurons working whereas specific abilities would require fewer neurons. Maxwell et al. studied 150 different 7-year-old schoolchildren. All completed the WPPSI, a standard test of intelligence designed for children, which is comprised of 10 subscales. The factor structure of the test was studied with a factor analysis, but the factor analysis was done separately in one group of children defined by having good reading scores and in another group defined by poor reading scores. The same three-factor structure was found in each group—a general factor, a verbal factor, and a performance factor. However, the factor loadings on the general factor were consistently higher in the poor reading group whereas the loadings for the verbal factor were lower. Maxwell et al. (1974) interpreted this interesting observation as consistent with Thompson's view that more neurons would be used for general cognitive performance and fewer neurons were used for the specific ability. Maxwell et al. (1974) thought this implied that poor readers made less efficient use of neurons than good readers. They noted that their data, "proposes the surprising hypothesis that high test scores or efficient cognitive functioning require fewer neurons for its elaboration than is otherwise the case" (p. 280; see also Maxwell, 1976). This, of course, is consistent with the subsequent findings of inverse correlations between GMR and scores on the RAPM. Other researchers have replicated and extended Maxwell's psychometric finding of higher factor loadings in lower ability subjects (Detterman & Daniel, 1989).

The pioneering work of Huttenlocher (1979) and colleagues also drew our interest. They had shown with autopsy work that the course of human brain development from birth to age 5 years included a sharp rise in the density of synapses; from age 5 through age 20 years, however, there is an equally sharp decline in synaptic density. The decline, sometimes called neural pruning, is attributed to mechanisms of programmed cell death. A similar developmental pattern is seen in GMR for most brain areas (Chugani et al., 1987). We speculated that a failure of these mechanisms could result in the failure of the normal pruning process (Haier, 1993). This would result in a brain where there were more synaptic connections than normal with the possible interpretation of an inefficient brain due to having too many redundant circuits with resulting higher cerebral GMR. Poor problem-solving ability may result and relatively low scores on psychometric tests of intelligence could be expected. By contrast, a particularly strong pruning process might result in a more efficient brain development where fewer synaptic connections would allow the better problem solving associated with higher performance on intelligence testing and lower cerebral GMR. This train of thought would be consistent with inverse correlations between intelligence test scores and GMR. In fact, a few early reports of synaptic density found higher than normal densities in some cases of mental retardation (Cragg, 1975; Huttenlocher, 1974).

Recent UCI PET/FDG Studies

Mental Retardation. We therefore chose our next study to test whether people with mild mental retardation (IQ between 50 and 75; $n = 10$) of unknown etiology had higher cerebral GMR than normal controls ($n = 10$; Haier et al., 1995). We also tested a group with Down Syndrome ($n = 7$) as a comparison. Each subject completed a CPT of attention during the 32-minute FDG uptake period; the digit stimuli were degraded for the normal controls but not for the retarded or Down groups to help match performance. At the time, no other PET studies of mild mental retardation had been done, and the expectation of most researchers was that we would find lower GMR, suggesting brain damage. The inefficiency hypothesis (i.e., low-IQ people would show high GMR because they had too many circuits) was counterintuitive.

The mean whole brain GMR for the normal group was 36.3 μmol glucose/100g brain tissue/minute. For the retarded group, the mean was 47.4, and for the Down Syndrome group the mean was 46.1. By ANOVA, the group effect was not quite statistically significant (df 2,23; F = 3.27, p = .056) although t tests showed the normals were lower than either of the other groups. These results were consistent with the brain inefficiency prediction.

Each subject also completed structural MRI determinations of brain volume. The retarded and Down groups had smaller brains—about 80% the volume of controls (ANOVA, p = .0024). For the combined group of all subjects (N = 26), the correlation between brain volume and IQ was .65 (r = .36 corrected for extreme groups). This is consistent with a number of other similar studies. The correlation between GMR and IQ was −.58. Interestingly, the correlation between brain volume and GMR was −.69, suggesting that bigger brains use less glucose. A similar inverse correlation between brain size and GMR was reported by Hatazawa et al. (1987). This suggests that neural density or packing of neurons may be an important factor.

An analysis of these data subject-by-subject (Haier, Hazen, et al., 1998a) showed even more striking individual differences. We took each person in the retarded and Down groups and displayed each PET scan as a standard score image. Usually, each image is displayed based on a GMR value for each pixel. For the standard score image, we calculated the mean GMR and standard deviation for each pixel in every scan for the normal control group. For each person in the retarded and Down groups, we calculated a standard score for each pixel person-by-person. The resulting image was thresholded to highlight brain areas where the standard score of GMR exceeded two standard deviations of the normal control values. Thus, in the standard score images, we can see patterns of GMR unique to the individual compared to the normal control group.

Some individuals in both the retarded and the Down groups showed GMR activations in mostly limbic areas, others showed activations mostly in motor areas, others in mostly speech areas, and others mostly in limbic and speech areas. The small sample sizes in these subgroups made it difficult to relate the GMR patterns to clinical features of each case. Nonetheless, the heterogeneity of GMR patterns, even within the Down group, emphasizes the importance of using imaging analyses based on individual differences, especially in groups where brain size or morphology may not be normal. Moreover, this kind of standard-score image may be particularly useful for clinical diagnosis and, possibly, for identifying unique treatment approaches on an individual basis.

As far as the efficiency hypothesis, the study of people with mental retardation and Down syndrome revealed some evidence consistent with a relationship between poor cognitive performance and overactive brains. Other PET results in Down Syndrome are inconsistent (Schapiro et al., 1990; Schwartz et al., 1983) with respect to whether brain activity is increased or decreased; additional studies are planned to investigate this further. Delayed funding for this purpose, however, led us to try other approaches. We decided to expand our efforts with more elaborate research designs in normal groups.

Mathematical Reasoning.   In the next study (Haier & Benbow, 1995), we selected male and female subjects for high or average math ability using the SAT. Each subject then solved mathematical reasoning problems for 32 minutes during PET with FDG. On the basis of our previous work, we would expect that subjects selected for high math ability would have lower cerebral GMR than the subjects selected for average ability. We also took this opportunity to investigate whether there may be any sex differences in brain activation during mathematical reasoning.

A total of 44 right-handed college students volunteered for the study. Eleven males had SAT-Math scores (for college entrance) of 700 or higher (95th percentile of college-bound high school seniors) as did 11 females. Another 11 males and 11 females had SAT-Math scores between 410 and 540 (30th to 68th percentile). The men and women were matched on age as well as for corresponding SAT Math level.

Contrary to the prediction of brain efficiency, the subjects selected for high math ability did not show lower cerebral GMR. Also, contrary to other predictions, there was no evidence of right greater than left hemisphere activation. In the 22 males, there were, however, significant correlations between GMR in temporal lobe areas (bilaterally) and math score attained on the test given during the 32-minute FDG uptake period. These correlations ranged between .42 ($p < .05$) and .55 ($p < .01$) for the areas of middle, inferior, and posterior temporal cortex in left and in right hemispheres. Scatterplots revealed that these correlations were not due to outlier data. In the women, there were no significant correlations between GMR in any cortical area and math score. Thus, although failing to substantiate brain efficiency in the high ability group, this study showed a clear sex difference such that the higher temporal lobe GMR, the better the performance of a math reasoning test in men but not in women (see also Mansour et al., 1996). Following the arguments proposed by Maxwell et al. (1974) brain efficiency may be more apparent when general cognitive ability is studied rather than a more specific ability like mathematical reasoning.

Mental Effort.   About the same time, we completed a PET study specifically designed to compare high and average ability subjects during the performance of easy and hard versions of the same task (Larson et al.,

1995). This study addressed whether mental effort played a key role in GMR determinations.

We recruited 28 normal male, right-handed volunteers after screening potential subjects on the Raven's Advanced Progressive Matrices (RAPM). Half had high scores (range, 28–33; IQ equivalents, 119–131) and half had average scores (range, 14–22; IQ equivalents, 97–107). Each subject completed PET scanning on two occasions about a week apart. For each PET, the subject performed a visual digit-span backwards task presented by computer for the 32-minute FDG uptake period. On one occasion, the task was set so that the subject's performance self-corrected to maintain a correct response rate of 90% (easy condition). This was done by adjusting the number of digits in the string to be remembered. On the other occasion, the correct response rate was adjusted to maintain a 75% accuracy (hard condition).

The strongest prediction for the brain efficiency hypothesis was that subjects selected for high RAPM scores would show lower GMR, irrespective of easy or hard mental effort conditions, than the average subjects. However, ANOVA showed no significant main effect for group or for condition. In fact, there was a trend for higher cortical GMR in the high RAPM group during both the easy and hard conditions. There was a significant group by condition interaction. In the average RAPM group, GMR was lower in the hard condition; GMR was higher in the hard condition in the high RAPM group. As in the math study, the use of a cognitive test tapping a more specific rather than general ability may be an important reason that an efficiency finding was not prominent.

This study and the math study demonstrate the importance of including both ability and effort levels into the research design of functional brain imaging experiments, especially those that attempt to examine individual differences. Clearly, the studies completed so far fail to establish a simple brain efficiency theory (see also Goldberg et al., 1998; Holcomb et al., 1996 for additional evidence of decreased brain function associated with increased effort). We have a number of findings that suggest additional hypotheses to be tested with more elaborate research designs, different cognitive tasks (using more than one strategy), and larger sample sizes.

Emotional Memory. We have completed one more brain imaging study that demonstrates the importance of individual differences. Many animal and some human studies implicate the amygdala in emotional memory. We undertook a PET study (Cahill et al., 1996) to test whether emotional experiences, which subsequently evoke strong memories, activate the amygdala at the time of encoding. Eight right-handed male volunteers completed two PET sessions about a week apart. During each session, the subject watched a 32 minute video during the FDG uptake. On one occasion, the video contained 12 emotionally arousing (aversive) clips depicting, for example, animal mutilation or violent crime. On the other occasion, 12 emotionally neutral clips were viewed, including a court proceeding and a travelogue. Three weeks following the viewings and PET, each subject completed a surprise memory test to determine how many emotional and neutral clips were remembered. Consistent with a number of previous studies, more emotional clips were remembered than neutral clips ($M = 6.25$, $SD = 1.1$ and 2.75, $SD = 0.7$, respectively, $p < .05$, 2-tailed t test).

A standard subtraction analysis compared GMR by t test, pixel-by-pixel, between the emotional and neutral conditions. There was no difference in the amygdala. However, there was a range of memory scores for the emotional clips; not all subjects apparently showed the effect equally. We correlated each subject's emotional memory score (obtained 3 weeks after the PET) with GMR during the emotional viewing, pixel-by-pixel. Few significant correlations were found but these included a correlation of .93 ($p < .01$) in the right amygdala. The corresponding correlation in the neutral condition was .33 (ns). The scatterplot demonstrated that the $r = .93$ was not an artifact of any outlying point.

Thus, the more the amygdala was activated during the emotional experience, the more was remembered about the experience 3 weeks later. Recently, we reported that there is a similar correlation ($r = .91$, $p < 0.001$) in another sample of eight normals between GMR in the hippocampus/parahippocampal region during encoding of a list of nonemotional words and recall scores after 24 hours. There is no correlation with amygdala GMR (Alkire, Haier, et al., 1998).

As in our study of mathematical reasoning, these findings of individual differences, and the lack of a simple group difference in amygdala or parahippocampal activation, demonstrates the fundamental importance of this approach for human cognitive studies using functional brain imaging techniques. Task complexity and performance differences, effort, subject ability level, strategy, and gender/age effects are important variables. As shown in the following brief review of over 20 other PET/learning studies, these variables are not often incorporated into learning paradigms. As a result, there is not yet a strong empirical basis for answering basic questions about where learning takes place in the brain or how the degree of learning is related to brain function.

Review of Other Imaging Studies of Learning

A large number of learning and memory experiments using functional imaging have been reported in the last 10 years. We have listed many of these studies in Table 5.1 based on a review of citations and on Medline

TABLE 5.1

Glucose Metabolic Rate (GMR) and Regional Cerebral Blood Flow (rCBF) Studies of Learning

STUDY	N	GMR/ rCBF	Learning Task	Functional results related to learning/practice
Risberg et al., 1977	12	rCBF	Reasoning	Frontal decreases after a day of habituation using Xenon technique.
Seitz et al., 1990	9	rCBF	Motor	Increases in the cerebellum; decreases in all limbic and paralimbic structures, and striatal decreases that change to striatal increases as task was learned.
Friston et al., 1992	4	rCBF	Motor	Increases during task confined to pyramidal and extrapyramidal motor system. Right cerebellar cortex and cerebellar nuclei show a decline after practice.
Grafton et al., 1992	6	rCBF	Motor	Increases in left primary motor cortex, left supplementary motor cortex, left pulvinar thalamus; no decreases.
Haier et al., 1992a	8	GMR	Visuospatial/ Motor	After practice, GMR in all cortical surface regions decreased while performance increased. The better the performance, the bigger the GMR decrease. A few GMR increases seen in Area 18 of right occipital cortex, right precuneus, right hippocampus, and left cingulate cortex.
Schlaug et al., 1994	9	rCBF	Motor	Increases in left putamen/globus pallidus and left hand area after practice; decreases in right superior and anterior parietal areas and right Broca homologue.
Jenkins et al., 1994	12	rCBF	Motor	After learning, increases in cerebellar vermis cortex, bilaterally in medial thalamus, and anterior cingulate cortex; decreases in temporal cortex bilaterally, bilaterally in posterior insula.
Raichle et al., 1994	12	rCBF	Verbal	Increases after practice in sylvan-insular cortex bilaterally and left medial extrastriate cortex; decreases after practice in anterior cingulate, left prefrontal and posterior temporal cortices, and right cerebellar hemisphere;
Berman et al., 1995	9	rCBF	Reasoning	Activation in the dorsolateral prefrontal cortex remained significant even after training and practice during the Wisconsin Card Sorting Test.
Grafton et al., 1995	12	rCBF	Serial reaction time/spatial	Increases in motor cortex, supplementary motor area and putamen during learning. Greater sequence awareness showed increases in bilateral parietal, superior temporal, and right premotor cortex.

(Continued)

133

TABLE 5.1
(Continued)

STUDY	N	GMR/rCBF	Learning Task	Functional results related to learning/practice
Andreasen et al., 1995a	13	rCBF	Verbal	Practiced and novel story conditions both activated frontal, inferior temporal, thalamic, anterior cingulate and cerebellar regions. Practiced story showed less activation than novel story.
Andreasen et al., 1995b	13	rCBF	Verbal	Practiced and novel word lists activated frontal, parietal, and temporal cortices, thalamus, anterior and posterior cingulate, precuneus, and cerebellum. Practiced list showed less activation than novel list.
Vandenberghe et al., 1995	10	rCBF	Visuospatial	Familiar outline showed decreases in left lateral anterior temporal neocortex, left medial temporal pole, and rostral anterior cingulate compared to unfamiliar outline; no rCBF increases.
Kawashima et al., 1995	10	rCBF	Spatial	Two different types of functional fields in the posterior part of the superior parietal lobule—one active when reaching for target, the other a storage site of visual target in long-term memory.
Logan & Grafton, 1995	12	GMR	Eyeblink Condition	GMR negatively correlated with degree of learning in contralateral occipitotemporal fissure, superior and middle temporal gyri, and ipsilateral posterior inferior cerebellum.
Blaxton et al., 1996	7	rCBF	Eyeblink Condition	Increases as learning increased in left hemisphere in caudate, hippocampal formation, fusiform gyrus, cerebellum, and right temporal cortex and pons; decreases in left frontal cortex with increased learning.
Doyon et al., 1996	14	rCBF	Visuomotor	Striatum and cerebellum activation in implicit acquisition of visuomotor skill. Cerebellum activation also found after acquisition of explicit knowledge.

Study	N	Method	Task	Findings
Passingham, 1996	20	rCBF	Motor and Verbal	Decreases in prefrontal and anterior cingulate cortex with practice during both tasks. Learning two tasks simultaneously (interference/increased effort) is associated with decreases.
Iacoboni et al., 1996	6	rCBF	Spatial/ motor	Reaction times decreased with practice with linear rCBF increases in dorsolateral prefrontal, premotor, and primary cortex of left hemisphere.
Jueptner, Frith, et al., 1997	12	rCBF	Aural Motor	Learning new task showed increases in right prefrontal, caudate, cerebellar nuclei, vermis.
Jueptner, Stephan, et al., 1997	12	rCBF	Motor	Dorsal prefrontal cortex and anterior cingulate cortex were activated during new learning but not during automatic performance.
Shadmehr & Holcomb, 1997	16	rCBF	Motor	Shift from frontal cortex to premotor, posterior parietal, and cerebellar cortex after practice.
Schreurs et al., 1997	10	rCBF	Eyeblink Condition	Right superior temporal, left lateral temporoccipital, and right transverse temporal increases during conditioning. Right superior temporal pole, right and left cerebellar cortex, left inferior temporal pole, and left inferior prefrontal lobe decrease during conditioning.
Berns et al., 1997	10	rCBF	Visual sequences	Ventral striatum responsive to novel information. Decreases observed in right dorsolateral prefrontal and parietal areas.
Deiber et al., 1997	7	rCBF	Arbitrary Mapping	Dorsolateral prefrontal cortex and posterior parietal cortex showed decreases during learning, all in right hemisphere.
Koepp et al., 1998	8	Raclopride	Motor	Dopamine reduced in striatum during video game practice positively correlated with performance level and was greatest in ventral striatum.
Petersen et al., 1998	12, 32	rCBF	Verbal and Motor	Verbal practice produced shift shown by decreases in activity in left frontal, anterior cingulate, and right cerebellar hemisphere and increases in Sylvian-insular cortex. Motor practice showed decrease in activity in right premotor and parietal cortex and left cerebellar hemisphere; medial frontal cortex showed increase.

searches using *learning* as a key word. It is inherently difficult to separate learning from memory studies; here we limit our discussion mostly to those studies reporting at least two scans in the same subject before and after a learning paradigm (studies of encoding, short-term recall, and similar memory paradigms are not our focus). A major point of this review is that the studies listed in Table 5.1 demonstrate such a range of research design and image analysis differences that an integrative review of results is difficult. Most studies have small samples where data from men and women and for young and old subjects often are combined. Few studies use practice or learning beyond a few minutes. Individual differences in task performance are not often used in analyses. One focus for this limited review is whether activation increases or decreases with learning and whether there is a relationship between the change in task performance with practice and the change in brain functioning.

Our Tetris study (Haier, Siegel, MacLachlan, et al., 1992) was the first and only study to use PET imaging before and after a lengthy period of learning (4–8 weeks of practice). As predicted by our previous RAPM results, GMR decreased with learning and, moreover, the largest decreases were found in subjects with the highest scores on the RAPM (Haier, Siegel, Tang, et al., 1992), suggesting that the smartest people become most brain efficient with learning. Tetris is a complex visuospatial motor task. Subsequently, other researchers have studied learning of more simple tasks and a few have studied complex tasks. In most of the 27 studies shown in Table 5.1, rCBF was determined for 1 to 2 minutes in different conditions assessed during the same day scanning session. Only the Haier, Siegel, MacLachlan, et al. (1992) study used FDG for a longer experimental trial (32 minutes) assessed before and after 4 to 8 weeks of practice. In most of these studies, learning and practice are associated with decreases in brain activity in at least some brain areas and increases in other areas. The brain areas involved depend on the nature of the task and some studies show task activation switches from one area to another. Given the many differences among these studies in task selection, spatial resolution, duration of practice, and other methodology, we wish to emphasize in this brief overview only that brain deactivation with practice is now a general finding in functional imaging studies of learning.

Possibly the first regional cerebral blood flow (rCBF) study relevant to learning was done with the Xenon technique and reported by Risberg et al. (Risberg & Ingvar, 1973; Risberg, Maximillian, & Prohovnik, 1977). Twelve subjects were studied, each worked on the Raven's Advanced Progressive Matrices (RAPM) test of abstract reasoning during blood flow determinations; this was done on 2 consecutive days using parallel forms of the test. On the second day, habituation was seen in that frontal areas were less activated than on the first day. This decrease was not apparent in posterior cortex (see also Berman et al., 1995, who reported frontal decreases after short practice on the Wisconsin Card Sort Test in nine subjects).

Subsequent PET blood flow studies began with learning simple motor sequences with fingers. In one of the earliest, Seitz et al. (1990), reported in nine subjects that rCBF increased in the cerebellum and the striatum as the task was learned, but decreases were noted in all limbic and paralimbic areas. Friston et al. (1992) studied only four subjects with a motor task and found rCBF decreases in some areas and increases in others; the very small sample size limited interpretation. Grafton et al. (1992), Schlaug et al. (1994), Jenkins et al. (1994), Grafton et al. (1995), Kawashima et al. (1995), Passingham (1996), Jueptner, Stephan, et al. (1997), Shadmehr & Holcomb (1997), Deiber et al. (1997), and Petersen et al. (1998) all studied motor learning of one kind or another and found rCBF decreases after practice in some areas along with increases in other areas. The Schlaug et al. (1994) report ($n = 8$) reanalyzed data from Seitz & Roland (1992) specifically to examine individual patterns of rCBF related to learning a simple motor sequence. No simple quantitative relationship between individual differences in task performance (after learning and rCBF changes) was found although the pattern of rCBF for each person may be related to individual performance in a qualitative way.

Grafton et al. (1992) reported increases and no decreases in six subjects in a motor learning paradigm. Doyon et al. (1996), used a visuomotor task in 14 subjects and found a number of regional rCBF increases associated with learning and no decreases. Jueptner, Frith, et al. (1997) also reported comparing an aural/motor learned-task to a nonlearned task and found only rCBF increases in the learned task. Thach (1996) has reviewed imaging studies of motor learning and discussed patterns of rCBF increases and decreases after practice.

Using verbal tasks, Raichle et al. (1994) and Andreasen et al. (1995a, 1995b) reported rCBF decreases along with other increases. In the Raichle et al. (1994) report, 12 normal subjects were studied with rCBF determinations during a simple verbal response test before and after 15 minutes of practice. Brain areas activated in the naïve condition decreased with practice and new areas were activated after practice. They concluded that their results showed two distinct circuits are used for saying verbs appropriate to specific nouns and these circuits change with short practice. This was amplified by Petersen et al. (1998) for practice after both verb generation and motor maze tasks. Andreasen et al. (1995a) studied 13 normals as they recalled two complex narratives. One narrative had been practiced and well learned for perfect recall the week before the PET/rCBF determinations and the other was novel—given for the first time just 60 seconds before recall on the PET day. The rCBF patterns for both recall conditions were similar but smaller areas were activated during recall of the practiced story. This was interpreted as consistent with greater neural efficiency following practice. Similar findings from the same study were reported using a practiced or novel word list rather than narratives (Andreasen et al., 1995b).

In a spatial stimulus–response study using six subjects, practice-related increases but no decreases in rCBF were reported (Iacoboni et al., 1996). Vandenberghe et al. (1995) reported in 10 subjects that familiarity of stimuli in a visuospatial task was associated with rCBF decreases, but not with increases. Berns et al. (1997) reported a rCBF pattern of increases and decreases in 10 subjects as they practiced a serial reaction-time task.

Logan and Grafton (1995) is the only other GMR study of learning. They studied eyeblink conditioning in 12 subjects. More learning was correlated with lower GMR. Blaxton et al. (1996) and Schreurs et al. (1997) also studied eyeblink conditioning with rCBF; both reported areas of increase and decrease.

Koepp et al. (1998) reported a PET study designed to measure dopamine release during the performance of a computer game (shooting tanks rather than Tetris). Reduced dopamine in striatum correlated with better task performance.

fMRI and EEG. Our focus is on PET, but a few words on learning studied with EEG/evoked potentials and fMRI are appropriate. EEG measures brain function with a temporal resolution of 1 ms and may be the best imaging technique for studying elemental cognitive processes related to learning. Using high-resolution EEG topography, Gevins et al. (1997), for example, suggested that practice resulted in the activation of fewer cortical resources. The fMRI technique has excellent spatial resolution and can address questions about where in the brain activation or deactivation occurs during short periods of learning over a few seconds. However, given the way fMRI signals are generated, the amount of activation or deactivation cannot be quantified as with the FDG/PET technique. Nonetheless, the relatively wide availability of fMRI has resulted in a large number of interesting studies. Spitzer et al. (1996), for example, showed decreased cortical activation with practice of a word list and Sakai et al. (1998), reported different time courses of changes in four cortical areas during learning a visuomotor task. Several other fMRI findings indicate that larger areas of cortex are activated with learning or effort, suggesting that more neurons are recruited in performing a learned task (see for example, Just et al., 1996; Karni et al., 1995; Karni et al., 1998). If replication studies confirm this general observation, it might seem inconsistent with PET and other imaging studies showing less activation. However, this is not obvious. Larger areas of activation may show less mean activation than smaller areas; this cannot be determined easily with the fMRI technique.

Review Summary

We conclude from this review of brain imaging studies of learning that we are still at an early stage of research. Only a few generalizations about brain deactivation after learning are apparent. More well-designed studies are necessary to test specific hypotheses about brain efficiency and other changes after learning over extended periods of time with a variety of tasks in subjects selected over a range of ability.

Brief Overview of Our Newest PET Studies—The Use of Drug Challenges

We argued in this chapter that task complexity and strategy as well as subject ability, age, and gender are important design parameters for imaging experiments of learning. When future studies incorporate these elements, new hypotheses can be tested about specific brain areas and neurotransmitter systems that are related to learning. When this occurs, even more powerful imaging experiments can be designed using drug challenges. Individual differences are likely to be important at this stage as well.

For example, a PET study of FDG changes during a divided attention task on and off alcohol showed striking individual differences with exactly opposite alcohol effects in some subjects (Haier et al., 1999). We also used PET in a series of studies with anesthetic drugs to examine mechanism/circuitry issues, memory effects, and the neurobiological basis of consciousness. These studies also illustrate the potential of combining drug challenges with functional imaging and we review them briefly.

Consciousness, Memory, and Neurocircuitry. We studied the effects of anesthetic drugs on cerebral GMR in normal volunteers. Each subject has three PET/FDG sessions—one fully awake (conscious), one fully anesthetized (unconscious), and one partially anesthetized (barely conscious). During the FDG uptake period for each session, the subject heard a taped list of words repeated for the 32 minutes. Free recall and forced-choice memory tests were given the following day for words on the list. To date, we have studied three anesthetic drugs: propofol (Alkire et al., 1995), isoflurane (Alkire et al., 1997), and halothane (Alkire, Pomfrett, et al., 1999).

Overall, the unconscious state (defined by loss of responsiveness to verbal and tactile stimulation) shows about a 50% reduction in global cerebral GMR compared to the awake condition for each of the three drugs. There are some regional differences depending on the drug. Some brain areas show GMR decreases during unconsciousness irrespective of drug. These areas tend to be related to sleep. Moreover, the correlation between size of

GMR decrease in a specific area and the density of various receptors in that area suggest that propofol has its greatest effect in those areas richest in GABA receptors (Alkire & Haier, submitted). These data demonstrate the potential for testing hypotheses about the mechanisms of anesthesia and the neurobiological basis of consciousness with functional brain imaging, especially emphasizing individual differences.

The memory data also suggest intriguing possibilities. Only propofol was associated with a greater forced-choice recognition score in the unconscious condition than expected by chance (Alkire et al., 1996a, 1996b). We correlated each subject's memory recognition score with GMR during the unconscious condition, pixel-by-pixel. Then, we did the same for memory recall and conscious GMR. Significant relationships were found in verbal memory areas for both conditions. GMR in the mediodorsal thalamic nucleus was correlated to memory score in the conscious (but not the unconscious) condition. This suggested to us that the thalamus may play an important role in the neurobiology of consciousness.

Conclusion

Our original 1992 report of GMR decrease with performance increase after lengthy practice, was thought to be counterintuitive. Clearly, subsequent studies, over a wide range of tasks and methods, appear to confirm decreases in brain function as performance increases after learning. These findings need careful consideration from a perspective of individual differences. Whether the decreases reflect a shift from some brain areas to other areas or whether the decreases are related to increased efficiency of processing or something else remains to be determined. More than one explanation is likely depending on type of task, ability of subjects, duration of practice, and other variables. Research designs using functional imaging with PET or fMRI or EEG, should incorporate task complexity and level of difficulty, ability levels of subjects, strategy, age/gender comparisons, and the use of drug challenges. The identification of specific brain areas salient to performing a task is only the starting point for functional imaging research.

Analyses should include correlations between task performance and regional cerebral function. Identifying relevant brain systems and circuits is a major goal. Multidisciplinary collaboration is required. As functional imaging techniques become more accessible, especially to psychologists, there is every reason to believe that brain anatomy and brain function can be linked to cognitive and emotional processes of all kinds in new and interesting ways. Brain imaging studies of learning are at an early but exciting stage where many basic questions await research with innovative designs and imaginative hypotheses.

ACKNOWLEDGMENTS

The author thanks Christopher Lawrence for his assistance in preparing this chapter.

REFERENCES

Ahern, S., & Beatty, J. (1979). Pupillary responses during information processing vary with scholastic aptitude test scores. *Science, 205,* 1289–1292.

Alkire, M. T., Haier, R. J., Barker, S., Shah, K., & Kao, J. (1995). Cerebral metabolism during propofol anesthesia in human volunteers studied with positron emission tomography. *Anesthesiology, 82,* 393–403.

Alkire, M. T., Haier, R. J., Fallon, J., & Barker, S. J. (1996a). PET imaging of conscious and unconscious memory. *Journal of Consciousness Studies, 3*(5–6), 448–462.

Alkire, M. T., Haier, R. J., Fallon, J., Barker, S. J., & Shah, N. K. (1996b). Positron emission tomography suggests the functional neuroanatomy of implicit memory during Propofol anesthesia. In B. Bonke, J. G. Bovill, & N. Moerman (Eds.), *Memory and awareness in anesthesia* (pp. –). The Netherlands: Van Gorum Publishers.

Alkire, M. T., Haier, R. J., Shah, N. K., & Anderson, C. T. (1997). A positron emission tomography study of regional cerebral metabolism in humans during Isoflurane anesthesia. *Anesthesiology, 86*(3), 549–557.

Alkire, M. T., Haier, R. J., Fallon, J. H., & Cahill, L. (1998). Hippocampal, but not amygdala, activity at encoding correlates with long-term, free recall of non-emotional information. *Proceedings of the National Academy of Sciences, 95,* 14506–14510.

Alkire, M. T., Pomfrett, C., Haier, R. J., Gianzero, M. V., Chan, C., Jacobsen, B., & Fallon, J. H. (1999). Functional brain imaging during anesthesia in humans: Effects of halothane on global and regional cerebral glucose metabolism. *Anesthesiology, 90*(3), 701–709.

Alkire, M. T., & Haier, R. J. (Submitted). In vivo human evidence that regional cerebral metabolic effects of propofol but not isoflurane are related to benzodiezapine receptor density, submitted.

Andreasen, N. C., O'Leary, D. S., Arndt, S., Cizadlo, T., Rezai, K., Watkins, G. L., Boles Ponto, L. L., & Hichwa, R. D. (1995a). I. PET studies of memory: Novel and practiced free recall of complex narratives. *Neuroimage, 2,* 284–295.

Andreasen, N. C., O'Leary, D. S., Cizadlo, T., Arndt, S., Rezai, K., Watkins, G. L., Boles Ponto, L. L., & Hichwa, R. D. (1995b). II. PET studies of memory: Novel versus practiced free recall of word lists. *Neuroimage, 2,* 296–305.

Barrett, P. T., & Eysenck, H. J. (1994). The relationship between evoked potential component amplitude, latency, contour length, variability, zero-crossing, and psychometric intelligence. *Personality & Individual Differences, 16,* 3–32.

Berent, S., Giordani, B., Lehtinen, S., Markel, D., Penny, J. B., Buchtel, H. A., Starosta-Rubinstein, S., Hichwa, R., & Young, A. B. (1988). Positron emission tomographic scan investigations of Huntington's disease: Cerebral metabolic correlates of cognitive function. *Annals of Neurology, 23,* 541–546.

Berman, K. F., Ostrem, J. L., Randolf, C., Gold, J., Goldberg, T. E., Coppola, R., Carson, R. E., Herscovitch, P., & Weinberger, D. R. (1995). Physiological activation of a cortical network during performance of the Wisconsin Card Sorting Test: A positron emission tomography study. *Neuropsychologia, 33,* 1027–1046.

Berns, G. S., Cohen, J. D., & Mintun, M. A. (1997). Brain regions responsive to novelty in the absence of awareness. *Science, 276,* 1272–1275.

Blaxton, T. A., Zeffiro, T. A., Gabrieli, J. D. E., Bookheimer, S. Y., Carrillo, M. C., Theodore, W. H., & Disterhoft, J. F. (1996). Functional mapping of human learning: a positron emission tomography activation study of eyeblink conditioning. *Journal Neuroscience, 16,* 4032–4040.

Cahill, L., Haier, R. J., Fallon, J., Alkire, M., Tang, C., Keator, D., Wu, J., & McGaugh, J. (1996). Amygdala activity at encoding correlated with long-term, free recall of emotional information. *Proceedings of the National Academy of Sciences, 93,* 8016–8321.

Callaway, E. (1975). Evoked potential latencies and intelligence. In *Brain Electrical Potentials and Individual Psychological Differences* (pp. 43–62). New York: Grune & Stratton.

Chalke, F. C. R., & Ertl, J. P. (1965). Evoked potentials and intelligence. *Life Sciences, 4,* 1319–1322.

Chugani, H. T., Phelps, M. E., & Mazziotta, J. C. (1987). Positron emission tomography study of human brain functional development. *Annals Neurology, 22,* 487–497.

Cragg, B. G. (1975). The density of synapses and neurons in normal, mentally defective and aging brains. *Brain, 98,* 81–90.

Deiber, M. P., Wise, S. P., Honda, M., Catalan, M. J., Grafman, J., & Hallett, M. (1997). Frontal and parietal networks for conditional motor learning: a positron emission tomography study. *J. Neurophysiology, 78,* 977–991.

Detterman, D. K., & Daniel, M. H. (1989). Correlations of mental tests with each other and with cognitive variables are highest for low IQ groups. *Intelligence, 13,* 349–359.

Doyon, J., Owen, A. M., Petrides, M., Sziklas, V., & Evans, A. C. (1996). Functional anatomy of visuomotor skill learning in human subjects examined with positron emission tomography. *European Journal of Neuroscience, 8,* 637–648.

Ertl, J. (1969). *Neural efficiency and human intelligence.* Final report, U.S. Office of Education Project No. 9-0105.

Ertl, J. (1971). Fourier analysis of evoked potentials and human intelligence. *Nature, 230,* 525–526.

Ertl, J., & Schafer, E. W. P. (1969). Brain response correlates of psychometric intelligence. *Nature, 223,* 421–422.

Friston, K. J., Firth, C. D., Passingham, R. E., Liddle, P. F., & Frackowiak, R. S. J. (1992). Motor practice and neurophysiological adaptation in the cerebellum: A positron emission tomography study. *Proc. Royal Soc. London B, 1323,* 223–228.

Gevins, A., Smith, M. E., McEvoy, L., & Yu, D. (1997). High-resolution EEG mapping of cortical activation related to working memory: Effects of task difficulty, type of processing, and practice. *Cerebral Cortex, 7,* 374–385.

Goldberg, T. E., Berman, K. F., Fleming, K., Ostrem, J., Van Horn, J. D., Esposito, G., Mattay, V. S., Gold, J. M., & Weinberger, D. R. (1998). Uncoupling cognitive workload and prefrontal cortical physiology: A PET rCBF study. *Neuroimage, 7,* 296–303.

Grafton, S. T., Hazeltine, E., & Ivy, R. (1995). Functional Mapping of Sequence Learning in Normal Humans. *Journal of Cognitive Neuroscience, 7*(4), 497–510.

Grafton, S. T., Mazziotta, J. C., Presty, S., Friston, K. J., Frackowiak, R. S., & Phelps, M. E. (1992). Functional anatomy of human procedural learning determined with regional cerebral blood flow and PET. *Journal of Neuroscience, 12,* 2542–2548.

Haier R. J. (1990). The end of intelligence research. *Intelligence, 14,* 371–374.

Haier, R. J. (1993). Cerebral glucose metabolism and intelligence. In P. A. Vernon (Ed.), *Biological approaches to the study of human intelligence* (pp. 317–332). Norwood, NJ: Ablex.

Haier, R. J. (1998). Brain scanning and neuroimaging. In H. S. Friedman (Ed.), *Encyclopedia of mental health* (pp. 317–329). New York: Academic Press.

Haier, R. J., Alkire, M. T., Chan, C., & Anderson, C. T. (1997). Functional brain imaging for anesthesiology research: How PET works. *Current Anesthesia and Critical Care, 8,* 86–90.

Haier, R. J., & Benbow, C. (1995). Gender differences and lateralization in temporal lobe glucose metabolism during mathematical reasoning. *Developmental Neuropsychology, 11,* 405–414.

Haier, R. J., Chueh, D., Toychette, P., Lott, I., MacMillan, D., Sandman, C., Lacasse, L., & Sosa, E. (1995). Brain size and glucose metabolic rate in mental retardation and Down Syndrome. *Intelligence, 20*, 191–210.

Haier, R. J., Hazen, K., Fallon, J., Alkire, M. T., Schell, M., & Lott, I. (1998). Brain imaging and classification of mental retardation. In S. Soraci & W. McIlvane (Eds.), *Perspectives on fundamental processes in intellectual functioning* (pp. 115–130). Norwood, NJ: Ablex.

Haier, R. J., Schandler, S. L., MacLachlan, A., Soderling, E., Buchsbaum, M. S., & Cohen, M. (1999). Alcohol induced changes in regional cerebral glucose metabolic rate during divided attention. *Personality & Individual Differences, 26*, 425–439.

Haier, R. J., Siegel, B. V., Crinnella F., & Buchsbaum, M. S. (1993). Biological and psychometric intelligence: Testing an animal model in humans with Positron Emission Tomography. In D. Detterman (Ed.), *New trends in intelligence research* (pp. 157–170). Norwood, NJ: Ablex.

Haier, R. J., Siegel, B. V., MacLachlan A., Soderling E., Lottenberg S., & Buchsbaum, M. S. (1992). Regional glucose metabolic changes after learning a complex visuospatial/motor task: A PET study. *Brain Research, 570*, 134–143.

Haier, R. J., Siegel, B. V., Nuechterlein, K. H., Hazlett, E., Wu, J., Paek, J., Browning, H., & Buchsbaum, M. S. (1988). Cortical glucose metabolic rate correlates of abstract reasoning and attention studied with positron emission tomography. *Intelligence, 12*, 199–217.

Haier, R. J., Siegel, B. V., Tang, C., Abel, L., & Buchsbaum, M. S. (1992). Intelligence and changes in regional cerebral glucose metabolic rate following learning. *Intelligence, 16*, 415–426.

Hatazawa, J., Brooks, R. A., Di Chiro, G., & Bacharach, S. (1987). Glucose utilization rate versus brain size in humans. *Neurology, 37*, 583–588.

Holcomb, H. H., Gordon, B., Loats, H. L., Gastineau, E., Zhao, Z., Medoff, D., Dannals, R. F., Woods, R., & Tamminga, C. A. (1996). Brain metabolism patterns are sensitive to attentional effort associated with a tone recognition task. *Biological Psychiatry, 39*, 1013–1022.

Huttenlocher, P. R. (1974). Dendritic development in neocortex of children with mental defect and infantile spasms. *Neurology*, 203–210.

Huttenlocher, P. R. (1979). Synaptic density in human frontal cortex-developmental changes and effects of aging. *Brain Research, 163*, 195–205.

Iacoboni, M., Woods, R. P., & Mazziotta, J. C. (1996). Brain–behavior relationships: Evidence from practice effects in spatial stimulus–response compatibility. *Journal of Neurophysiology, 76*, 321–331.

Jenkins, I. H., Brooks, D. J., Nixon, P. D., Frackowiak, R. S. J., & Passingham, R. E. (1994). Motor sequence learning: A study with positron emission tomography. *Journal of Neuroscience, 14*, 3775–3790.

Jueptner, M., Frith, C. D., Brooks, D. J., Frackowiak, R. S. J., & Passingham, R. E. (1997). Anatomy of motor learning. II. Subcortical structures and learning by trial and error. *J. Neurophysiology, 77*, 1325–1337.

Jueptner, M., Stephan, K. M., Frith, C. D., Brooks, D. J., Frackowiak, R. S. J., & Passingham, R. E. (1997). Anatomy of motor learning. I. Frontal cortex and attention to action. *J. Neurophysiology, 77*, 1313–1324.

Just, M. A., Carpenter, P. A., Keller, T. A., Eddy, W. F., & Thulborn, K. R. (1996). Brain activation modulated by sentence comprehension. *Science, 274*, 114–116.

Karni, A., Meyer, G., Jezzard, P., Adams, M. M., Turner, R., & Ungerleider, L. G. (1995). Functional MRI evidence for adult motor cortex plasticity during motor skill learning. *Nature, 377*, 155–158.

Karni, A., Meyer, G., Rey-Hipolito, C., Jezzard, P., Adams, M. M., Turner, R., & Ungerleider, L. G. (1998). The acquisition of skilled motor performance: fast and slow experience-driven changes in primary motor cortex. *Proceedings of the National Academy of Sciences, 95,* 861–868.

Kawashima, R., Roland, P. E., & O'Sullivan, B. T. (1995). Functional anatomy of reaching and visuomotor learning: A positron emission tomography study. *Cerebral Cortex, 2,* 111–122.

Koepp, M. J., Gunn, R. N., Lawrence, A. D., Cunningham, V. J., Dagher, A., Jones, T., Brooks, D. J., Bench, C. J., & Grasby, P. M. (1998). Evidence for striatal dopamine release during a video game. *Nature, 393,* 266–268.

Larson, G., Haier, R. J., Lacasse, L., & Hazen, K. (1995). Evaluation of a "Mental Effort" hypothesis for correlations between cortical metabolism and intelligence. *Intelligence, 21,* 267–278.

Logan, C. G., & Grafton, S. T. (1995). Functional anatomy of human eyeblink conditioning determined with regional cerebral glucose metabolism and positron emission tomography. *Proceedings of the National Academy of Sciences, 92,* 7500–7504.

Mansour, C. S., Haier, R. J., & Buchsbaum, M. S. (1996). Gender comparisons of cerebral glucose metabolism during a cognitive task. *Personality and Individual Differences, 20,* 183–191.

Maxwell, A. E. (1976). The learning of motor movements: a neurostatistical approach. *Psychological Medicine, 6,* 643–648.

Maxwell, A. E., Fenwick, P. B. C., Fenton, G. W., & Dollimore, J. (1974). Reading ability and brain function: A simple statistical model. *Psychological Medicine, 4,* 274–280.

Parks, R. W., Crockett, D. J., Tuokko, H., Beattie, B. L., Ashford, J. W., Coburn, K. L., Zec, R. F., Becker, R. E., McGeer, P. L., & McGeer, E. G. (1989). Neuropsychological "systems efficiency" and positron emission tomography. *Journal of Neuropsychiatry, 1,* 269–282.

Parks, R. W., Loewenstein, D. A., Dodril, K. L., Barker, W. W., Toshii, F., Chang, J. Y., Emran, A., Apicella, A., Sheramata, W., & Duara, R. (1988). Cerebral metabolic effects of a verbal fluency test: A PET scan study. *Journal of Clinical & Experimental Neuropsychology, 10,* 565–575.

Passingham, R. E. (1996). Attention to action. *Phil. Trans. R. Soc. Lond. B,* 1473–1479.

Petersen, S. E., van Mier, H., Fiez, J. A., & Raichle, M. E. (1998). The effects of practice on the functional anatomy of task performance. *Proc. Natl. Acad. Sci., 95,* 853–860.

Prabhakaran, V., Smith, J. A. L., Desmond, J. E., Glover, G. H., & Gabrieli, J. D. (1997). Neural substrates of fluid reasoning: An fMRI study of neocortical activation during performance of the Raven's Progressive Matrices Test. *Cognitive Psychology, 33,* 43–63.

Raichle, M. E., Fiez, J. A., Videen, T. O., MacLeod, A. K., Pardo, J. V., Fox, P. T., & Petersen, S. E. (1994). Practice-related changes in human brain functional anatomy during nonmotor learning. *Cerebral Cortex, 4,* 8–26.

Risberg, J., & Ingvar, D. H. (1973). Patterns of activation in the grey matter of the dominant hemisphere during memorizing and reasoning. *Brain, 96,* 737–756.

Risberg, J., Maximillian, A. V., & Prohovnik, I. (1977). Changes of cortical activity patterns during habituation to a reasoning task. *Neuropsychologia, 15,* 793–798.

Sakai, K., Hikosaka, O., Miyauchi, S., Takino, R., Sasaki, Y., & Putz, B. (1998). Transition of brain activation from frontal to parietal areas in visuomotor sequence learning. *Journal of Neuroscience, 18,* 1827–1840.

Schapiro, M. B., Grady, C. L., Kumar, A., Herscovitch, P., Haxby, J. V., Moore, A. M., White, B., Friedland, R. P., & Rapoport, S. I. (1990). Regional glucose metabolism is normal in young adults with Down Syndrome. *J. Cerebral Blood Flow & Metabolism, 10,* 199–206.

Schlaug, G., Knorr, U., & Seitz, R. J. (1994). Inter-subject variability of cerebral activations in acquiring a motor skill: A study with positron emission tomography. *Experimental Brain Research, 98,* 523–534.

Schreurs, B. G., McIntosh, A. R., Bahro, M., Herscovitch, P., Sunderland, T., & Molchan, S. E. (1997). Lateralization and behavioral correlation of changes in regional cerebral blood flow with classical conditioning of the human eyeblink response. *Journal of Neurophysiology, 77,* 2153–2163.

Schwartz, M., Duara, R., Haxby, J., Grady, C., White, B. J., Kessler, R. M., Kay, A. D., Cutler, N. R., & Rapoport, S. I. (1983). Down's syndrome in adults: Brain metabolism. *Science, 221,* 781–783.

Seitz, R. J., & Roland, P. E. (1992). Learning of sequential finger movements in man: a combined kinematic and positron emission tomography (PET) study. *European Journal of Neuroscience, 4,* 154–165,

Seitz, R. J., Roland, P. E., Bohm, C., Grietz, T., & Stone-Elander, S. (1990). Motor learning in man: a positron emission tomographic study. *Neuroreport, 1,* 17–20.

Shadmehr, R., & Holcomb, H. H. (1997). Neural correlates of motor memory consolidation. *Science, 277,* 821–825.

Spitzer, M., Bellemann, M. E., Kammer, T., Guckel, F., Kischka, U., Maier, S., Schwartz, A., & Brix, G. (1996). Functional MR imaging of semantic information processing and learning-related effects using psychometrically controlled stimulation paradigms. *Brain Research. Cognitive Brain Research, 4,* 149–161.

Thach, W. T. (1996). On the specific role of the cerebellum in motor learning and cognition: Clues from PET activation and lesion studies. *Behavioral and Brain Sciences, 19,* 411–433.

Thomson, G. H. (1939). *The factorial analysis of human ability.* London, England: University of London Press.

Thompson, R., Crinella, F. M., & Yu, J. (1990). *Brain mechanisms in problem solving and intelligence.* New York: Plenum.

Vernon, P. A. (1993). Intelligence and Neural Efficiency. In D. Detterman (Ed.), *Current topics in human intelligence: Individual differences and cognition* (pp. 171–187). Norwood, NJ: Ablex.

Vernon, P. A., Nador, S., & Kantor, L. (1985). Reaction times and speed-of-processing: Their relationship to timed and untimed measures of intelligence. *Intelligence, 9,* 357–374.

Vandenberghe, R., Dupont, P., Bormans, G., Mortelmans, L., & Orban, G. (1995). Blood flow in human anterior temporal cortex decreases with stimulus familiarity. *Neuroimage, 2,* 306–313.

COMMENTARY

Breakthroughs in Using Individual Differences to Study Learning: Comments on Goldin-Meadow, Haier, McClelland, Merzenich, and Siegler

James W. Stigler
UCLA

The previous chapters by Goldin-Meadow, Haier, McClelland, Merzenich, and Siegler reflect a surprising convergence in ideas regarding change of neuroscientists, cognitive psychologists, and developmental psychologists. An impressive unity emerges as we watch these scientists struggle to understand the basic processes that underlie learning and cognitive change. As Merzenich points out, the "mind is expressed in physical form as the brain, [which] can in principle be studied to any level of detail." Ten years ago, such a statement might have been interpreted as politeness, equivalent to Chairman Mao's famous exhortation to "let a hundred flowers bloom." However, these five papers represent more than just the blooming of different research traditions. They show, instead, that the study of mind has become genuinely interdisciplinary, with each field contributing in direct ways to the others. Rather than researchers from different disciplines seeming to describe different minds, the portrayals from the three disciplines seem to be evolving into an integrated description of one mind. In this chapter, I focus on this integration, as I see it.

CULTURAL DESCRIPTIONS OF MIND

Individual differences have long been used as a lever for researching how people learn, and the authors discussed here follow in this tradition. Goldin-Meadow seeks to understand the difference between children who are ready to learn a new concept and those who are not; McClelland wonders why adult Japanese speakers have such difficulty learning to distinguish /r/ from /l/; Siegler asks why some children, at some times, invent new strategies, whereas others do not; Merzenich and Haier study the brain correlates of individual learning experiences, and of group differences in intelligence. Significantly, however, each of these researchers departs from the traditional way that psychologists have theorized about individual differences. They do focus on the cognitive, and in that sense they resemble psychologists of the past. Yet, their descriptions of mind are not in terms of abstract traits. Instead, they explain cognitive variation among individuals in cultural terms, using concrete domain-specific language to describe mental representations.

Goldin-Meadow, for example, does not see gesture–speech mismatch as an individual trait, but as a transitory state that individuals pass through in relation to a specific concept within a specific domain of knowledge. McClelland, similarly, explains the deficit shown by native Japanese speakers not as an example of a loss of plasticity but as a product of their specific learning history. Siegler paints a picture of the mind as a collection of specific strategies that, whether individually constructed or passed down through culture, are still specific to, and describable in terms of, the domain in which they apply. Haier, who seeks to relate individual differences to measures of regional glucose metabolic rate (GMR), finds that such measures make sense only when linked to the specific content of his subjects' cognition. Merzenich rejects attempts to understand mental representation in terms of individual neurons, arguing that the mind must be described in terms of neuronal ensembles that can be labeled and described in terms of behaviorally important stimuli.

In each case, the language of explanation is concrete, domain specific, and connected to the specific histories of the learners.

The decision to describe mental representations in cultural terms reminds me of an old paper by the anthropologist Leslie White (1947). White's paper is on first glance somewhat contentious. He contrasts what he calls culturological and psychological explanations of behavior, and basically dismisses the psychological ones, albeit politely, as transitory and insignificant. For White, human behavior "is not a single homogeneous substance like copper or gold, but a compound like water or table salt. Human behavior is made up of two separate and distinct elements, the one biological, the other cultural." In White's view, the biological basis of behavior varies little within our species. Individual and group differences, therefore, are best explained in cultural terms.

Culture, in White's formulation, is "an organization of phenomena— acts (patterns of behavior), objects (tools; things made with tools), ideas (belief, knowledge), and sentiments (attitudes, values)—that is dependent on the use of symbols." Created by man, culture is the environment to which man adapts behaviorally. It is "erected over the species like a great architectural edifice." Like biology, culture is long-lasting and evolves slowly. Although culture is constrained by biology, it cannot be explained in biological terms. Once humans create it, culture "has a life of its own," and "is to be explained in terms of the science of culture, of culturology, not in terms of psychology."

White uses language as an example.

> "A symbolic language would, of course, have no existence were it not for human organisms. But once the linguistic process gets under way it proceeds along its own lines, in terms of its own principles and in accordance with its own laws. The linguistic process is composed of phonetic elements. These interact with one another forming various kinds of combinations and patterns—phonetic, syntactic, grammatical, lexical, and so forth. The language acquires form and structure and uniformities of behavior. In other words, it develops certain principles upon which it rests and in terms of which it functions." (p. 693)

According to White, these principles will be understood by the study of philology, not biology or psychology. Returning to the work of Goldin-Meadow, Haier, McClelland, Merzenich, and Siegler, each appears to recognize the fact that learning, in humans, is largely a process of acquiring culture. Just as culture itself must be described in cultural terms, so too must the process by which culture is learned. In the past, psychologists might have considered the study of how Japanese speakers can acquire the /r/-/l/ distinction to be merely applied psychology. Quite the contrary, nothing could be more basic to the study of human learning than the process by which cultural forms are acquired. These scientists, working to find what is most general about human learning, have made breakthroughs by describing mind in cultural terms.

THREE GENERALIZATIONS ABOUT LEARNING

Having decided to describe the mind in cultural, domain-specific ways, these researchers go on to paint a picture of learning that is highly consistent among themselves. Some of these consistencies include:

1. *Learning is, first and foremost, the construction of new representations within specific content domains.*

Although increasingly efficient use of old representations is also important, these investigators all stress the (re)creation of new representations as the most critical part of learning. This makes sense given what I have argued previously: If mind is best described in terms of specific cultural representations, then the key problem facing the theorist of learning is that of how cultural representations are acquired by individuals.

McClelland, in his discussion of how native Japanese speakers learn to differentiate /l/ and /r/, sees the problem as one of how to replace a unitary representation that covers both phonemes with two specific representations that are different for the two. (Merzenich's view of the dynamic nature of the internal, represented world provides a convincing backdrop against which the processes studied by McClelland take place.) For Siegler, learning in the mathematical domain consists largely of constructing new strategies. The strategies that children generate are mostly determined by the domain itself, along with the children's understanding of the principles that govern legal strategies in the domain. Goldin-Meadow sees learning as the construction of new principles, and the increasing use of these principles to inform understanding of concepts.

In general, learning for these researchers involves the ability to envision new alternatives, whether on the neuronal, perceptual, strategic, or conceptual level. In their research, they have identified some of the conditions that facilitate the envisioning of alternatives, and also some of the obstacles that make this process more difficult.

2. *Exposure to variation is a fundamental requirement for learning.*

Of course, variation is in the eye of the beholder. Learning is enhanced by variation, but variation alone will not produce learning unless it is seen as distinct by the learner. Where the problem lies is: How can we increase the likelihood that variation in input will lead to construction of new representations?

It seems that this is a key concern of all of the researchers. McClelland starts with the problem that native Japanese speakers, when presented with /l/s and /r/s simply cannot hear them as different. Repeated exposure to the two sounds, even in the face of feedback, not only does not help but actually makes things worse, by reinforcing the connection that both /l/ and /r/ have to the single, underlying representation. The solution is to stop this automatic mapping by exaggerating the differences between /l/ and /r/ so much that it becomes impossible to not see the two as distinct. Once two representations are formed, the exaggerations can be gradually lessened.

Siegler and Goldin-Meadow also identify failure to consider variation as a key obstacle to learning, but they conceptualize the problem somewhat differently than does McClelland. They see the problem as one of attention: It takes resources to consider alternatives. Novices tend to have fewer extra resources to spare than do those who are more competent in a domain.

Siegler finds two mechanisms to increase attention to variation in the invention of new strategies. The first of these is conceptual knowledge, or principles, that keep the learner from considering unlikely alternatives, thus leaving the learner free to focus on the likely ones. The second mechanism is automatization of strategy execution. Siegler finds that the invention of new strategies tends to occur on the easy problems rather than the hard ones, where learners are most effective at using their old strategies. The reasoning is that automatization frees up cognitive resources to search for alternative strategies, or construct new ones.

Goldin-Meadow, similarly, zeroes in on gesture as a tool that enables learners to consider more information than they are able to represent in speech alone. Goldin-Meadow and her colleagues have shown that the mismatch of speech and gesture predicts learning, and that the new information represented by gesture plays a causal role in the construction of new, more sophisticated concepts. Merzenich, interestingly, provides some insight as to the brain basis for the mechanisms proposed by Goldin-Meadow: "Temporally coincident, coherent inputs are highly effective in driving representational changes." (p. 68) Gesture allows learners to represent new ideas *simultaneous* with old ones. It is this simultaneous activation of variations in ideas that appears to enable learning.

3. *Learning takes time.*

Variation can come from a number of sources. It can be self-generated or can be stimulated by an instructor. In either case, it is important that the learner be able to consider variation over time, not just once. In fact, in none of these lines of research do we see support for the "Eureka!" type of learning that amounts to a sudden "Aha!" In Siegler's research, there is little fanfare to accompany the first appearance of a new strategy, and the new strategy comes to be applied only gradually. In Goldin-Meadow's research, there is quite a long time after information is indicated gesturally before it finds its way into students' verbal explanations. In McClelland's work, one could argue that the exaggeration in input acts to slow down a process that has been happening too fast. Time alone does not produce learning, yet neither does simple exposure to new strategies, new ideas, or new phonemes. There must be time to incorporate the new into the evolving cognitive system.

PRACTICAL CONSIDERATIONS

Finally, I want to speculate on what some of the educational implications might be of the work done by these three researchers.

The first issue I consider is the role of errors in the educational process. There has been some disagreement over the years in what the proper role of errors should be. Behaviorist psychologists have implied that errors are bad, because they reinforce incorrect or undesirable behaviors. More cognitively oriented psychologists have taken a different view: Their goal, in an educational setting, is to change thinking, not simply behavior. And, they see the learner as a thinker, and not just a behaver. For them, errors are an inducement to thinking and problem solving, and analysis of errors can lead to learning.

Although all of the researchers discussed here come from a cognitive tradition, they nevertheless contradict each other in the value they apparently assign to errors. Siegler, on the face of it, takes a negative view of errors, in that innovation and new strategies tend to follow on success with easy problems, not struggle with difficult ones. McClelland, also, paints a negative picture of errors. Errors, in McClelland's view, cause learners to strengthen activation of incorrect representations, thus making it more difficult to construct new pathways to new, more desirable representations. Goldin-Meadow, on the other hand, sees a more positive role for errors in the learning process. Indeed, it is students' explanations of incorrect responses that provide the context for construction of new understandings.

Embedded within this apparent contradiction is a way of thinking through the role that errors might play in the learning process. In all three of these researchers' views, learning requires the opportunity to focus on new ideas, be they perceptual representations, strategies, or concepts. Interestingly, each researcher's view on errors derives from this principle. In McClelland's work, a rapid incorrect mapping of perceptual input onto a unified /r/-/l/ representation is not bad because it is an error but because it forecloses the opportunity to focus on the distinctiveness of the input. For Siegler, similarly, it is not that errors themselves are bad, but that automaticity is good, freeing up cognitive/resources for generating new solution strategies. Goldin-Meadow's more positive view of errors can be explained using the same principle: in the case of conceptual learning, focusing on errors provides learners with the time, opportunity, and motivation to search for alternative ways of understanding the domain in question.

Learning is a dynamic process in which some things are best sped up, others slowed down. For Siegler, errors slow you down where you need to speed up, in order to save time for constructing new strategies. (But, Siegler also believes that when the task is to choose among several existing strategies, the slowing down that results from using error-prone strategies is a positive thing, in that it promotes greater use of more accurate approaches.) For McClelland, errors speed you up where you need to slow down, right at the point where a new representation needs to be considered. For Goldin-Meadow, errors slow you down right where you need to be slowed down in order to consider the contradictions that can induce new conceptual understandings. It is not sufficient to conclude that learning takes time; it does, but at specific points in time, learners need to focus on specific things. As Merzenich points out, a complete understanding of mind will depend critically on understanding how "neural processes take place in context and through time."

A second implication is really just a half-baked thought based on McClelland's work. McClelland finds that the way to foreclose the hasty mapping of /r/ and /l/ onto the same underlying representation is to present the learner with exaggerated input that highlights and amplifies the key differences between /r/ and /l/. It strikes me that this approach to the enhancement of learning might be a justification for case studies as a way to advance research. Case studies are limited in some ways, for example, small samples, nonrandom selection of cases, and so forth. However, if cases are chosen because they are unusual or distinctive in some way, they may advance research in the same way exaggerated input helps the native Japanese speaker learn English. Slowing down to take a hard look at an unusual case may enable researchers to construct a new, more differentiated representation of the processes they are trying to study. Especially in education, where true variation is hard to find, case studies of real innovation might help the field to envision potentially quite valuable alternatives.

REFERENCE

White, L. A. (1947). Culturological vs. psychological interpretations of human behavior. *American Sociological Review, 12*, 686–698.

Part II

STUDIES OF CHANGE
OVER LONGER
TIME SCALES

BEHAVIORAL APPROACHES

Chapter 7

Dynamic Mechanisms of Change in Early Perceptual–Motor Development

Esther Thelen
Indiana University

This chapter takes a detailed look at a single developmental transition in infancy. The transition, the onset of successful goal-directed reaching, is important in itself. The ability to grasp and manipulate objects is a major milestone in the development of human cognition. Control over our arms and hands is central to our lives as human beings, allowing us to use tools, communicate with others, provide caregiving, and create music and art. The very nature of our cognition is shaped by the fact that we have agile, highly differentiated appendages. Much research has documented the intimate connections between thought and manual actions of all kinds (Jeannerod, 1997).

I focus on this transition for a second, equally important reason: To describe the processes of developmental change in a way that begins to capture its complexity and its richness. By necessity, we must simplify and isolate in order to understand. Sometimes it is also useful to step back and try to recapture some of the intricacies as well. To do this, I report on research in my own laboratory as well as work done by others, as there is a long history of work on this important sensorimotor behavior. The picture drawn here is not simple, but even in its complexity, it is far from complete, and much work remains to be done.

A DYNAMIC SYSTEMS APPROACH
TO DEVELOPMENT

Although my goal is description, it is description strongly grounded in the principles of dynamic systems theory. The outlines of this theory are now fairly well known, so I focus first on some core assumptions (see Thelen, 1989; Thelen & Smith, 1994, 1997).

The first assumption is that developing humans, like all living organisms, are composed of multiple, heterogeneous parts that cooperate together to produce coherent and patterned behavior. This coordination of the parts happens without a central executive giving commands, but rather through the interactions of the components, such that the system is truly self-organizing. The patterns produced in such systems depend critically on the organism itself, but also on the constraints of the environment in which the organism resides. In this sense, it is theoretically impossible to isolate any one component as a single cause of developmental change. System coherence means that the participating elements are always coupled, and causality is always circular. This is the *systems* part.

Second, the patterns that self-organize can themselves be complex, but they are always continuous in time. This is the *dynamic* part. Continuity of process applies to components at many different levels of organization ranging from ion channels in cell membranes, through network properties of populations of neurons, through the time scales of synaptic change and consolidation, and into longer-term development. The critical point here is that processes occurring on these different time scales are completely nested within one another, as well as being coupled to one another. For our thinking, therefore, the major challenge—at whatever level—is to understand how activity on the short time scales, say on the order of milliseconds, or seconds, becomes transduced to the behavioral changes we are most interested in, which happen in days, weeks, and months.

The third major assumption is about *stability*. When patterns self-organize, they may be variously stable or unstable. In complex dynamic systems, stable patterns must lose stability in order to change. The system may be destabilized by a variety of means. Sometimes the coherence of a pattern is threatened by new input, but just as often, by changes in one of the organic components as well. Instability allows the components to self-organize in new ways. The goal of developmental research therefore, is to describe patterns, characterize their stability, and then ask what parts of the system or the environment are important in engendering the loss of old patterns and the emergence of new ones. The appearance of entirely new forms of behavior, such as reaching for objects, appear as reorganizations of underlying processes that themselves are continuous in time. These ideas, of course, are not unique to dynamic systems theories. Many, if not most, developmental theories propose change as a succession of equilibrium states arising from disequilibria (cf. Siegler, Case, this volume). All recognize that patterns that are stable do not change, and that the system must open up the possibilities for learning by selecting new combinations of components. How do these principles, then, help us to understand the universal behavior of learning to reach? The first step is a task analysis to identify the contributing components and processes.

THE TASK OF LEARNING TO REACH

Babies first reach out and grab objects placed in front of them at about 4 months of age. There is considerable individual variability in this milestone, however. Among the 10 infants we have followed extensively in my laboratory, age of reach onset varied from 12 to 24 weeks. Piaget (1952) saw a similar range in his own children, with Laurent reaching at 3 months and Jacqueline at 6 months. Whatever is involved in this new skill, it takes some infants twice as long as others to acquire it. This should also serve to remind us what we already know: Age is an imperfect indicator of level of skill.

To begin to understand how multiple, continuous processes can lead to the emergence of the novel behavior of reaching and grasping, it is useful to identify, as best we can, just what infants must do to reach and grasp objects.

1. Infants must have some motivation to obtain an object in the world around them. We know that this is normally provided by vision. Infants universally visually fixate objects before reaching for them. Blind infants, with normal motor functions, are severely delayed in this milestone as they are denied the usual visual motivation (Fraiberg, 1977).

2. Infants must thus have the visual capacity to detect and localize a graspable object in reachable space. If the baby cannot accurately see the target, they cannot control their movements toward it (see review in Kellman & Arterberry, 1998).

3. Related to Items 1 and 2, infants must have, or soon learn, what objects afford reaching and grasping and what constitutes a *reachable space*, that is, what they can successfully obtain by extending their arms or moving arms and torso (Bruner & Koslowski, 1972; Field, 1977).

4. Infants must have some means by which to transduce the coordinates of the object in external, Cartesian space, into a frame of reference suitable to moving the arm on the shoulder, which must be internally referenced (e.g., Flanders, Tillery, & Soechting, 1992).

5. Infants must have some, even rudimentary, ability to plan their movements ahead of actually executing them. This requires recognizing what muscle activations and what forces move the limbs into what positions in space ahead of actually moving them. Otherwise, the movement will be badly misdirected right from the start and require large corrections (Hofsten, 1993).

6. Infants must have some—perhaps also rather crude—ability to correct movements on the basis of sensory feedback once they are executed. No movement is ever planned in an entirely feedforward manner. We know that adults, children, and even older infants have a good ability to rapidly correct reach trajectories online on the basis of visual or proprioceptive feedback (proprioception is how the arm *feels*) (see, for example, Desmurget et al., 1995).

7. Infants must have enough control of their muscles to actually lift the arm against gravity and keep it sufficiently stable while the hand moves forward and then while the object is grasped. This is biomechanically an extremely complex task, and indeed an engineering nightmare. Muscles and tendons are highly nonlinear—that is, they have very different properties depending on how much they are currently stretched and how fast they are stretching. Moreover, because the arm is a linked segment, forces delivered to one segment are transmitted to the other segments (Hollerbach & Flash, 1982). Thus, the arm must be stabilized to avoid the problem of the marionette—pulling only one string activates the whole puppet and not always in desired ways.

8. For the same reason of linked segments, infants must also stabilize their heads and trunks so that when the arm is lifted, the reactive forces generated do not cause them to slump over. These processes are sometimes called *anticipatory postural reactions* because such adjustments are often done before the focal movement begins. In reality, they may be planned and executed as part of a single movement synergy (Savelsbergh & van der Kamp, 1993).

9. Finally, infants must have some ability to remember the action that they just performed in relation to their success in obtaining the goal. This is critical for any improvement to take place. The networks' underlying movements that are more successful and less effortful should be competitively strengthened in comparison with others less good (Sporns & Edelman, 1993).

Thus, the emergence of even that first awkward and inaccurate reach requires the cooperative interactions of every part of the infant's body and nervous system. Here, it is appropriate to mention that there is a vigorous debate among researchers in motor neuroscience on just how the nervous system solves the problems of controlling the arm. From a computational point of view, the complexity is nearly intractable. Indeed, despite intense effort, and highly sophisticated models, engineers have still not built a robot arm with the immense flexibility, gracefulness, adaptibility, and speed of the human prototype. Although any one robot arm can do some pretty fancy things, the robots are brittle. They can do what they are programmed to do, but no more. Even young babies are more flexible. The complexity inherent in an engineering solution is not such a problem for a biological system, however, even a quite immature one.

How Infants Learn to Reach

How, then, do babies do it? The most complete and developmentally elegant theory of this sensorimotor transformation comes, not surprisingly, from Piaget (1952). Piaget spends a considerable portion of his *Origins of Intelligence* providing rich descriptions of the construction of this skill in his own children. Piaget's account is dynamic and epigenetic, and should be required reading of every student of development. According to Piaget, visually elicited reaching and grasping evolve gradually from the progressive modification and melding of several more elementary schemata, especially the schema for visually interesting objects, vision (and feel) of the hand, and the reflexes of grasping and sucking. As is well known, Piaget believed that with each repetition, a schemata changes by assimilating the environment and that this progressing activity itself leads to change in the schemata, the process of accomodation. The critical, rate-limiting step is the integration of the sight of the object and the sight of the grasping hand. In Piaget's account, this coordination is acquired as infants first link these two visual components of reaching by chance encounter of hand and object within the visual field. Then, by repetition, infants succeed in getting the hand and eye schemata together. The Piagetian position was echoed by White, Castle, and Held (1964), who described a period of intense hand-regard as a prerequisite for reaching. However, it should be noted that White et al. (1964) observed mainly institutionalized infants who spent large portions of their day supine and with little stimulation other than their own hands.

Thus, Piaget's position is essentially constructionist, or putting a skill together from initially independent parts. More recently, Trevarthen (1984) offered a different account. Using stop-framed movie films, he described visually elicited smooth arm extension movements in newborn infants, so-called "prereaching." This was about the same time that Hofsten's (1982) careful and pioneering studies demonstrated similar movements—infrequent, and unsuccessful, but apparently well-aimed, reach-like movements in newborn and very young infants. Both authors reported that these early movements decline in frequency, to be replaced some months later by the familiar, clumsy first attempts at reaching.

These authors differ from Piaget, therefore, in suggesting that vision and hand do not need to be coupled in development, but that, as Trevarthen (1984) noted, ". . . infants are born with a considerable part of the neural structures that will coordinate these functional patterns . . . in adults" (p. 247). Development, in this view, consists of refining and remodeling of patterns that are already there. In particular, both Trevarthen and von Hofsten propose a major neural reorganization at about 4 months, whereby cortical centers become functional and coordination between the more cortical control of grasping and the subcortical innate connections is established.

The issue of whether hand–eye coordination is present at birth or is acquired by experience matching object and hand is still somewhat unresolved. Recently, van der Meer, van der Weel, and Lee (1995) reported remarkable results of newborn infants actively "working" to bring the hand into the visual field, suggesting a similar strong innate coupling to facilitate hand–object visual matching. On the other hand, an important study by Clifton, Muir, Ashmead, and Clarkson (1993) showed that at the time of their very first reaches, infants were equally adept at hitting the target with or without vision of their hands. If visual matching did occur in the early months, it was clearly not necessary for early reaches because the "feel" of the arms and the hands sufficed.

These accounts all capture aspects of early reaching, and my purpose here is not to choose among them. Rather, I prefer another look at this milestone from a dynamic perspective, where task, motivation, vision, proprioception, and trajectory control are coequal contributors.

A Dynamic Account

A common theme among previous accounts of reach emergence is that infants come into the world with some existing organization that facilitates the developmental cascade. The nature of the preorganization is debated—Piaget believed in "reflexes" and others talk of "amodal matching of hand and eye." All agree that something is there beforehand, whether or not it actually looks like reaching. It is useful to recast these classic issues in the terms of dynamic systems. What patterns of behavior are stable and easily elicited at birth? What destabilizes these patterns? What new patterns emerge? In this way, we need not look for continuity of form—the elements that coalesce to produce reaching need not look like reaching to begin with. Nor do we need to propose any sort of hard-wiring. Rather, we look for patterns that may assemble under some circumstances and dissolve under others. In particular, I suggest that the stable perinatal patterns work to establish a motivational matrix that allows reaching to truly self-organize from other nonreaching components.

Motivation: A Dynamic Cascade.   The first question I address is: Why reach at all? Learning to reach is hard work. What keeps infants trying, despite many unsuccessful efforts? Do they need a "reach icon" in their brains to motivate this behavior?

Following Piaget, we can now construct a story where reaching emerges from very simple, nonreaching precursors. Following Edelman (1987) and others, I suggest that what is necessary is a rather general, biologically plausible, and quite nonspecific bias in the system, or in Edelman's words, "value." *Value* means that certain neural connections, when activated, are intrinsically favored and their activation strengths are increased. We know from many neural net simulations that systems need such biases to reach a stable solution, biases either present in the initial connections or acquired through differential input. For example, the autonomous robot built by Almássy, Edelman, and Sporns (1998) learned to visually discriminate real world objects with different patterns (blobs and stripes). Initially, these objects differed to the robot only in their value—some tasted good and others were aversive. However, by repeated and varied movements, the robot came to recognize the favored objects by vision alone. The robot appeared to develop the motivation to seek out objects with a particular pattern.

Likewise, motivation to reach can itself emerge dynamically from a system with a few simple biases. Here are my candidates for what starts this cascade in human infants:

1. A bias to look at visually complex parts of the environment. In Piaget's terms, interesting sights act as *aliments* to the vision schema and they are sought actively by babies.
2. A bias to stimulate the tactile receptors in the mouth. "Anything in the mouth feels good." Again in Piaget's terms, this is the reflexive sucking system, which seeks its own activation.
3. A bias to grasp when the tactile receptors in the hand are stimulated.
4. Linkages between these system such that activation of any one also activates one or more of the others.

Reaching itself need not be specifically prefigured: These initial biases set up the motivational substrate for reaching to emerge in the context of other general changes in the system.

There is good evidence for such initial biases and linkages. The first three patterns have been noted for many years. First, since the groundbreaking work of Fanz (1958, 1963) it is well established that infants' attention is captured by high-contrast patterns, and moderately complex stimuli, including faces and bull's-eyes, and that this bias can be detected, however painstakingly, in newborns. Preference for complexity increases with age, although there is likely an optimal range that maximizes attentiveness. Babies like to look at interesting visual displays.

Second, the drive to suck and to suck on objects placed in the mouth is of course, extremely strong at birth. Sucking movements are established early in fetal development, appear as the first patterned movements after birth, and remain powerfully reinforcing throughout infancy. Spontaneous mouthing movements are also extremely common (Wolff, 1987). Third, the grasp response to tactile stimulation is also elicited easily at birth (Prechtl & Beintema, 1964).

However, there is also strong evidence for coupling among these systems (Rochat, 1993). Several modalities are linked to movements of the mouth. For instance, stimulation of the hand, as well as fluid in the mouth, leads to mouthing movements (Blass, Fillion, Rochat, Hoffmeyer, & Metzger, 1989; Korner & Beason, 1972; Rochat, 1993; Rochat, Blass, & Hoffmeyer, 1988). Rochat (1993) provided evidence that this hand–mouth linkage is established early in prenatal development. Very young infants also respond with oral activity to interesting *visual* sights. Jones (1996) showed increases in mouthing and tongue protrusion movements in 6-week-old infants when they were shown flashing lights.

There is a strong tendency to put the hand in the mouth as well. Spontaneous hand-to-mouth contacts are seen at birth (Butterworth & Hopkins, 1988) and throughout the early months (Lew & Butterworth, 1997). Moreover, Korner & Beason (1972) reported a concurrence between hand-to-mouth contact and visual alerting in newborn infants. Beginning at about 2 months, when experimenters put objects in infants' hands, the contacts with the mouth increased (Lew & Butterworth, 1997; Rochat, 1989). Overall, there appears to be a tight coupling between vision, tactile input to the hands and mouth, and movements of the hands and mouth.

Thus, in the first months of life, the mouth is activated by both visual and tactile stimuli, and is the target for both the hands and for objects placed in the hands, whatever is available. Early hand-to-mouth movements look like infants are "reaching" for the mouth with their hands (Bower, Broughton, & Moore, 1970), although the presumed target is felt, not seen (Butterworth & Hopkins, 1998). Our own data further illustrate these patterns: Figure 7.1 shows the frequency of different types of hand-to-face and hand-to-mouth contacts in the first half of the year in four individual infants (G. Smith, 1998). The infants were seated and offered an object at midline and shoulder height. Consider first the time before reach onset, which is indicated by the vertical line in the graph. At first, hand-to-face contacts were most common. As infants' arm control improved, more of their movements actually hit the mouth target. The near-misses decreased as successful hand-to-mouth movements increased. The first step in reaching is to reach for one's own mouth because stimulating oral receptors is intrinsically pleasurable.

Here, then, is how a few of these simple biases might initiate a motivational cascade that culminates in goal-directed reaching for objects: Stim-

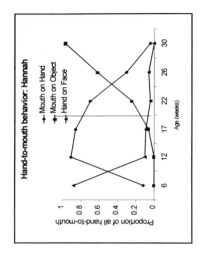

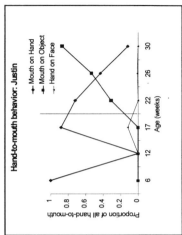

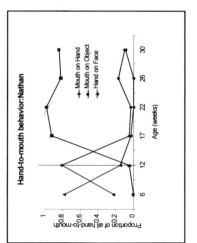

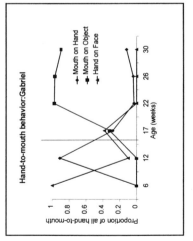

FIG. 7.1.   Hand-and-object to mouth behavior in four infants followed longitudinally, showing the shifting proportions of hand contacts with the face and mouth and object contacts with the mouth.

ulation of oral receptors by nipples, objects, and their own hands is presumably pleasurable for infants and they repeat the actions that get things into the mouth. This bias is perfectly plausible as a selective advantage in a feeding system. This leads to considerable practice in bringing hands and objects held in their hands to their mouths. When infants are given toys to hold, they explore them visually, haptically, and orally. What is left for infants to discover is how to actually grab the objects themselves by extending their arms. Reaching is a natural solution for obtaining a distal object to place in the mouth (Bruner, 1969; Rochat, 1993). Object reaching and grasping need not be acquired directly through evolutionary design, but rather indirectly as a consequence of biases for feeding, which probably do have a strong selective advantage.

Several other lines of evidence support the view that reaching is an emergent online solution for oral exploration. First, when infants could reach out and grasp toys, they substituted these toys for their hands in their mouths. As seen in Fig. 7.1, infants switched from hands-in-the-mouth to objects-in-the-mouth at the time of reach onset. Oral activity continued with what was available. Second, the "empty" oral exploration movements of tongue protrusions in the presence of visually interesting sights dramatically declined just at the time of reach onset (Jones, 1996). Mouth openings remained relatively stable. After reach onset, infants still opened their mouths in anticipation of the grasped object, but the tonguing movements were now presumably used in service of oral exploration and could no longer be seen. Finally, is the intriguing phenomenon described by Rochat (1993) of "oral capture." Here, infants unable to reach nonetheless attempted to grab the offered toy with their mouths directly. This is an excellent example of "motor equivalence," or using whatever motor solution is available to attain the object in the mouth. Infants are flexible and problem-solving: When objects are unavailable for oral exploration, hands will do. When hands are not fully skilled in reaching, mouths will do. This opportunistic flexibility is further illustrated in Fig. 7.2. Nathan, in his first week of successful reaching, has reached out with both hands to grab the toy, offered on a device in front of him. He missed the toy, but grasped the other hand instead, and both hands are immediately transported to his mouth.

Moving Hands to a Distal Target. The desire for oral exploration motivates infants to solve the problems that remain: How to extend their arms to a specific location in space; hold them up against gravity; keep the head and torso steady; and eventually, grasp the offered toy. Again, following Edelman (1987; Sporns & Edelman, 1993), I suggest that this motor skill is acquired through selection—generation of many movement possibilities and retention of movement parameters that satisfy the goal

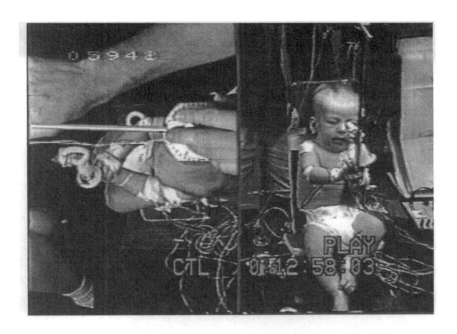

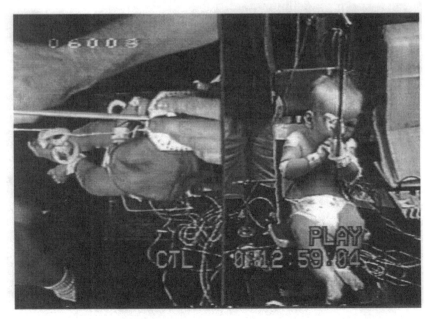

FIG. 7.2.   The mouth as motivation to reach. Nathan at 12 weeks, the first week of successful reaching. In this example, Nathan has reached bilaterally for a toy presented to him at midline (top photo). He missed the toy but grabbed his other hand and brought both hands to his mouth.

state. When the baby is successful in capturing the toy and bringing it to the mouth, the neural pathways that produced the desired end are selectively strengthened, and the movement parameters associated with that reach are more likely to be activated again. Many repetitions in many situations result in a class of movements that have been favored in the past with the object capture and oral reward. Although we do not yet know what movement parameters are most important at particular stages in the process, we can suggest some likely candidates of what infants must learn as they attempt to reach.

Generating Task-Appropriate Forces. An essential component of goal-directed reaching is the ability to generate forces in the arm appropriate to carry the hand to the target and to position the hand for a grasp. Adults, children, and even older infants are very adept at gauging what forces are sufficient to move the hand to the target, but not excessive, and in timing those forces to accelerate and brake their movements efficiently. Newly reaching infants must learn these parameters.

We have reason to believe that the process of learning to modulate arm forces may be occurring many weeks before infants are actually successful reachers. In our longitudinal studies, we observe infants visually fixating the toy and moving their arms (albeit not clearly aimed toward the toy) many weeks before reach onset. More importantly, infants are continually moving their arms spontaneously since their day of birth. Thus, by 3 or 4 months, infants may well have had hundreds of hours of moving and learning about the *feel* of their arms.

Although all infants move in their first months, the style of their movements can vary greatly. In our studies, some infants were especially active and produced extended, stiff, fast, forceful limb actions. Others were more flexed, quiet, and slow (Thelen et al., 1993). These movement characteristics were quite stable within an infant. Figure 7.3 provides an illustration of the wide differences in the speeds of early nonreaching movements and the corresponding differences in preferred arm postures for a subset of those infants (Spencer et al., 2000).

What this means for the process of controlling forces is that infants must individually discover the appropriate speeds from the background of their characteristic styles. Our infant Gabriel, for example, had to damp down his very vigorous movements in order to successfully reach, and he did. In contrast, Hannah, who moved slowly and spent considerable time with her hands flexed near her face, had to activate her arms more to extend them out in front of her. She indeed increased her characteristic movement speed around the onset of reaching (Thelen et al., 1993). Reaching is thus sculpted from ongoing movements of the arms, through a process of modulating what is in place and presumably remembering

Speed differences in pre-reaching movements

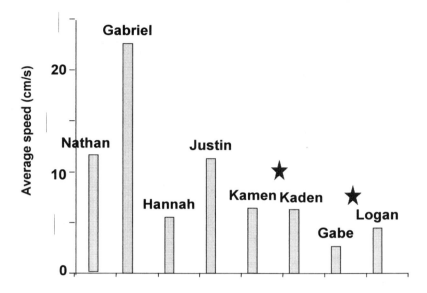

Individual differences in preferred arm postures

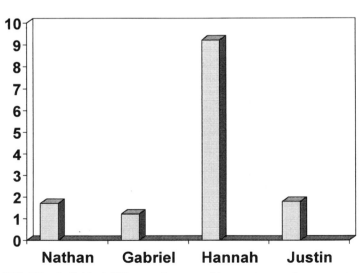

FIG. 7.3. Individual differences in prereaching postures and movements. Top panel: Mean 3-dimensional speed of all prereaching movements in six infants followed longitudinally. The first four infants were observed from 3 weeks. Kamen and Kaden were identical twins followed from 12 weeks. Gabe and Logan were also identical twins followed from 18 weeks. Bottom panel: Percent of time spent in a flexed arm postures before reach onset.

force parameters that lead to successful approaches to the toy. The process of discovering task-appropriate forces and modulating their timing and amplitude continues at least throughout the first year, and into the second (Konczak et al., 1995, 1997; Thelen, Corbetta, & Spencer, 1996).

Associated with the problem of generating appropriate arm forces to move the arm forward is the issue of lifting the arm up against gravity and holding it sufficiently steady to grasp the toy. Because of the gravity vector, holding the arms straight and extended when supine is especially effortful (try reading when lying flat on your back), and indeed, upright posture appears to facilitate reaching in babies (Out et al., 1998; Spencer et al., 2000). In either posture, lifting the arm requires strength and control in the muscles of the shoulder and neck.

Our longitudinal electromyographic (EMG) recordings of shoulder, neck, and arm muscles revealed a dramatic change in the activation patterns of these muscles associated with the onset of reaching. Spencer & Thelen (2000) developed a method to determine what proportion of time each of the four monitored muscles (biceps, triceps, deltoid, and trapezius) and all possible combinations were active during all reaching and nonreaching movements. As shown in Fig. 7.4, prereaching movements were largely generated by activation in the upper arm muscles, biceps and

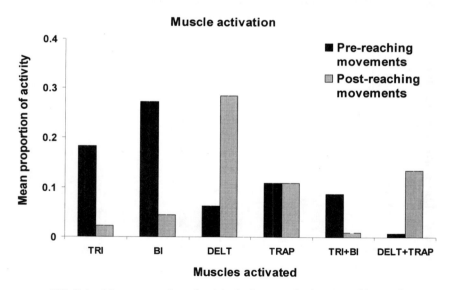

FIG. 7.4.   Mean proportion of activity in four muscles in prereaching and postreaching movements for the four singleton infants. TRI = triceps, BI = biceps, DELT = anterior deltoid; TRAP = trapezius. TRI + BI; DELT + TRAP refer to coactivity, i.e., when muscles were both active at the same time.

triceps, which serve to flex and extend the elbow. In contrast, the proportion of time with the deltoid (extension at the shoulder) and trapezius (neck and shoulder) muscle activation was significantly increased in the weeks following reach onset.

One possible explanation of these results is that infants do not use neck and shoulder muscles before about 3 months of age because they have no reason to extend and lift their arms: They could occupy the spaces that these muscles bring the arm to, but they do not. Further analysis revealed, however, that muscle pattern differences are not simply a function of a change in the postures and movements of the arms between the pre-reaching and postreaching periods, but are an actual shift in the muscles used. Spencer and Thelen (2000) began the analyis by partitioning infants' three-dimensional reaching space into 11 sectors. They then calculated, on the basis of the three-dimensional coordinates of infants' hands during reaching and nonreaching movements, the relative time spent in each sector. This revealed infants' predominant movement configurations, that is, how close or far away from the body and whether the hand was more likely to be high or low. As expected, infants did spend more time in spaces close to the body in the early months. However, when Spencer and Thelen asked what muscle combinations infants used to get to each sector in space, the analysis showed that in the prereaching period, biceps and triceps, alone or coactivated, predominated, even for areas far from the body and high up (Fig. 7.5). After reaching onset, the deltoids and trapezius muscle groups were much more frequently used for spaces that required lifting and extending the arms.

A possible explanation is that during the first 3 or 4 postnatal months, infants' movements are influenced by the predominantly flexor posture assumed in utero, and that shoulder and neck antigravity muscles would have comparatively little activation compared to the elbow flexors. Movements to all areas in the reaching space are accomplished using these distal muscles proportionately more often. Only after several months of exercise in a gravitational environment—with practice supporting the head and lifting the arms, do infants gain sufficient strength and control of shoulder and neck muscles to use them consistently for arm lifting and extension. This strength and control is necessary for voluntary reaching and holding the arms steady. Again, I suggest that shoulder and neck strength are not specific to reaching, but are gained through infants' more general movements—waving, wiggling, grasping, swiping, and so on. The association with reach onset means that such changes are necessary, but not in themselves sufficient for reaching to occur. When these muscles gain such strength, they are used for holding the head steady, supporting the torso in sitting posture, and lifting the head and supporting the torso on the arms when in prone, all of which co-occur around the time of reach onset

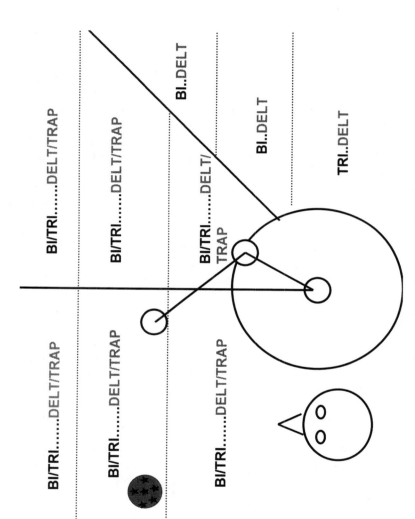

FIG. 7.5. Predominant muscle activity when infants' hands were in different areas of the potential reaching space. The first muscle or muscle pair is during the prereach period; the postreach period follows after the three dots. Adapted from Spencer & Thelen (2000).

(Spencer et al., 2000). Indeed, in the four infants tracked longitudinally, there was a strong association with head control and reach onset such that being able to hold the head centered in midline when supine was a necessary (but again, not sufficient) prerequisite for reach onset (Spencer et al., 2000).

More on Arm Control

I have shown thus far how motivation to reach out and grab objects emerges from the convergence of a few simple, biologically plausible biases, such that infants are highly interested in distal objects in order to put them in their mouths. I have further suggested that one rate-limiter in the actual emergence of reaching is the ability to lift and steady the arm against gravity. Finally, I proposed that infants gain this strength, as well as the control of muscle force amplitude and timing, as they move their arms freely and experience a range of movement speeds and consequences. Here, I show how this process is aided by constraints seemingly inherent in the neuromuscular architecture from early in life.

The issue here involves another problem in arm trajectory control—the so-called "degrees of freedom" problem (Bernstein, 1967). In general form, the question is how the brain coordinates all the possible combinations of many body segments, muscle fibers, neurons, and so on to produce coordinated movement because the number of these combinations taken independently is nearly limitless. One particular instance of the degrees of freedom problem concerns the mechanical interactions of the body segments. When people produce forces at any one joint in the articulated limbs and torso, forces are also transmitted to the other connected segments. For instance, if one produces a shaking movement at the shoulder while the arm is relaxed, the lower arm and the hand will flop around. Only when the elbow and wrist joints are stiffened can oscillations at the shoulder be fully controlled, say to hammer a nail. Voluntary movement requires that these mechanical interaction forces be precisely counterbalanced to prevent unwanted inertial effects (Hollerbach & Flash, 1982). Another example is pushing a heavy door: Postural muscles must stabilize the trunk to counteract the forward push of the extended arm to prevent falling over backward. So in the realm of forces for movement, too many degrees of freedom mean lack of control.

Bernstein suggested that the degrees of freedom of the neuromotor system was constrained by parts naturally working together as synergies, such that fewer elements need be independently controlled. Recently, Gottlieb and his colleagues (Gottlieb et al., 1996a, 1996b) discovered such a constraint in the realm of forces that move the joints. In particular, Gottlieb and colleagues found that in adults' natural and comfortable reaching movements, the forces moving the shoulder and elbow are not independently controlled. Rather, these forces (torques rotating the joints) are allocated in a proportional manner, such that forces moving the elbow are always a stable proportion of those moving the shoulder. This is illustrated for an adult movement in Fig. 7.6, showing the two joints controlled as a *linear synergy*. Presumably, then, the neural control is significantly simplified in a way that takes account of inertial interactions. This is not a hard-wired link: Adults can violate this synergy, and they do, as when

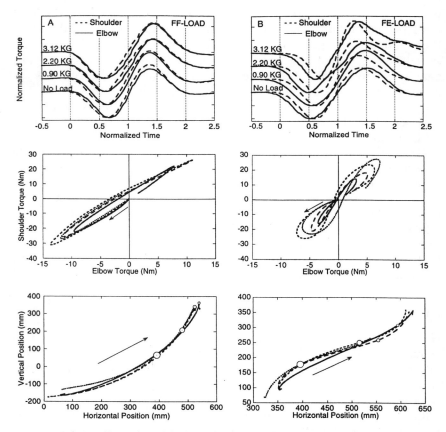

FIG. 7.6.   Illustration of the linear synergy in adult reaching. The top panels show the torques (forces) rotating the shoulder and elbow joints when a participant had different weights attached to his wrist. In the left panels, the participant was instructed to flex the shoulder and the elbow. In the right panel, he was told to flex the shoulder and extend the elbow. The paths of his hand are shown in the bottom panels. The center panels illustrate the linear synergy, when the elbow torque is plotted as a function of the shoulder torque. From Gottlieb et al. (1996b).

performing certain movements, such as reaching across their midline or to a position far to the left or right. It is not rigid or inevitable.

The adults tested by Gottlieb and colleagues had been reaching hundreds of times each day for many years. But what about infants? If such a natural constraint were available early in life, it would also vastly simplify the problem of learning to reach, because infants would not have to control each joint independently. Indeed, this appears to be the case: Zaal et al. (1999), discovered that even in early non-goal-directed and seemingly uncontrolled prereaching movements, infants displayed a nearly perfect linear synergy. This relationship persisted throughout the first year as arm control greatly improved and as infants transversed from no accurate reaching, to early, poorly executed attempts through the more adult-like reaches seen at the end of the first year. This is illustrated in Fig. 7.7, which shows examples of trajectories taken from three epochs: prereaching, early reaching, and late, more stable trajectory control. Remarkably, despite many changes at the level of the hand path and at the joint angles, the constraints at the level of the linear synergy were evident.

This means that, by reducing the problem space, the neural architecture in place early in life facilitates the exploration and selection of appropriate forces to move the hand to a desired location in space. Again, I must emphasize that this constraint is not an icon of reaching nor does it prefigure reaching. Infants do not have knowledge of reaching built-in. Rather, this constraint makes learning easier and optimizes infants' own discoveries of what works to get the toy into the mouth and what does not. In addition, while adults easily break the synergy by voluntarily moving in different ways than comfortable reaching, infants seem to be rather tightly constrained as we found virtually no violations of the principle (Zaal et al., 1999). This suggests that infants at first reach using movements that are simplified by the linear synergy constraint, and only later, learn to break the synergy with more unusual movements. In short, early reaching movements are carved out from a background of more limited possibilities, but they are possibilities that work.

CONCLUSION

An infant's first goal-directed reach is a new behavior. I have argued here that this new form emerges from the confluence of many processes that have been occuring over the first months of life, none of which have reaching itself prefigured in them. It is only when all of these come together that we can identify the behavior as an arm extension to a visually fixated target that results in grasping an object and bringing it to the mouth. I have suggested mechanisms for several of these converging processes:

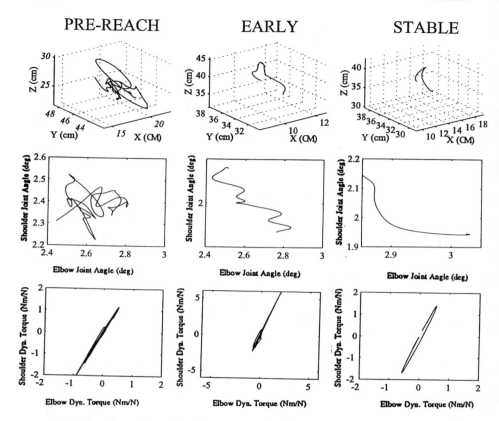

FIG. 7.7. Illustration of the linear synergy in infants. The three panels show the same infant sampled in the prereach period, in the first month of reaching, and in the stable last quarter of the first year. The top panels are the three-dimensional hand paths; the middle panels plot the elbow joint angle versus the shoulder joint angle. The bottom panel is the shoulder and elbow torques, calculated as in Gottlieb et al., 1996b.

motivation, control of movement speed, reducing the degrees of freedom of segmented limbs, and the use of proximal muscle groups. All of these components are critically important, but they are no more or less important than those I did not address such as visual localization, coordinate transformation, and movement planning, learning, and memory. Future research must address both the mechanisms of the individual components and how they coalesce and mutually influence each other.

I had two, related goals for this chapter: (a) to describe a single developmental transition and (b) to try and capture the dynamic, complex, contingent nature of any ontogenetic change. Describing complexity can be an alien exercise for most of us in developmental studies. We are trained by our models of science to isolate and manipulate a single variable and to minimize complexity by holding other variables as constant as we can. In addition, most of our studies capture but a snapshot of processes that are continuous in time. The take-home message of a dynamic systems approach, however, is that can be seductive to assign the manipulated variable as the single cause of developmental change, or to assume that a snapshot picks out a time-independent characteristic of the system. Sometimes contributions to change are nonobvious and often components themselves change in individual and nonlinear ways, but the system is always multiply causal, and it is always enduring in time.

Motor development provides a particularly transparent entry into these complexities of change because the contributions of the nonmental components are real, observable, and cannot be overlooked. The child's tasks seems apparent and the solutions he or she generates to solve them can be overtly measured over time. The issues with what are considered more traditional cognitive tasks are harder, because many of the changes must be inferred indirectly and cannot be read out as a trajectory over time. Yet, I believe the principles are identical. When children learn to solve a cognitive problem, it is an open question as to what underlying mechanisms engendered the change. Every task requires attention, perceptual processing, memory, planning, idiosyncratic experience, and a rich context, including the child's postures and movements, which may be variously familiar or novel. It is not even clear that individual children of the same age find the same routes to solving similar problems.

One method of investigating such multiple causality within the framework of conventionally good experimental design is to manipulate not just one variable, but to systematically vary several of the components that contribute to performance. For instance, Smith, Thelen, and colleagues have used this approach to explain infant perseverative reaching, or the tendency for 8- to 12-month-old infants to reach persistently one of two targets even when the cue changed. This is the behavior described by Piaget as his well-known A-not-B error, traditionally believed to be a measure of infant knowledge of hidden objects. Rather than focus on the static, and single property of what infants know, Smith, Thelen, and colleagues deconstructed the task into one involving multiple processes, in particular: looking, planning, reaching, and remembering (Smith et al., 1999; Thelen et al., in press). Using the single task, they systematically varied aspects of the visual task environment (Diedrich et al., in press a), locus of visual attention (Smith et al., in press), the number, order, and location of the trials (Smith et al., 1999), the memory load (Smith et al., 1999), the infants' posture (Smith et al., 1999), and the proprioceptive input to the limbs (Diedrich et al., 2000). The results were that performance could be affected by all of these manipulations. Infants perseverated or not in a predictable fashion, depending on the relative influence of these multiple factors and how they interacted over time (see Thelen et al., in press). The message for development is that if all of these factors could influence change in real-time performance, then any or all of them may contribute to longer-term changes as well. As infants become older, their attention becomes more focused, *and* their perceptual discrimination improves, *and* their memories get better, *and* their movements become more skilled. A rich, complex, and realistic account of change must include this dynamic interplay.

ACKNOWLEDGMENTS

I thank Melissa Clearfield, Daniela Corbetta, Kristin Daigle, Fred Diedrich, Gerald Gottlieb, Christian Scheier, Gregor Schöner, Gregory Smith, Linda Smith, John Spencer, Beatrix Vereijken, and Frank Zaal for invaluable contributions to the research reported here, and Bob Siegler for his helpful editorial suggestions. This research was supported by NIH RO1 HD22830 and an NIMH Research Scientist Award.

REFERENCES

Almássy, N., Edelman, G. M., & Sporns, O. (1998). Behavioral constraints in the development of neuronal properties: A cortical model embedded in a real world device. *Cerebral Cortex, 8*, 346–361.

Bernstein, N. (1967). *Coordination and regulation of movements.* New York: Pergamon Press.

Blass, E. M., Fillion, T. J., Rochat, P., Hoffmeyer, L. B., & Metzger, M. A. (1989). Sensorimotor and motivational determinants of hand–mouth coordination in 1–3-day old human infants. *Developmental Psychology, 25*, 963–975.

Bower, T. G. R., Broughton, J., & Moore, M. K. (1970). Demonstration of intention in the reaching behavior of the neonate. *Nature, 228*, 679–681.

Bruner, J. S. (1969). Eye, hand, & mind. In D. Elkind & J. Flavell (Eds.). *Studies in cognitive development: Essays in honor of Jean Piaget* (pp. 223–236) New York: Oxford University Press.

Bruner, J. S., & Koslowski, B. (1972). Visually prepared constituents of manipulatory action. *Perception, 1*, 3–14.

Butterworth, G., & Hopkins, B. (1988). Hand–mouth coordination in the newborn baby. *British Journal of Developmental Psychology, 6*, 303–314.

Clifton, R. K., Muir, D. W., Ashmead, D. H., & Clarkson, M. G. (1993). Is visually guided reaching in infancy a myth? *Child Development, 64*, 1099–1100.

Desmurget, M., Prablanc, C., Rossetti, Y. Arzi, M., Paulignan, Y., Urquizar, C., & Mignot, J. C. (1995). Postural and synergic control for three-dimensional movements of reaching and grasping. *Journal of Neurophysiology, 74*, 905–910.

Diedrich, F. J., Highlands, T., Spahr, K., Thelen, E., & Smith, L. B. (in press a).The role of target distinctiveness in infant perseverative search errors. *Journal of Experimental Child Psychology.*

Diedrich, F. J., Thelen, E., & Smith, L. B. (2000). Infant spatial location memories include perceptions of limb mass. Manuscript submitted for publication.

Diedrich, F. J., Thelen, E., Corbetta, D., & Smith, L. B. (in press b). Motor memory is a factor in infant perseverative errors. *Developmental Science.*

Edelman, G. M. (1987). *Neural Darwinism.* New York: Basic Books.

Fanz, R. L. (1958). Pattern vision in young infants. *Psychological Record, 8,* 43–47.

Fanz, R. L. (1963). Pattern vision in newborn infants. *Science, 140,* 296–297.

Field, J. (1977). Coordination of vision and prehension in young infants. *Child Development, 103,* 48–57.

Flanders, M., Tillery, S. I. H., & Soechting, J. F. (1992). Early states in a sensorimotor transformation. *Behavioral and Brain Sciences, 15,* 309–362.

Fraiberg, S. (1977). *Insights from the blind.* New York: Basic Books.

Gottlieb, G. L., Song, Q., Hong, D., Almeida, G. L., & Corcos, D. M. (1996a). Coordinating movement at two joints: A principal of linear covariance. *Journal of Neurophysiology, 75,* 1760–1764.

Gottlieb, G. L., Song, Q., Hong, D., & Corcos, D. M. (1996b). Coordinating two degrees of freedom during human arm movement: Load and speed invariance of relative joint torques. *Journal of Neurophysiology, 76,* 3196–3206.

Hofsten, C. von (1982). Eye-hand coordination in the newborn. *Developmental Psychology, 18,* 450–461.

Hofsten, C. von (1993). Prospective control: A basic aspect of action development. *Human Development, 36,* 253–270.

Hollerbach, J. M., & Flash, T. (1982). Dynamic interactions between limb segments during planar arm movements. *Biological Cybernetics, 44,* 67–77.

Jeannerod, M. (1997). *The cognitive neuroscience of action.* Cambridge MA: Blackwell.

Jones, S. J. (1996). Imitation or exploration? Young infants matching of adults' oral gestures. *Child Development, 67,* 1952–1969.

Kellman, P. J., & Arterberry, M. E. (1998). *The cradle of knowledge: Development of perception in infancy.* Cambridge, MA: MIT Press.

Konczak, J. Borutta, M., Topka, H., & Dichgans, J. (1995). Development of goal-directed reaching in infants: Hand trajectory formation and joint torque control. *Experimental Brain Research, 106,* 156–168.

Konczak, J., Borutta, M., & Dichgans, J. (1997). The development of goal-directed reaching in infants. II. Learning to produce task-adequate patterns of joint torque. *Experimental Brain Research, 113,* 465–474.

Korner, A. F., & Beason, L. M. (1972). Association of two congenital organized behavior patterns in the newborn: Hand–mouth coordination and looking. *Perceptual and Motor Skills, 35,* 115–118.

Lew, A. R., & Butterworth, G. (1997). The development of hand–mouth coordination in 2- to 5-month old infants: Similarities with reaching and grasping. *Infant Behavior and Development, 20,* 59–69.

Out, L., van Soest, A. J., Savelsbergh, G. J. P., & Hopkins, B. (1998). The effect of posture on early reaching movements. *Journal of Motor Behavior, 30,* 260–272.

Piaget, J. (1952). *The origins of intelligence in children.* New York: International Universities Press.

Prechtl, H. F. R., & Beintema, D. J. (1964). The neurological examination of the full-term newborn infant. *Clinics in Developmental Medicine (No. 12)* London: Heinemann Medical Books.

Rochat, P. (1989). Object manipulations and exploration in 2- to 5-month old infants. *Developmental Psychology, 25,* 871–884.

Rochat, P. (1993). Hand–mouth coordination in the newborn: Morphology, determinants, and early development of a basic act. In G. J. P. Savelsbergh (Ed.), *The development of coordination in infancy* (pp. 265–288). Amsterdam: North Holland.

Rochat, P., Blass, E. M., & Hoffmeyer, L. B. (1988). Oropharyngeal control of hand–mouth coordination in newborn infants. *Developmental Psychology, 24,* 459–463.

Savelsbergh, G. J. P, & van der Kamp, J. (1993). The coordinatin of infant's reaching, grasping, catching, and posture: A natural physical approach. In G. J. P. Savelsbergh (Ed.), *The development of coordination in infancy* (pp. 289–317). Amsterdam: North Holland.

Smith, G. (1998). Hand-to-mouth behavior in the first 6 months. Unpublished Masters' Thesis, Indiana University.

Smith, L. B., Thelen, E., Titzer, R., & McLin, D. (1999). Knowing in the context of acting: The task dynamics of the A-not-B error. *Psychological Review, 106,* 235–260.

Spencer, J. P., & Thelen, E. (2000). Spatially specific changes in infants' muscle co-activity as they learn to reach. *Infancy, 1,* 275–302.

Spencer, J. P., Vereijken, B., Diedrich, F., & Thelen, E. (2000). A dynamic systems study of posture and the emergence of manual skills. *Developmental Science, 3,* 216–233.

Sporns, O., & Edelman, G. M. (1993). Solving Bernstein's problem: A proposal for the development of coordinated movement by selection. *Child Development, 64,* 960–981.

Thelen, E. (1989). Self-organization in developmental processes: Can systems approaches work? In M. Gunnar & E. Thelen (Eds.), *Systems and development: The Minnesota symposia on child psychology, Vol. 22* (pp. 77–117). Hillsdale, NJ: Lawrence Erlbaum Associates.

Thelen, E.. Corbetta, D., Kamm, K., Spencer, J. P., Schneider, K., & Zernicke, R. F. (1993). The transition to reaching: mapping intention and intrinsic dynamics. *Child Development, 64,* 1058–1098.

Thelen, E., Corbetta, D., & Spencer, J. P. (1996). The development of reaching during the first year: The role of movement speed. *Journal of Experimental Psychology: Human Perception and Performance, 22,* 1059–1076.

Thelen, E., Schöner, G., Scheier, C., & Smith, L. B. (in press). The dynamics of embodiment: A field theory of infant perseverative reaching. *Behavioral and Brain Sciences.*

Thelen, E., & Smith, L. B. (1994). *A dynamic systems approach to the development of cognition and action.* Cambridge, MA: Bradford Books/MIT Press.

Thelen, E., & Smith, L. B. (1997). Dynamic systems theories. In R. M. Lerner (Ed.) *Theoretical models of human development. Handbook of child psychology. Vol. 1.* (5th edition; pp. 563–634). Editor-in-chief: William Damon. New York: Wiley.

Trevarthen, C. (1984). How control of movement develops. In H. T. A. Whiting (Ed.), *Human motor actions: Bernstein reassessed* (pp. 223–261). Amsterdam: North-Holland.

van der Meer, A. L. H., van der Weel, F. R., & Lee, D. N. (1995). The functional significance of arm movements in neonates. *Science, 267,* 693–695.

White, B. L., Castle, P., & Held, R. (1964). Observations on the development of visually-directed reaching. *Child Development, 35,* 349–364.

Wolff, P. H. (1987). *The development of behavioral states and the expressions of emotions in infancy.* Chicago: University of Chicago Press.

Zaal, F. T. J. M., Daigle, K., Gottlieb, G., & Thelen, E. (1999). An unlearned principle for controlling natural movements. *Journal of Neurophysiology, 82,* 255–259.

Chapter 8

Differentiation, Integration, and Covariance Mapping as Fundamental Processes in Cognitive and Neurological Growth

Robbie Case
School of Education, Stanford University
and
Institute of Child Study, University of Toronto

Michael P. Mueller
Institute of Child Study, University of Toronto

The general goal of the present volume is to lay the groundwork for a more integrated understanding of behavioral and neurological change. Achieving this goal is complicated by the fact that each of these two types of change has been studied in three different theoretical traditions, each with its own methods, its own language, and its own implicit model of the relationship between brain and behavior. In this chapter, we first consider each of these three traditions separately; we then go on to suggest how work from the three traditions can be integrated, and used to help us understand the neurological and cognitive changes that take place in the development of a specific competence—namely, the competence that is involved in comparing two numbers, and deciding which of the two is larger.

DIFFERENT THEORETICAL PERSPECTIVES ON THE RELATIONSHIP BETWEEN BRAIN AND BEHAVIOR

The Empiricist Perspective

The epistemological roots of the first tradition lie in British empiricism, as articulated by Locke (1690/1975) and Hume (1748–51/1975). According to the empiricist approach, knowledge of the world is acquired by a process in

185

which the sensory organs first detect stimuli in the external world, and the mind then detects the customary patterns or conjunctions in these stimuli. Developmental psychologists who have been influenced by this view have tended to view the goals of developmental psychology as describing: the process by which new stimuli are discriminated and encoded (perceptual learning); the way in which correlations or associations among these stimuli are detected (cognitive learning); and the process by which new knowledge is accessed, tested, and/or used in other contexts (transfer). The general methodology that has been favored is one drawn from experimental physics, namely: make detailed empirical observations of children's learning under carefully specified conditions; generate explanations for these observations that are clear and testable; and conduct experiments to test these hypotheses—experiments that manipulate the variable that is hypothesized to affect performance, and rule out any possible rivals.

Under the influence of behaviorism, early work in this tradition concentrated on developing a small number of paradigms in which children's behavior could be recorded in a clear fashion, and the variables that produced changes in this behavior could be carefully manipulated (e.g., see Kendler & Kendler, 1967). After the cognitive revolution, experimenters began looking at children's thinking on a broader array of tasks and developing more detailed theoretical models of the cognitive processes that underpin children's performance—models that included such internal constructs as goals, strategies, and knowledge bases. Computer simulations of these models were also developed, first using production systems (Klahr & Wallace, 1976) then connectionist networks, (McClelland, Rumelhart, & Hinton, 1987) and, more recently, a combination of these two approaches (Siegler, this volume).

Work on the neurological substrate of children's cognition in this tradition has proceeded independently from the work on children's thinking but has been grounded in the same basic assumptions. Among these assumptions are the following: (1) different types of external stimuli trigger the activation of different types of neurons or neuronal groups in the cortex, (2) lawful connections or patterns that are obtained among these stimuli are represented by physiological associations among the corresponding neurons or neuronal groups; and (3) the physiological processes that produce these associations are time-sensitive and incremental. Thus, every time two neuronal groups are activated at the same time, or in immediate succession, a physiological change takes place in the synapses that connect them. The result is that firing of the first element is more likely to stimulate firing of the second element on a subsequent exposure, and less likely to stimulate firing of the various alternatives (Hebb, 1949). The more general consequence is that, with repeated exposure, the spatiotemporal structure of neuronal firing comes to mirror the spatiotemporal structure of the events in the external world.

The foregoing (Hebbian) conception of learning has generated a number of interesting lines of empirical work. At the sensory level, the emphasis has been on determining what kinds of cells are responsive to what kind of input. At higher levels, the emphasis has been on mapping the higher-order areas to which these receptors project and determining which particular circuits are involved in executing which kinds of higher-order decisions and processes. Finally, using animal models and single-cell recordings, advances have been made in charting the synaptic changes that take place while learning actually takes place. Results from all three lines of work have been broadly congruent with Hebb's position.

Developmental work in this tradition has focused on a number of different questions. Among the most important have been: (a) What kind of changes take place, with normal experience, in the responsiveness of the cortical circuits that are known to be crucial for particular types of adult competence?, and (b) Do critical windows exist within which particular areas and circuits are particularly sensitive to particular types of experience? In response to the first question, early results suggest that—once children can succeed at a particular task—the same general circuit tends to be operative as for adults. However, the speed with which the components of the circuit are activated is usually slower and the amplitude of activation greater for quite a prolonged period (McCandliss, Posner, & Givon, 1997; Taylor, 1988, 1995; Temple & Posner, 1988).

In response to the second question (i.e., sensitive periods), early findings suggested that, if an animal is completely deprived of early sensory experience, then the areas of cortex that would normally be used to encode that experience may be used for other purposes, and will not develop normally even when normal sensory conditions are restored (Cynader & Frost, 1999). In keeping with this finding, imaging techniques have shown that naturally occurring differences in the early linguistic experience of young children are associated with different patterns of cortical functioning when the children are fully grown, even when later linguistic experience is equivalent (see Neville, this volume). Taken together, these findings support the notion that sensitive periods exist in development, and that early experience should receive particular attention (Keating & Hertzman, 1999). Although this conclusion has been of great interest to early childhood educators, it has not gone unchallenged. Thus, in recent years, there has been an attempt to show, both theoretically and practically, that a good deal of neural plasticity remains, providing the nature of the early learning is understood and appropriate remedial experience is crafted (see McClelland, this volume, and Merzenich, this volume).

The Rationalist (Structuralist) Perspective

A second theoretical tradition in which children's cognitive development has been studied is quite different from the first, and has been associated with a different set of behavioral and neurological investigations. This tradition has drawn its inspiration from Continental rationalism and its further development by Kant in his Transcendental Philosophy. Kant (1796/1961) suggested that knowledge is acquired by a process in which organization is imposed by the human mind on the data that the senses provide, not merely *registered* as the empiricists had suggested. Examples of concepts that played this organizational role in Kant's system were space, time, and causality. Kant argued that, without some preexisting concept in each of these categories, it would be impossible to make any sense of the data of sensory experience, that is, to see events as taking place in space, as unfolding through time, or as exerting a causal influence on each other. He therefore believed that these categories must exist in some a priori form rather than being induced from experience.

Developmental psychologists who were influenced by Kant's view tended to see the study of children's cognitive development differently from those influenced by empiricism. They presumed that one should begin by exploring the foundational concepts with which children come equipped at birth; then, go on to document any change that may take place in these concepts with age. The first developmental theorist to apply this approach was Baldwin (1894/1968), who proposed that children's conceptual schemata progress through a sequence of four universal stages, which he termed the stages of sensorimotor, quasilogical, logical, and hyperlogical thought. In any given stage, Baldwin believed that new experience was assimilated into the existing set of schemata, in a manner similar to that in which the body assimilates food. He saw transition from one form of thought to the next as driven by accommodation, a process by which existing schemata are broken down and then reorganized into new and more adaptive patterns. Finally, he saw children's conceptual understanding in each of Kant's categories as something that children construct, not something that is inborn, or that is simply written in by experience.

Baldwin believed that the process by which these patterns were constructed required the active intervention of the child's attention. For this reason, he suggested that two criteria had to be met for the reorganization process to take place. First, children had to be exposed to experience that demonstrated the limits of their existing schemata. Second, they needed to develop a certain minimum span and power of attention, so that they could attend to all the relevant schematic elements, and assemble them into a new configuration. He believed that the span and power of attention, in turn, were dependent on the maturation and integration of the cortex. Neurological variables that he felt contributed to cortical integration (and hence to the increase in attentional span) included the myelinization of neurons and the formation of new dendritic connections among different regions of the brain.

Baldwin called on subsequent generations of biologists to chart out the changes that take place with age in children's understanding of Kant's categories, and to map the cortical events on which these changes are dependent. The person who took up the first of these challenges was Piaget (1960, 1970). Using modern biology (not physics) as his model, he developed a set of methods designed to map out the psychological structures that children develop at each major stage of their lives, and the processes by which these structures are transformed as children move from one stage to the next. The basic elements of his procedure were:

1. Present children with a wide (indeed, a vast) range of simple problems or questions in each of Kant's categories, at each of the major stages outlined by Baldwin
2. Interview children to determine the reasoning on which their response is based
3. Hypothesize both an age-typical pattern of responding and an underlying structure (analogous to biological structure, but based on logic) that might generate this pattern
4. Invent new tasks that would serve to test and refine the formulation concerning the nature of this underlying logical structure
5. Hypothesize a set of adaptive processes that would serve to transform early logical structures into later ones.

In addition to the processes of differentiation and integration that Baldwin had mentioned, Piaget included *reflexive abstraction* in this latter category. In Piaget's lexicon, reflexive abstraction was a superordinate process in which the subject takes his or her own internal structures as objects and abstracts a new pattern in them.

Given Piaget's heavy emphasis on logic, it is not surprising that his work did not have much immediate impact on neurological investigations. In recent years, however, neo-Piagetian theorists have sought to reintroduce many of Baldwin's earlier notions, and to integrate them with Piaget's (Case, 1985; Fischer, 1980; Halford, 1993; Pascual-Leone, 1970). Thatcher (1992) developed a line of research that is broadly congruent with Piagetian and neo-Piagetian theory and that explicitly draws on Baldwin's notion regarding the underlying neurological basis of these stages: namely, the degree of cortical integration. Using second order methods that correlate one pattern of EEG activation with another, Thatcher has focused on the coherence between activation in different parts of the brain.

Because his interest is in charting structural changes in the overall connectivity of the cortex, he does not ask children to engage in any task that might engage one particular circuit or another differentially. Rather, he simply asks them to relax and shut their eyes. Under these conditions, what he has shown is that—with age—there is decreasing local coherence across adjacent areas of the cortex (which he interprets as indicating increasing differentiation due to the formation of inhibitory connections) and increasing coherence across areas that are more widely separated (which he interprets as indicating increasing general integration due to the formation of long-distance connections of a facilitory sort).

These changes appear to take place in waves, which Thatcher has attempted to relate to the general stages described by Piaget and his followers on one hand, and to the growth of long-distance fibers connecting different cortical regions on the other. Thatcher's model does not imply that the brain is unresponsive to incoming empirical experience. On the contrary, he presumes, along with empiricists, that the brain is highly responsive to such experience and encodes it as a series of specific neural connections. However, in the rationalist tradition, Thatcher sees responsiveness to specific experience as highly constrained by the general state of development of the cortex, which, in turn, is driven both by maturation and general (as opposed to task-specific) experience.

Thatcher's work is congruent with work in the rationalist tradition in one other respect. He sees higher-order regulation and reworking of experience (Piaget's reflexive abstraction) as being just as important as experience itself in the process of development. In his view, development is a cyclic process. During certain phases, growth is led by experience; while during other phases, growth is led by the higher order reworking of this experience. He suggests that the frontal lobes are particularly important in this reworking, and developed a mathematical model in which frontal and posterior growth are linked to each other, much the way that the growth of predators and prey are linked in a dynamic ecosystem (Thatcher, 1997).

The Sociohistoric Perspective

The third epistemological tradition within which children's development has been studied has its roots in the sociohistoric interpretation of Hegel's epistemology, developed by Marx and further expanded by modern continental philosophers (Kaufmann, 1980). According to the sociohistoric view, conceptual knowledge does not have its primary origin in the structure of the objective world (as empiricist philosophers suggested). Nor does it have its origin in the structure of the subject and his or her spontaneous cogitation (as rationalist philosophers suggested). It does not even have its primary origin in the interaction between the structure of the subject and the structure of the objective world (as both Baldwin and Piaget maintained). Rather, conceptual knowledge has its primary origin in the social and material history of the culture of which the subject is a part, and the tools, concepts, and symbol systems that the culture has developed for interacting with its environment.

Developmental psychologists who have been influenced by the sociohistoric perspective have viewed the study of children's conceptual understanding in a different fashion from empiricists or rationalists. They believe that one should begin a study of children's thought by analyzing the social, cultural, and physical contexts in which human cultures find themselves, and the social, linguistic, and material tools that they have developed over the years for coping with these contexts. One should then proceed to examine the way in which these intellectual and physical tools are passed on from one generation to the next in different cultures and different time periods.

The best known of the early sociohistoric theories was Vygotsky's (1934/ 1962). According to Vygotsky, children's thought must be seen in a context that includes both its biological and its cultural evolution, and the study of this thought must combine the sorts of ethnographic methods used by anthropologists with those used by physicists and evolutionary biologists. According to Vygotsky, three of the most important features of human beings, as a species, are: (1) that they have developed language, (2) that they fashion their own tools, and (3) that they transmit the discoveries and inventions of one generation to the next, via institutions such as schooling. From the perspective of Vygotsky's theory, the most important aspect of children's development is not their exposure to the "stimulation" that the world provides. Nor is it their construction of universal logical (or neurological) structures for organizing that experience. Rather, it is their acquisition of language (circa 2 years of age) and their use of language for internal regulatory purposes (4–7 years of age), and for the purpose of acquiring the basic intellectual technology of their culture.

Given this view, it was natural that most developmental work in this tradition would attempt to explore the effects of such variables as schooling, literacy, numeracy, and the teaching of higher-order concepts and forms of representation in different cultures (Cole, 1997; Cole, Gay, Glick, & Sharp, 1971; Greenfield, 1966; Olson, 1994; Rogoff, 1990). Under the guidance of Luria, however, an extensive series of neurological investigations was also initiated. In his early work, Luria (Luria & Judowitsch, 1959) focused on establishing the increasing dominance of words over external signals as stimuli for controlling children's behavior. In his later work, Luria (1966, 1973) utilized a unique data base to which he had access, which included information on all the gunshot wound patients in the Soviet Union during World War II. By combining the assessment of normal subjects with those who had experienced focused neurological insults, he attempted to isolate the neurological system that had the primary responsibility for self-regulation, which, in his view, involved both regulation of affect as well as cognition.

The area that Luria saw as exercising this self-regulatory function was the frontal region. By the time he finished his career, Luria had proposed a view of the cortex that was differentiated along a left–right, and top–down as well as a front–back axis. In his model, the frontal lobes were specialized for planning and self-regulation, the posterior lobes were specialized for the processing of information, and the limbic system for affective and emotional processing. Within the posterior lobes, the left hemisphere was seen as being specialized for digital, sequential aspects of cognitive processing; the right hemisphere was specialized for analogical, parallel aspects. Applying this general framework to the study of children's development, a number of investigators have interpreted the shift in children's cognition that takes place just prior to, or concurrent with, the introduction of schooling as being potentiated by the development of lateralization on the one hand and of frontal development and dominance on the other (Witelson, 1983).

Neo-Piagetian researchers have also drawn heavily on Luria's ideas, and have seen the frontal system as the mediator of the sort of reflexive abstraction postulated by Piaget. Thatcher's work has already been mentioned in this regard. Pascual-Leone's work is also worth mentioning. He has suggested that one of the mechanisms by which the frontal lobes have their effect is by "boosting" the activation strength of cortical units that are localized in the posterior lobes, thus permitting integrations and differentiations to take place rapidly that would otherwise take place slowly or not at all (Pascual-Leone, Hamstra, Benson, Khan, & Englund, 1990). In our own work, we have stressed that the reflexive or executive requirements of the integration process may also be dependent on frontal input; hence frontal development may strongly affect the development of these functions as well as the development of working memory (Case, 1992). Fischer and Rose (1994) have proposed a similar view.

Dialogue Among the Three Traditions

Relationships among the empiricist tradition and the other two traditions have not been close, and indeed have often been actively hostile. In part, this appears to be because the three traditions take such different stances on the underlying nature of knowledge and its acquisition in the course of ontogenesis. Another factor that has kept the three traditions apart is that they have adapted such different models of what a mature science of cognitive development should look like, and the methods on which it should rely. Still another possible factor that has kept the traditions apart is that each tradition was conceived in a different cultural and intellectual milieu, with its own unique values and forms of expression.

In spite of (or perhaps, because of) the fact that the three traditions have proceeded in such different directions, we believe that each one has an important contribution to make as we work toward a more integrated view of behavioral and neurological change. On the basis of the evidence gathered in the empiricist tradition, it seems to us to be indisputable that experience produces changes in local neurological circuits in the general fashion described by Hebb—by the formation of local inhibitory and facilitory connections. Following Baldwin and Thatcher, it seems to us equally likely that these changes take place in the context of a general cortical system in which different local regions are increasingly differentiated from their immediate neighbors, and are increasingly connected to regions that are quite distant, as a function of general experience and maturation as well as specific experience. Finally, following Vygotsky and Luria, it seems to us that the role of language and self-reflection in change—however these are embodied neurologically—must be understood, if we are to move toward models of change that do justice to the full range of phenomena that human children exhibit in the course of their ontogenesis. We see the work of Carpenter and Just (this volume), as one way of moving toward a model that includes all three of the previous elements and points of view, and the work that we describe in the next few sections as another way.

NEUROLOGICAL CIRCUITRY REQUIRED
FOR COMPARISONS OF NUMERICAL MAGNITUDE

We begin by summarizing a program of research that has been conducted in the empiricist tradition, and that focuses on the neurological circuitry that is involved in making comparisons of numerical magnitude. A typical task given to subjects is: "Is the number on the screen bigger or smaller than five?" Moyer and Landauer (1967) were the first to demonstrate that the time required to answer this question increases as the numerical distance between the two numbers decreases. Since their original study, this effect has been replicated many times and shown to hold true for different kinds of stimuli such as written number words, Arabic numerals, or arrays of dots of different numerosity (Buckley & Gillman, 1974; Dehaene, 1996; Dehaene, Dupoux, & Mehler, 1990; van Oeffelen & Vos, 1982). Over the past two decades, researchers have started to probe the neurological underpinnings of this effect. Because ERPs (Event-related potentials) provide information about the precise timing of mental events, as well as the events' locations, they have proven to be particularly useful for measuring the subtle and rapid processes that are involved in this task.

ERP recordings made while adults perform a number comparison task reveal stable patterns (Dehaene, 1996; Grune, Mecklinger, & Ullsperger, 1993; Grune, Ullsperger, Moelle, & Mecklinger, 1994; Mecklinger, Ullsperger, Moelle, & Grune, 1994; Schwarz & Heinze, 1998; Ullsperger & Grune, 1995). The early components found in the ERP depend on the different modalities in which the stimuli are presented. For visual stimuli, Dehaene (1996) describes a first component (P1) 104 ms after stimulus onset at electrode sites surrounding the parieto-occipito-temporal junction. At this point in time, he finds no impact of varying the type of stimulus that is presented: Written number words (e.g., 'six') and Arabic numerals (e.g.,'6') produce the same effect. The second major event is a negative component (N1) over symmetrical temporo-parietal sites, showing a peak at 148 ms for written number words and 160 ms for Arabic numerals, respectively. The difference between the pattern of activation for these two sorts of stimuli becomes significant at around 110 ms after stimulus onset. He also found a significant effect on the symmetry of the N1 component, with larger activations at the left sites for number words, but no difference for Arabic numbers. These effects of the nature of the stimulus still continue when a third component begins to appear (P2). At this point in time (162 ms), Dehaene observes an effect of word length for verbal stimuli. Three-letter words produce a greater negative response than four-letter words over symmetrical parieto-temporal localizations.

A few milliseconds later (174 ms for numerals, 190 ms for written number words) he finds the first differential effect for numbers at different distances from the target, with larger amplitudes for numbers close to the target. This difference is apparent at the parieto-occipito-temporal junction of both hemispheres, but more pronounced on the right side. A second major effect is found on the morphology of the P3 component at parietal sites: The amplitude is smaller, and the latency is longer, if the number presented is close to the target (e.g., 4 vs. 5) than if it is distant (e.g., 9 vs. 5), regardless of the nature of the stimulus (e.g., word vs. Arabic number; Grune, Mecklinger, & Ullsperger, 1993; Schwarz & Heinze, 1998; Ullsperger & Grune, 1995).

The next electrophysiological event is the emergence of activation over motor sites contralateral to the response key. Finally, a medial superior prefrontal activation is found on trials where the subject makes an error, about 150 ms after the motor response (Dehaene, 1996).

In modeling these data, Dehaene and his colleagues (Dehaene, 1992, 1996; Dehaene & Cohen, 1995, 1998) have proposed that a three-step process is involved: (1) stimulus identification (which is responsive to the differences between written number words and Arabic numerals), (2) comparison (which is responsive to the size of the difference between the target number and the number presented) and (3) response (which generates response-side effects). Figure 8.1 illustrates the areas, the connecting tracts, and functional units that they postulate as being involved in the three processing stages. Note that many of the areas indicated (especially the left temporal area and frontal areas) are considerably more developed in humans than in other primates. Note also that other primates have great difficulty in mastering a response of this nature (Boysen & Berntson, 1995; Boysen, Berntson, Shayer, & Hannan, 1995), and that humans exhibit a distinctive pattern of differential impairment in numerical processing if one of the above areas is impaired by a tumor or external injury (Dehaene & Cohen, 1995; Deloche & Seron, 1987; McCarthy & Warrington, 1990).

Taken together, the foregoing data make a strong case for the existence of a neurological circuit that plays a foundational role in numerical cognition, and that involves components that may possibly be less well developed and integrated in lower species. Although this circuit primarily involves areas that are localized in the posterior lobes, there is reason to believe that prior to the automization of the process—or when the automized response leads to an error that the subject detects—the frontal lobes are essential for making similar judgments in a slower and more voluntary basis (Dehaene, 1996; Pauli et al., 1994).

CHILDREN'S EARLY ACQUISTION OF NUMBER CONCEPTS AND THE ABILITY TO MAKE COMPARISONS OF NUMERICAL MAGNITUDE

What do we know about children's mastery of the task that Dehaene has studied, and the sort of changes that must take place before the circuitry that he has documented can become functional? We begin with the behavioral competencies. Children's first success on Dehaene's task typically occurs somewhere between 4 and 6 years of age. Prior to this, children have two clear precursor cognitive capabilities, which appear to be vital for their subsequent success and that are illustrated in Fig. 8.2. The top panel in the figure illustrates a competence that has to do with counting. By the age of 4 to 5, most preschoolers can count a small set of objects without error. They no longer miss out items in the string of counting words (e.g., 1–2–3–4–6), and they no longer miss out or double count items in the array that they are asked to count. Although their counting is much slower and more effortful than that of adults, they tag each object once and only once, and cite the last number counted as the one that answers the question "How many?" (Gelman, 1978).

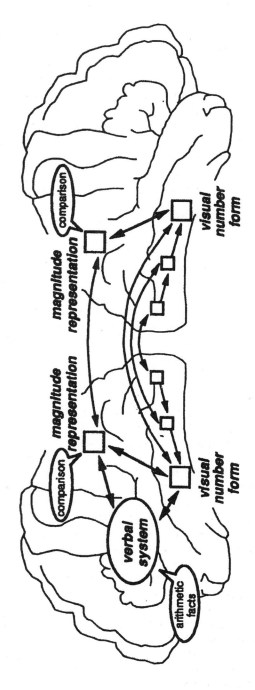

FIG. 8.1. Dehaene's diagram of the brain areas and pathways that are involved in making comparisons of magnitude. Representations of written numerals (visual number forms) are localized in the ventral visual pathways in the inferior occipito-temporal areas. Number words (verbal system) are localized in the language areas of the brain, for example the left temporal lobe. The analogical magnitude representations are localized in the vicinity of the occipito-temporo-parietal junction (Dehaene & Cohen, 1995).

196

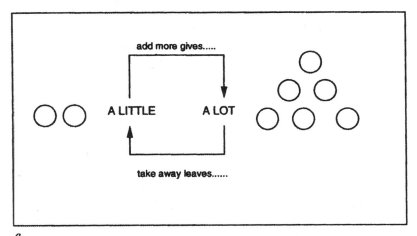

a

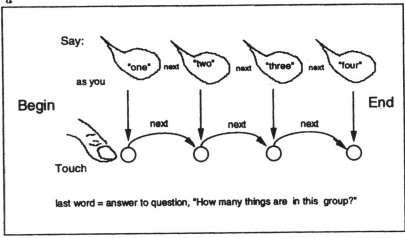

b

FIG. 8.2.   a) The global quantity schema that permits children to answer questions about "more" and "less". b) The counting schema that permits children to state how many objects are in a set.

Children's ability to make relative judgments of numerosity, without counting, is also quite well developed by this age. From birth, children show a predisposition to orient to the number of objects in a small array, regardless of the spatial distribution of these objects (Starkey, 1992; see also Wynn, 1992). By the age of 4 years, they cannot only make such judgments for a larger number of objects, but they can also make verbal predictions about changes in numerosity as a function of the sort of transformation that is made to them. Their general competence is represented visually in the bottom panel of the second figure.

Several changes take place in these two competencies as children enter the elementary school years and become capable of succeeding on Dehaene's tasks. The most obvious change is the one that takes place in their counting. Children learn more words in the number sequence: Typically, they can now count from 1 to 20—and often on to 100—instead of just 1 to 5. A less obvious change—and one that is a great help to them when they encounter subtraction problems in first grade—is that they learn to count backwards, at least for small numbers. This new counting knowledge is indicated in the top panel of Fig. 8.3. Note that the diagram includes a more detailed set of verbal labels for the connections among elements, as well as more elements themselves.

During the same time period, children's schema for nonverbal quantification also becomes more elaborated and differentiated. As children make the transition to formal schooling, they begin to distinguish between adding and taking away—as global operations—and additions and subtractions that have a particular magnitude associated with them. Numbers 4 and 5 not only become numbers that are associated with particular perceptual appearances, they also become a set sizes that can be derived from each other. Thus 5 can be derived from 4 by adding 1, and 4 can be derived from 5 by taking away 1. The bottom panel of Fig. 8.3 is intended to illustrate the complexity of the new knowledge that children acquire in this area in the same fashion as the top panel indicates the increased complexity of their knowledge about counting.

A third change that takes place during the same age range is that the two representations gradually merge into a single knowledge network, of the sort indicated in Fig. 8.4 (Griffin & Case, 1996, 1997). Although questions about counting and questions about magnitude yield two discrete factors during the preschool years, by the first grade they yield a clear, single factor (Y. Okamoto, personal communication, September, 1998). Several other behavioral changes accompany this transition, including: (1) Children can now answer questions about relative magnitude when they are presented in verbal form—without any concrete props (e.g., Which is more, 5 or 4?) (Griffin & Case, 1997; Okamoto & Case, 1996; Siegler & Robinson, 1982); (2) Children can now answer questions about quantity transformation when they are verbally presented, once again, without any concrete props (e.g., "If you have 4 chocolates, and someone gives you 3 more, how many will you have altogether?"); (3) Children can now answer questions that take their own counting words as objects that can be quantified in their own right (e.g., "What number comes 2 numbers after the number 5?"); (4) Children can use information derived from verbal counting to draw inferences about relative magnitude along a number of dimensions such as weight, length, musical tonality, distributive justice, and so forth (Marini, 1992); (5) Children begin to master the visual symbols

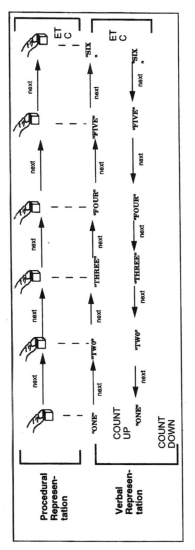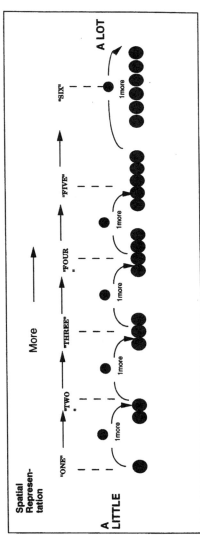

FIG. 8.3. Elaborated schema for counting and addition / subtraction (roughly 6 years of age).

for referring to numbers, and become capable of answering questions about the relative magnitude of two numbers when they are presented in a symbolic format (Griffin & Case, 1996; Okamoto & Case, 1996).

Figure 8.4 is a model of the structure that we have hypothesized children acquire during this age range, one that is responsible for producing this correlated set of behavioral changes. By extension, we see it as constituting the psychological competence that must be present, in order for the circuit that Dehaene has studied to become functional. Because we see the structure as playing this foundational role, it is worthwhile to say a word or two about its basic components, and the way they are represented in the figure. The top two rows of the figure are an attempt to represent children's increasingly elaborate and bidirectional counting competence. The bottom two rows of the figure attempt to represent children's increasingly elaborate and bidirectional magnitude knowledge, and its representation by second-order symbols. The edge of the figure attempts to capture children's ability to take this structure and use it to make a variety of dimensional judgments. Finally, the vertical lines in the figure attempt to capture the fact that all these various types of knowledge have become tightly linked to each other in a relationship of one-to-one correspondence.

There is one final element of interest in the figure. In the middle row is a representation of the standard pattern in which quantities are represented with one's fingers. This row has been placed in the middle of the figure because we believe that the representations it entails play a central role in integrating the two precursor representations that have already been mentioned, ones that appear in the rows directly above and below it. The advantage of the fingers is not just that they are universal and highly portable. It is that they can be used for two distinct purposes namely, counting (by moving one finger at a time, in sequence) and indicating a set size (by forming the standard finger pattern for a particular pattern). We believe that because the fingers can serve both these representational functions, they play a foundational role in helping young children integrate their previous competencies.

Fingers may also support the integration process in other ways. First, they may provide a pointer to, and hence carry with them, the child's first kinesthetic/motoric representation of counting, thus permitting the intuitions developed when counting in this fashion to be integrated with the more abstract insights and intuitions being developed at this later time. Second, they may allow children to play events backwards and forwards, thus reviewing them and understanding how they are related more explicitly. In effect, they may constitute a sort of slow motion camera children can play forwards and backwards at will. Finally, they may help children link numerical transformation processes as well as representations of numerical quantity, per se.

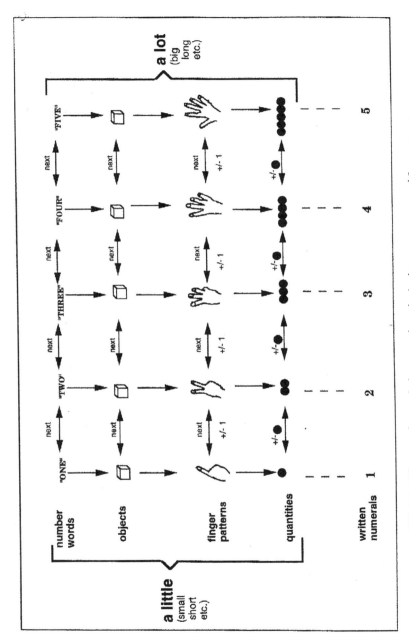

FIG. 8.4. The central numerical structure hypothesized to emerge at around 6 years as a result of the differentiation, elaboration, and integration of the two earlier schemas shown in Fig. 2 above.

201

This last function is subtle but potentially important, particularly in a neurological context. One of the most important challenges for children to grapple with—repeatedly stressed by Piaget—is that of linking different but formally equivalent types of transformation. It is one thing to connect different canonical representations of number and thus to realize (for example) that the word *four* is equivalent to the canonical pattern for four that is found on dice, or the Arabic numeral *4*. It is quite another thing to link the operations by which four is generated in these different cases: that is, (in the first case) by counting forward one word from three, and (in the second case) by adding one unit to a set that has three objects in it. We believe that it is only if children make the latter connection that they will ever realize that verbal counting can be used as a reliable model for addition or subtraction—and it is only if they encounter a context in which these two operations are brought into one-to-one correspondence (as they are in finger counting) that this realization will be ever be achieved.

How do children come to this insight? Fuson (1982) conducted a detailed study of the steps that are involved in mastering this equivalence principle for one-unit transformations, and generalizing it to transformations of more than one unit. Our own belief is that progress through this sequence is greatly facilitated by using the fingers, for all the reasons mentioned before. The fingers represent both the cardinal and ordinal aspects of number simultaneously; the fingers provide an object of contemplation that can be played forward or backward at will; and the fingers serve to harmonize the rate at which the two event sequences take place, and thus emphasize (or permit children to reflect on) the equivalence of the underlying processes that generate them. Finally, and typical of the reflection process, we believe that children need a certain minimum attentional capacity—of the sort stressed by those in the rationalist tradition from Baldwin through Piaget to Pascual-Leone—in order to engage in this reflection process. As a minimum, they need the attentional capacity that is required to activate both lower schemata at the same time, so that higher-order mappings can be made between them (Halford, 1982, 1993). An attentional capacity of this magnitude is typically acquired somewhere between the ages of 4 and 6 years (Case, 1985).

The foregoing model is a rather complex one for explaining the emergence of such a simple competence as pushing one button if a symbol on a screen is bigger than five, and another if it is smaller. As we will indicate next, however, it has been quite productive in generating a number of novel predictions, most of which have been confirmed. We believe that it has done so, in large measure, because it incorporates elements from all three of the developmental traditions mentioned earlier. The basic content of the structure hypothesized in Fig. 8.4, and its specification as a set of associatively linked nodes and relations, derives from the empiricist tradition. The notion that the structure is assembled by a process of differentiation and integration, in which conscious attention plays a critical role, derives from the rationalist tradition. And the notion that the structure contains critical, culturally derived representations, both linguistic and nonlinguistic—and hence depends on specific sociolinguistic experience—derives from the sociohistoric tradition. Next, we consider the way in which these different elements may be neurologically embodied.

POSSIBLE NEUROLOGICAL UNDERPINNINGS OF CHILDREN'S GROWING COMPETENCE AT NUMERICAL COMPARISON

What sort of neurological model of development might one propose that would complement the cognitive–developmental model that was sketched out in the preceding section on the one hand, and Dehaene's model of adult number comparison, on the other? In addressing this question, it is worthwhile to consider each of the general perspectives described previously, in sequence.

The Empiricist Perspective

Consider first the specific circuitry that might be involved, in light of the model proposed by Dehaene and his colleagues. Several of the rows in Fig. 8.4 correspond quite directly to the elements in Dehaene's model. The written numerals in the bottom row correspond to his visual number forms, are very probably localized in the ventral visual pathways in the inferior occipito-temporal areas (Ungerleider, 1995; Ungerleider & Mishkin, 1982). The number words in the top row correspond to his verbal system, and are very probably localized in the language areas of the brain, for example the left temporal lobe. The quantity representations in the fourth row correspond to Dehaene's analogical representations, and are very probably localized in the vicinity of the parieto-occipito-temporal junction. Dehaene has no manual component in his model to correspond to the pointing gestures whose mastery is such an important feature of the preschooler's competence, and which are very probably localized in the motor and premotor cortex. However, it must be remembered that his subjects are adults, for whom the task of comparing one number to another is highly automatized. It seems quite likely that for children, who are in the early stages of assembling the structure, the motor component may be a good deal more important, indeed critical (see next). Once the overall structure has become automatized and re-represented symbolically, this component may become unnecessary and in many situations may drop out entirely.

Given the close correspondence between the cognitive model in Fig. 8.4 and the neurological model in Fig. 8.1, might there also not be a close correspondence between the mechanisms that we have postulated for the assembly of the cognitive structure (e.g., differentiation and integration), and those that are necessary to put the neurological substrate in place? Almost certainly. As children learn more counting words and practice counting with their existing counting words, the neurological units and subunits (e.g., in the left temporal lobes) that represent each counting word must become more numerous and more differentiated from each other (by the formation of inhibitory connections), while the connections due to temporal adjacency should become stronger, and more bidirectional (by the formation of facilitory connections).

The same should apply to the representations of numerical magnitude and the representations of written numerals. Specific experience and some sort of Hebbian learning process very probably play an important part in making these neural networks more elaborate and differentiated. One can imagine that the areas of cortex in which the representations are housed might well be in the process of becoming more elaborate and differentiated anatomically and that this might play an important role as well. Under influences that might include both biological and experiential variables of a more general sort, there might be a proliferation of dendrites and synapses, which would make the establishment of the previous facilitative and inhibitory connections (and the pruning away of connections not used for either purpose) a good deal easier. In short, the increasing differentiation and elaboration of children's existing networks might have both hardware and software components, which might be under the control of both specific and general influences.

What about the integration of the different neural networks? Once again, one would expect that the long-distance pathways connecting the different neural networks would undergo a substantial process of development, as the discrete schemas whose components were outlined in Fig. 8.3 gradually differentiated and were integrated into the more elaborate structure in Fig. 8.4. There would be more positive connections (e.g., between the units representing visual symbol *4* in the visual pathways, and those representing the analogical value of *4* at the parieto-occipito-temporal junction, and the units representing verbal number word *four* in the language areas), and there would be a corresponding growth of inhibitory connections among the possible alternatives (e.g., the number word *four* and the visual symbol *5*). The same would apply for the elements of the neuronal network representing transformations. Once again, specific experience should play a big role in effecting these changes. Because the underlying system that is responsible for making such connections works on principles of temporal and spatial association, one might also hypothesize that the sort of specific experience that would have the greatest impact would be experience in which the different representations of number were placed in close temporal correspondence.

In this regard, it is worthwhile to mention one other way in which the fingers—or other representational devices that play a linking role in cognitive development—can have their effect. This is by providing a model of two different cognitive event sequences in which the rate of change is normally quite different, but which in this context can be harmonized. In the normal course of events, verbal counting is something that proceeds quite rapidly. If it is done orally, as a simple demonstration of verbal competence, it can occur at the rate of several words per second. If it occurs in the context of counting objects, it still occurs at the rate of one to two units per second in children at this age level (Case, Kurland, & Goldberg, 1982). By contrast, picking up several objects and adding them to a set, one at a time, is something that occurs much more slowly. As children look at their fingers, however, and count them one at a time, examining the set size after each finger movement—they bring the rate of these two event sequences into temporal harmony. In effect, then, the fingers may play a sort of pulsing or rhythmic role; that is, they may set up a sort of series of slow motion shots in which the movement from one canonical set representation to the next, and the movement from one counting word to the next takes place at precisely the same rate. As Hebbian learning processes are designed to strengthen connections among cortical units that are activated simultaneously, the fact that these two event sequences are brought into this sort of one-to-one correspondence should greatly increase the likelihood that the two different sorts of operator (counting by ones, and adding items one at a time), will start to be associated in children's minds, and become the object of further reflection.

The Rationalist (Structuralist) Perspective

Although structuralists have never denied the possibility of learning due to Hebbian processes, they have always taken pains to emphasize the more general context in which this learning takes place, and the importance of the active intervention of the child's own attentional processes. Recall Baldwin's interest in the overall connectivity of the cortex (not just certain specific circuits) and his emphasis on the importance of changing attentional limits, a point that has been echoed by neo-Piagetian theorists with their emphasis on the growth of attentional resources, and/or the growth of working memory (Case, 1995; Halford, 1993; Pascual-Leone, 1970). A frequent theme in modern neuropsychological writing is that the working memory function depends on a set of frontally localized circuits, and which—by their projection to the posterior lobes—influence both the speed and the content with which learning takes place (Goldman-Rakic, 1987, 1994; Pascual-Leone et al., 1990). To use a computer metaphor, if one wanted to establish connections between two different files, it would be a great help to be able to keep both open simultaneously. Conversely, it would be a great handicap not to be able to do so. Within a file as well, if one wanted to establish stronger (positive or inhibitory) connections between any two units, it would be a great help to be able to display both units simultaneously.

Continuing the analogy, one might expect that the key general requirements for assembling a network of the sort Dehaene proposed would not just be the availability of potential connections, within and between the different (posterior) cortical areas cited in his model, but the availability of long-distance projections from the frontal lobes, and a well-developed frontal system. Dehaene & Cohen (1995) suggest that frontal input is crucial in establishing original task performance, a suggestion that is in line with much contemporary theorizing (Krasnegor, Lyon, & Goldman-Rakic, 1997; Stuss & Benson, 1986). Thatcher's observation of increased coherence among EEG patterns in the frontal and posterior cortex during this age range is also of relevance. Recall that this increase was a general one, which was observable even when subjects were not engaged in specific task activity. It seems likely that this increase, and the additional planning/working memory capability that it provides, may be crucial to the more specific developments that are seen during this time period.

A final possibility that the rationalist perspective might alert us to is that a new level of "reflexive abstraction" may be developing during this age range, and that this may also impact on the child's capability for assembling a central conceptual structure such as that indicated in Fig. 8.4. It is not just that the child has a greater working memory for existing units—can keep more units in a high state of activation simultaneously—it is also that the child is better able to form second-order representations in which his or her own first-order representations are explicitly modeled. From a neurological view, the implication would be that the frontal lobes may do more than provide a context within which particular posterior networks may be brought to full activation, given appropriate level of executive functioning and working memory. They may also provide a higher-order area, to which posterior lobes project, which maps higher-order covariances that obtain among different types of posterior lobe activity. In effect, the frontal lobes may possibly house the self-referential "address" system, or "master directory," which makes the cognitive structure in Fig. 8.4 more than the sum of its parts. This would correspond to the reflexive abstraction, which both Baldwin and Piaget stressed in their writing, that continues to play a central role in modern theories in the rationalist tradition.

The Sociohistoric Perspective

There would be no disagreement from investigators in the sociohistoric tradition regarding the existence of experience-dependent circuitry of a local sort, or the assembly of local circuits into more general ones—as theorists in the empiricist tradition have suggested. Nor would there be any disagreement on the more general attentional requirements for assembling such circuits, including those for the general executive function, frontal–posterior priming, working memory, and/or the self-mapping function. Indeed, on this latter point, there would no doubt be a strong resonance. At least four additional suggestions might be ventured, however, which would not be as salient within the other two frameworks.

Interiorization of Language. In Vygotsky's theory, one of the most important general developments that takes place between the ages of 4 and 7 is the interiorization of language, and its use for the self-regulation of behavior. Whatever else children may need to do in order to assemble the structure in Fig. 8.4, it is clear that they must learn the counting word sequence from their parents or peers, that they must practice and gradually internalize it, and that they must finally begin to use it mentally in problems for which it is relevant. This being the case, one might expect that this process would be facilitated by the differentiation of the left temporal cortex, and the increased coherence of temporal and frontal activity, which Thatcher has documented during this age range.

Interleaving of Digital and Analog Representations. One of Luria's contributions to the tradition inaugurated by Vygotsky was the suggestion that linguistic processing is just one example of the sort of digital/sequential processing for which the left hemisphere is specialized, while spatial processing is just one example of the sort of analog/parallel activity for which the right hemisphere is specialized. To assemble the structure in Fig. 8.4, subjects must interleave these two types of processing, in a precisely timed fashion. Thus, one might suggest that the increased differentiation and integration of the two hemispheres that takes place during this age range (Witelson, 1983) might be another general cortical change that would potentiate and/or constrain the development of the specific circuits on which the numerical comparison process is dependent.

Importance of Culturally Created Signs and Symbols. A third possibility to which the sociohistoric tradition might alert one is that the visually represented number symbols—because they are significant cultural inventions and utilize a different modality housed separately in the cortex—may provide more than a convenient way of representing the two meanings of numbers (i.e., numbers as counting tags and numbers as set size labels). In addition, these visual symbols may facilitate the re-representation process, and play a vital role in permitting higher-level development. In effect, they may actually represent the higher-order meaning of numbers, formed by the coordination of the two lower-order meanings.

Role of Instructional Scaffolding. In the sociohistoric tradition, as in the empiricist tradition, instruction is seen as playing a potentially vital role in the development of high-level cognitive functions. Two ways that it can do so are: by the provision of temporary social *scaffolding* as a new competence is being acquired, and by the provision of an external form of representation—a sort of intellectual technology—that supplements and extends naturally emerging intellectual competencies. Putting these two possibilities together, one can suggest that the representation of numbers by the hands may be a form of cultural representation that can facilitate an integration of the different meanings of numbers, and bring the operation of counting and the operation of addition into temporal harmony. One of the major accomplishments of academic disciplines, and of those who teach them, may be to provide additional types of representation that play this sort of role. One example of a representation that might play this sort of role in the present case is the sort that appears on board games. On a board game, numerals are written on the squares that tokens travel over, thus making explicit the connection between the digital and analogical aspects of number. In addition, when one plays such games, one often counts forward or backward aloud, which should facilitate the realization that adding or subtracting X units to an existing array yields the same result as counting forward or backward by the same amount. In effect, such devices may provide a temporary external scaffold that facilitates the formation of internal, neurologically based connections across spatially distributed neural circuits and pathways. In effect, instruction may have significant neurological as well as cognitive consequences.

FURTHER BEHAVIORAL AND NEUROLOGICAL DATA OF RELEVANCE TO THE INTEGRATED MODEL

Before concluding, we would like to mention several additional types of data we see as broadly supportive of the cognitive and neurological model presented, which may lead to its refinement or modification.

Sociocultural Differences in Numerical Competence

Because language and other forms of cultural symbols (numerals, canonical finger patterns) play such an important role in the previous model, it follows that social and cultural factors should have some influence on the nature and the rate of formation of the mental counting schema, which permits numbers to be compared in magnitude on the basis of verbal input alone (i.e., even in the absence of analog representations). In fact, this is the case. Studies of early mathematics achievement have consistently shown large differences across different socioeconomic groups (Case, Griffin, & Kelly, 1999). Although these results are sometimes attributed to the differential effectiveness of schooling for different socioeconomic status (SES) groups, our own data suggest that this cannot be the entire story, as differences in numerical competence sometimes exist before children ever enter the school system. In one study that we conducted, for example, we found differences of 1 to 2 years between very high and very low SES children in the age at which they first solved Dehaene's task, and all the other tasks with which it is associated. High SES groups often pass these tasks by 4 to 4½ years of age, while low SES groups often did not pass them until 6 years of age or older (Griffin & Case, 1997). Although some might prefer a genetic explanation for socioeconomic effects such as these, our own interpretation is that these differences must be experientially produced, because the size of the differences varies from one country to the next as a function of social policy, even when the diversity of the populations are controlled for (Willms, 1999).

Results of Targeted Interventions

A second set of data concerns the possibility of modifying the early differences mentioned before. Drawing on our analysis of what the central conceptual structure for number entails, Griffin and Case (1997) designed a preschool program that provides children with a wide number of activities that involve the different sorts of representations indicated in Fig. 8.4, coupled with opportunities to master the language of each, and to move back and forth among the different representations (learning, for example, that three moves forward on a board game corresponds to the addition of three dots to a die, or to moving up three units on a thermometer). The results have been clear. Young children exposed to extensive experience with this sort of program (6 months to 2 years) have made large gains relative to controls on tasks such as those used by Dehaene, and on the full range of other numerical competencies associated with it. The children who have had these programs also compare very favorably with high SES groups from China and Japan that have been tested on the same measures (Griffin & Case, 1997).

**Data on the Limiting and/or Potentiating Effects
of Working Memory**

A third set of data are those having to do with general attentional resources, and/or working memory. If the model we presented in the previous section is correct, then frontal development and working memory should be factors that potentiate and/or limit the more specific and experientially based changes that take place as a result of such intervention programs. In this regard, it has been found that working memory for the products of counting operations shows a dramatic growth spurt during this general age range, which has the same general time course as the growth spurt in frontal/posterior coherence documented by Thatcher (Case, 1992). Further, that children who are exposed to opportunities for learning quantitative principles show a large difference in their responsiveness to these opportunities as a function of their working memory. In an early study on number conservation, for example, we showed that children with a working memory of 3 learned the conservation principle spontaneously and in a few trials, whereas those who had a working memory of 1 showed little signs of spontaneous learning, even after hundreds of trials, or when an explicit teaching condition was added (Case, 1977). In a more recent study using the preschool curriculum described previously, Griffin (1994) showed that working memory itself showed no increase as a function of the targeted intervention but did predict the magnitude of gain that individual children would derive from participation in the program.

Changes in Neuropsychological Functioning as Increased Numerical Competence Is Acquired

A fourth set of data are those that have resulted from ERP recordings. In a recent study, Temple and Posner (1998) gave Dehaene's task to a group of bright, middle-class 5- and 6-year-olds who were capable of solving it. They describe their results as remarkably similar to those seen in adults in the general form of waves that are recorded at different sites, but with the children showing slower latencies and larger amplitudes.

Although these data are consistent with our model, they are also consistent with many other models, including ones based entirely on empiricist principles. As our own model focuses on the differentiation and integration of existing capabilities that must take place if such circuits are to develop in the first place, we have begun to look at the change that takes place as children make the transition from not being able to pass the task to passing it. This shift in focus has necessitated using a slightly different paradigm, because children who cannot yet pass the number-comparison task typically cannot yet read written number words or numerals, and so cannot respond to either sort of visual stimulus.

To get around this problem, we first developed a task in which the numbers were presented auditorally rather than visually. Thus, subjects heard one of four numbers (1, 4, 6, or 9) and were asked to indicate for each whether it was bigger than 5 (in which case, they had to push the right-hand button) or smaller than 5 (in which case, they had to push the left-hand button). Before we asked our subjects to do the number-comparison task, we gave them a simpler task, in which they were simply asked to listen to the same digits (1, 4, 6, and 9) without any response whatever. For adults, as might be expected, the N1/P2 components showed the classic morphology elicited by auditory stimuli, which are different from the ones obtained by visual stimuli but reliable nonetheless. Interestingly, these components were strikingly similar for the listening task and the numeral-comparison task. However, a classical P3 response (from the central-parietal region) discriminated the listening and comparison conditions very clearly, and showed an amplitude and latency that varied with the difference between the stimulus presented and the target stimulus. There was also a negative response at frontal electrode sites that was identical in form and precisely synchronous with the posterior P3 component.

With these data in hand, the next question we asked was whether children who had just acquired the ability to compare orally presented numbers would show the same ERP pattern. We expected the same wave patterns for these children as in adults but longer latencies and larger amplitudes. The latter expectation was based on the results from many other EEG and ERP studies, which have shown that a decrease in amplitude and latency is a typical pattern observed during the school-age years until adulthood. It is also well known that these changes are of general nature and not related to any particular task (Friedman, 1991; Niedermeyer, 1999; Taylor, 1995).

The data collected from the children supported our hypothesis. The morphology of the components at central-parietal locations was similar to that of adults, although the responses were larger in their amplitude and—in particular for the later components—longer in latency. As for adults, also, the P3 was accompanied by a synchronous negative wave at frontal locations. Both the similarity in morphology and the differences in amplitude and latency were results that Temple and Posner had obtained using the visual version of the number comparison task. Thus, the final question we asked was how children who had *not* yet acquired the ability to pass the task would respond.

The behavioral results from the ERP experiment for these children showed that they had significantly more problems comparing the numbers close to the target (4 vs. 5 and 6 vs. 5) than the distant numbers (1 vs. 5 and 9 vs. 5). In fact, their performance for the close numbers was almost at chance level. This result suggested that they were attending to the stimuli but had not yet acquired the ability to solve the task. In the ERPs, they also showed a N1/P2 pattern. The most striking result, however, was a lack of a clear P3 component, as is shown in Fig. 8.5. Basically, the P3 component, and the corresponding negative wave in the frontal lobes, both remained completely flat. Finally we found a subgroup of children who

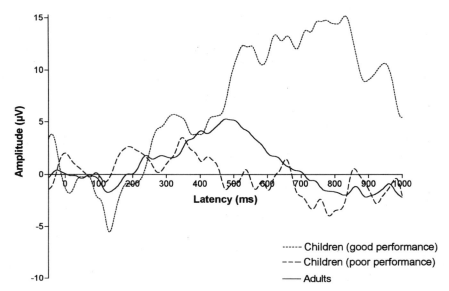

FIG. 8.5. Grand averaged ERPs for comparing the number "1" to the standard of "5" at electrode site Pz (see text for details).

judged the number 1 as smaller than the target, and the numbers 4, 5, and 9 as bigger. Their ERPs showed a P3 component similar to that of children who succeeded on all trials.

Although several different interpretations of this pattern might be possible, the one that we favor is that children who showed the intermediate pattern were attending to the problem and were attempting to solve it, but did not yet have sufficiently differentiated numerical schemata to resolve the discrimination that they were being asked to make. Although 9 was perceived as a "very big" number and 1 a "small" number, the numbers 4, 5, and 6 were all just perceived as being "big." For the children who were completely unsuccessful, we believe that these discriminations were possible for arrays of objects, but not yet on the basis of the number words alone, due to the absence of connections that directly map these representations on to each other, and coordinate their activation under frontal supervision. The question we are now examining is whether these EEG patterns will change as a function of exposure to the early intervention program designed by Griffin and Case (1996, 1997).

Parallel Modelling Efforts in Other Domains

A final support for the model of early numerical development that we have proposed comes from attempts of cognitive neuroscientists to solve the so-called binding problem, mostly investigated in the field of visual perception. The problem here is how to understand the way in which the brain combines the many different attributes of objects and events in space and time in order to construct a coherent representation. One possible solution to this problem is a hierarchical one, in which feedforward connections converge selectively on cortical *gnostic-cells*, which are capable of integrating the different information into a coherent representation. The finding of single neurons responding selectively to particular objects like hands or faces (Desimone, 1991; Desimone, Albright, Gross, & Bruce, 1984) supports this notion. However, if this were the only mechanism by which such coherence is achieved neurologically, one would have to postulate a combinatorial explosion in the number of gnostic-cells required to represent every aspect of reality. One would also have to postulate some sort of mastermind, which finally puts together the pieces of information. To avoid these problems, an increasing number of researchers have suggested that a dynamical binding mechanism is required as well, which is based on reciprocal connections between distributed neural assemblies.

Following this approach, coding of information is accomplished by coherent firing patterns of spatially distributed patterns of neurons. Thus, the integration of information is reflected in the correlation of activity in time. This notion is also supported by the neuroanatomical fact that in the cortex most functionally segregated neural assemblies are connected by reciprocal tracts. Because representations in this approach are based on functional rather than anatomical connections, it allows a dynamical regrouping that makes coherent and integrated processing very flexible. This new approach to the binding problem fits well into our own theory, which assumes that advances in understanding are related to the increasing capability of the child to synchronize different functional units of the brain. Discontinuous shifts in cognitive development could reflect access to newly acquired functional networks and qualitatively new features in reasoning can be understood as functional higher-order combinations of lower subsystems. Note that this approach does not exclude the notion of some neural systems working in the hierarchical mode, perhaps under frontal guidance.

Recently, Tononi and colleagues (Tononi, 1996; Tononi, Edelman, & Sporns, 1998; Tononi, Sporns, & Edelman, 1994) investigated the functional properties of neural systems acting in the way just described. They were interested in the relationship between functional segregation and functional integration in neural systems. Using concepts and measures from information theory, they ran large-scale computer simulations and applied the model to neurophysiological data. They found that the complexity (as understood in terms of information theory) of a neural system was high when the number of intra-area connections was high and the functional subunits were connected specifically. Decreasing the amount of intra-area connections in this model resulted in a significantly lower complexity. Finally, when they connected the neuronal subunits to many other subunits in a uniform way, the system became overconnected and showed a lowered complexity, again, as a result. These results suggest that a complex neural system depends on strong and rich set of connections within an area and a specific set of connections across areas. Again, this is very much in line with our model.

SUMMARY AND CONCLUSION

The way in which one models behavioral and neurological change depends strongly on the sort of change one is interested in and one's philosophical assumptions regarding the nature of the human mind. Here, we have taken as our focus the general change that takes place from ages 4 to 6 years in children's conceptual understanding of number. For a number of years, now, our research team has been developing a model of children's numerical competence in which the perspectives of three major traditions in developmental psychology are integrated (Case, Griffin, & Kelly, 1999; Griffin & Case, 1996, 1997; Okamoto & Case, 1996).

According to this model, 4-year-old children possess two types of number schema, one of which is primarily analogical, and deals with relative numerosity and its transformation via addition or subtraction; the other is primarily digital and deals with the sequential assignment of number-tags, via counting. What happens during development, according to the model, is that both these schemata become more differentiated and elaborated, and each newly differentiated element in the relative numerosity schema is mapped on to one, and only one, corresponding element in the counting schema. Because the mapping process is not straightforward, and requires children to make connections between operations (or transformations) as well as state-schemes, we believe that this mapping requires active executive mediation, and attentional priming and/or working memory that is sufficient to keep the to-be-mapped elements active simultaneously. We also believe that it requires—or is at least greatly facilitated by—the use of external forms of representation such as one's fingers, or the context provided by board games. Finally, we conjecture that the unified schema may be more than simply the sum of the previous two schemata and their connections. It may include a separate and higher-order map (an *addressing* schema, as it were) that explicitly represents the covariances among the two prior schemata.

What we have tried to do in the present chapter is to refine this model, and articulate a neuropsychological counterpart that is congruent with it. According to the neurological model, the verbal (counting) schemas that must be elaborated and differentiated are housed in the language areas of the brain (e.g., the left temporal lobe), while the analogical schemas are housed in the vicinity of the parieto-occipito-temporal junction, with the right hemisphere probably being superior to the left one. What happens with development is the neural elements within each of these regions become more numerous, more differentiated, and more tightly sequenced, by the formation of local positive and inhibitory connections. At the same time, the connections across these regions also become more extensive, and a second-order map is formed of the covariance between them. Formation of this covariance map requires active input from the frontal lobes, which (in conjunction with input from the fingers) serves to harmonize the firing of neuronal sequences in the posterior regions, thus rendering their correspondence more apparent. Frontal input also helps maintain key elements of these sequences in working memory, thus speeding the formation of connections between them, and perhaps contributing to the formation of a second-order map or directory that records and labels these covariances. Because the competence that emerges depends on general as well as local circuitry, one would expect that it would be open both to local, experiential effects (e.g., exposure to cultural devices and contexts that facilitate cross-regional mapping) as well as influences of a more epigenetic nature.

Throughout the chapter, we attempted to indicate which elements of the model were derived from which tradition. Our general parting message is simply that the model contains elements from all three of the classical developmental traditions, and that it has its own internal structure and coherence. Although the model is of course speculative, we see it as being congruent with most of the existing data that have been gathered so far in different laboratories, as well as our own preliminary data from ERP experiments and early intervention experiments that are ongoing. Our hope is that the model will be seen as laying out the general sorts of parameters that future, improved models must possess, if we are to take advantage of the insights of the pioneers in all three developmental traditions and move toward a deeper and more general model of the change process.

ACKNOWLEDGMENTS

This research was supported by grants from the James S. McDonnell Foundation. We are grateful to Cheryl Zimmerman for her help in preparing the manuscript.

REFERENCES

Baldwin, J. M. (1968). *The development of the child and of the race.* New York: Augustus M. Kelly. (Original work published 1894).

Boysen, S. T., & Berntson, G. G. (1995). Responses to quantity: Perceptual versus cognitive mechanism in chimpanzees. *Journal of Experimental Psychology: Animal Behaviour Processes, 21,* 23–31.

Boysen, S. T., Berntson, G. G., Shayer, T. A., & Hannan, M. B. (1995). Indicating acts during counting by a chimpanzee. *Journal of Comparative Psychology, 109,* 47–51.

Buckley, P. B., & Gillman, C. B. (1974). Comparison of digits and dot patterns. *Journal of Experimental Psychology, 103,* 1131–1136.

Case, R. (1977). Responsiveness to conservation training as a function of induced subjective uncertainty, M-space, and cognitive style. *Canadian Journal of Behavioral Science, 9,* 12–25.

Case, R. (1985). *Intellectual development. Birth to adulthood.* Orlando, FL: Academic Press.

Case, R. (1992). The role of the frontal lobes in the regulation of cognitive development. *Brain and Cognition, 20,* 51–73.

Case, R. (1995). Capacity based explanation of working memory growth: A brief history and a reevaluation. In F. M. Weinert & W. Schneider (Eds.), *Research in memory development and competencies* (pp. 23–44). Hillsdale, NJ: Lawrence Erlbaum Associates.

Case, R., Griffin, S., & Kelly, W. (1999). Socioeconomic gradients in mathematical ability and their responsiveness to intervention during early childhood. In D. P. Keating & C. Hertzman (Eds.), *The developmental health and wealth of nations: Social, Biological and Educational Dynamics* (pp. 125–149). New York: Guilford.

Case, R., Kurland, M., & Goldberg, J. (1982). Operational efficiency and the growth of short term memory. *Journal of Experimental Child Psychology, 33,* 386–404.

Case, R., & Okamoto, Y. (Eds.). (1996). The role of central conceptual structures in the development of children's thought. *Monographs of the Society for Research in Child Development, 61* (1–2, Serial No. 246).

Cole, M. (1997). *Cultural psychology.* Cambridge, MA: Harvard University Press.

Cole, M., Gay, J., Glick, J. A., & Sharp, D. D. (1971). *The cultural context of learning and thinking.* New York: Basic Books.

Cynader, M. S., & Frost, B. S. (1999). Mechanism of brain development: Neuronal sculpting by the physical and social environment. In D. P. Keating & C. Hertzman (Eds.), *The developmental health and wealth of nations: Social, Biological and Educational Dynamics* (pp. 153–184). New York: Guilford.

Dehaene, S. (1992). Varieties of numerical abilities. *Cognition, 44,* 1–42.

Dehaene, S. (1996). The organization of brain activations in number comparison: Event-related potentials and the additive-factors method. *Journal of Cognitive Neuroscience, 8,* 47–68.

Dehaene, S., & Cohen, L. (1995). Towards an anatomical and functional model of number processing. *Mathematical Cognition, 1,* 83–120.

Dehaene, S., & Cohen, L. (1998). Levels of representation in number processing. In B. Stemmer & H. A. Whitaker (Eds.), *Handbook of neurolinguistics* (pp. 331–341). San Diego: Academic Press.

Dehaene, S., Dupoux, E., & Mehler, J. (1990). Is numerical comparison digital: Analogical and symbolic effects in two-digit number comparison. *Journal of Experimental Psychology: Human Perception and Performance, 16,* 626–641.

Deloche, G., & Seron, X. (Eds.). (1987). *Mathematical disabilities. A cognitive neuropsychological perspective.* Hillsdale, NJ: Lawrence Erlbaum Associates.

Desimone, R. (1991). Face-selective cells in the temporal cortex of monkeys. *Journal of Cognitive Neuroscience, 3,* 1–8.

Desimone, R., Albright, T. D., Gross, C. G., & Bruce, C. (1984). Stimulus-selective properties of inferior temporal neurons in the macaque. *Journal of Neuroscience, 4,* 2051–2062.

Fischer, K. W. (1980). A theory of cognitive development: The control and construction of hierarchies of skills. *Psychological Review, 87,* 477–531.

Fischer, K. W., & Rose, S. P. (1994). Dynamic development of coordination of components in brain and behavior. In G. Dawson & K. W. Fischer (Eds.), *Human behavior and the developing brain* (pp. 3–66). New York: Guilford.

Friedman, D. (1991). The endogenous scalp-recorded brain potentials and their relationship to cognitive development. In J. R. Jennings & M. G. H. Coles (Eds.), *Handbook of cognitive psychophysiology: Central and autonomic nervous system approaches* (pp. 621–656). Chichester: Wiley.

Fuson, K. C. (1982). An analysis of the counting-on solution procedure in addition. In T. P. Carpenter, J. M. Moser, & T. A. Romberg (Eds.), *Addition and subtraction: A cognitive perspective* (pp. 67–82). Hillsdale, NJ: Lawrence Erlbaum Associates.

Gelman, R. (1978). Counting in the preschooler: What does and what does not develop? In R. Siegler (Ed.), *Children's thinking: What develops?* (pp. 213–242). Hillsdale, NJ: Lawrence Erlbaum Associates.

Goldman-Rakic, P. S. (1987). Development of cortical circuitry and cognitive functions. *Child Development, 58*, 642–691.

Goldman-Rakic, P. S. (1994). Specification of higher cortical functions. In S. H. Broman & J. Grafman (Eds.), *Atypical cognitive deficits in developmental disorders. Implications for brain function* (pp. 3–17). Hillsdale, NJ: Lawrence Erlbaum Associates.

Greenfield, P. M. (1966). On culture and conservation. In J. S. Bruner, R. R. Oliver, & P. M. Greenfield (Eds.), *Studies in cognitive growth* (pp. 225–256). New York: Wiley.

Griffin, S. (1994, June). *Working memory capacity and the acquisition of mathematical knowledge*. Paper presented at the biennial meeting of the International Society for the Study of Behavioral Development, Amsterdam, Netherlands.

Griffin, S., & Case, R. (1996). Evaluating the breadth and depth of training effects, when central conceptual structures are taught. In R. Case & Y. Okamoto (Eds.), The role of central conceptual structures in the development of children's thought. *Monographs of the Society for Research in Child Development*, pp. 83–102 (Serial Number 246).

Griffin, S., & Case, R. (1997). Rethinking the primary school math curriculum: An approach based on cognitive science. *Issues in Education, 3*, 1–65.

Grune, K., Mecklinger, A., & Ullsperger, P. (1993). Mental comparison: P300 component of the ERP reflects the symbolic distance effect. *NeuroReport, 4*, 1272–1274.

Grune, K., Ullsperger, P., Moelle, M., & Mecklinger, A. (1994). Mental comparison of visually presented two-digit numbers: A P300 study. *International Journal of Psychophysiology, 17*, 47–56.

Halford, G. S. (1982). *The development of thought*. Hillsdale, NJ: Lawrence Erlbaum Associates.

Halford, G. S. (1993). *Children's understanding: The development of mental models*. Hillsdale, NJ: Lawrence Erlbaum Associates.

Halford, G. S., Maybery, M. T., O'Hare, A. W., & Grant, P. (1994). The development of memory and processing capacity. *Child Development, 65*, 1338–1356.

Hebb, D. O. (1949). *The organization of behavior*. New York: Wiley.

Hume, D. (1975). *Enquiries concerning human understanding and concerning the principles of morals*. Oxford, England: Clarendon Press. (Original work published 1748–51)

Kant, I. (1961). *Critique of pure reason*. New York: Doubleday Anchor. (First published 1796).

Kaufmann, W. (1980). *Discovering the mind: Goethe, Kant and Hegel*. New York: McGraw-Hill.

Keating, D. P., & Hertzman, C. (Eds.) (1999). *The developmental health and wealth of nations: Social, biological and educational dynamics*. New York: Guilford.

Kendler, T. S., & Kendler, H. H. (1967). Experimental analysis of inferential behavior in children. *Advances in Child Development and Behavior, 3*, 157–190.

Klahr, D., & Wallace, J. G. (1976). *Cognitive development: An information-processing view*. Hillsdale, NJ: Lawrence Erlbaum Associates.

Krasnegor, N. A., Lyon, G. R., & Goldman-Rakic, P. S. (Eds.). (1997). *Development of the prefrontal cortex. Evolution, neurobiology, and behavior.* Baltimore: Paul H. Brookes.

Locke, J. (1975). *Essay concerning human understanding.* Oxford, England: Clarendon Press. (Original work published 1690)

Luria, A. R. (1966). *Higher cortical functions in man.* New York: Basic Books.

Luria, A. R. (1973). *The working brain.* London: Penguin Books.

Luria, A. R., & Judowitsch, F. J. (1959). *Speech and the development of mental processes in the child.* London: Staples Press.

Marini, Z. (1992). Synchrony and asynchrony in the development of children's scientific reasoning. In R. Case (Ed.), *The mind's staircase* (pp. 55–73). Hillsdale, NJ: Lawrence Erlbaum Associates.

McCandliss, B. D., Posner, M. I., & Givon, T. (1997). Brain plasticity in learning visual words. *Cognitive Psychology, 29,* 88–110.

McCarthy, R. A., & Warrington, E. K. (1990). *Cognitive neuropsychology: A clinical introduction.* San Diego: Academic Press.

McClelland, J. L., Rumelhart, D. E., & Hinton, G. E. (1987). The appeal of parallel distributed processing. In D. E. Rumelhart & J. L. McClelland (Eds.), *Parallel distributed processing. Explorations in the microstructure of cognition. Vol. 1: Foundations* (pp. 3–44). Cambridge, MA: MIT Press.

Mecklinger, A., Ullsperger, P., Moelle, M., & Grune, K. (1994). Event-related potentials indicate information extraction in a comparative judgement task. *Psychophysiology, 31,* 23–28.

Moyer, R. S., & Landauer, T. K. (1967). Time required for judgements of numerical inequality. *Nature, 215,* 1519–1520.

Niedermeyer, E. (1999). Maturation of the EEG: Development of waking and sleep patterns. In E. Niedermeyer & F. Lopes Da Silva (Eds.), *Electroencephalography. Basic principles, clinical applications, and related fields* (pp. 189–214). Baltimore: Williams & Wilkins.

Okamoto, Y., & Case, R. (1996). Exploring the microstructure of children's central conceptual structures in the domain of number. In R. Case & Y. Okamoto (Eds.), The role of central conceptual structures in the development of children's thought (pp. 27–58). *Monographs of the Society for Research in Child Development, 61,* (1–2, Serial No. 246).

Olson, D. R. (1994). *The world on paper: The conceptual and cognitive implications of writing and reading.* New York: Cambridge University Press.

Pascual-Leone, J. (1970). A mathematical model for the transition rule in Piaget's development stages. *Acta Psychologica, 32,* 301–345.

Pascual-Leone, J., Hamstra, N., Benson, N., Khan, I., & Englund, R. (1990). *The P300 event related potential and mental capacity.* Paper presented at the fourth international Evoked Potentials Symposium, Toronto. (Available in conference proceedings).

Pauli, P., Lutzenberger, W., Rau, H., Birbaumer, N., Rickard, T. C., Yaroush, R. A., & Bourne, L. E. (1994). Brain potentials during mental arithmetic: Effects of extensive practice and problem difficulty. *Cognitive Brain Research, 2,* 21–29.

Piaget, J. (1960). *The psychology of intelligence.* Totowa, NJ: Littlefield Adams.

Piaget, J. (1970). Piaget's theory. In P. H. Mussen (Ed.), *Carmichael's handbook of child psychology: Vol. 1* (pp. 703–732). New York: Wiley.

Rogoff, B. (1990). *Apprenticeship in thinking: Cognitive development in social context.* New York: Oxford University Press

Schwarz, W., & Heinze, H.-J. (1998). On the interaction of numerical and size information in digit comparison: A behavioral and event-related potential study. *Neuropsychologia, 36,* 1167–1179.

Siegler, R. S., & Robinson, M. (1982). The development of numerical understanding. In H. W. Reese & L. P. Lipsitt (Eds.), *Advances in child development and behavior, 16* (pp. 241–312). New York: Academic.

Starkey, P. (1992). The early development of numerical reasoning. *Cognition, 43,* 93–126.

Stuss, D. T., & Benson, D. F. (1986). *The frontal lobes.* New York: Oxford University Press.

Taylor, M. J. (1988). Developmental changes in ERPs to visual language stimuli. *Biological Psychology, 26,* 321–338.

Taylor, M. J. (1995). The role of event-related potentials in the study of normal and abnormal cognitive development. In F. Boller & J. Grafman (Series Eds.) & R. Johnson, Jr., & J. C. V. E. Baron (Vol. Eds.), *Handbook of neuropsychology: Vol. 10. Section 14: Event-related brain potentials and cognition. Section 15: Positron emission tomography and neurobehavior* (pp. 187–211). Amsterdam: Elsevier.

Temple, E., & Posner, M. I. (1998). Brain mechanisms of quantity are similar in 5-year-old children and adults. *Proceedings of the National Academy of Sciences USA, 95,* 7836–7841.

Thatcher, R. W. (1992). Cyclical cortical reorganization during early childhood. *Brain and Cognition, 20,* 24–50.

Thatcher, R. W. (1997). Human frontal lobe development: A theory of cyclical cortical reorganization. In N. A. Krasnegor, G. R. Lyon, & P. S. Goldman-Rakic (Eds.), *Development of the prefrontal cortex: Evolution, neurobiology, and behavior* (pp. 85–113). Baltimore: Paul H. Brookes.

Tononi, G. (1996). Specialization, integration, and complexity. *Journal of Psychophysiology, 10,* 273–274.

Tononi, G., Edelman, G. M., & Sporns, O. (1998). Complexity and coherence: Integrating information in the brain. *Trends in Cognitive Sciences, 2,* 474–484.

Tononi, G., Sporns, O., & Edelman, G. M. (1994). A measure for brain complexity—Relating functional segregation and integration in the nervous-system. *Proceedings of the National Academy of Sciences USA, 91,* 5033–5037.

Ullsperger, P., & Grune, K. (1995). Processing of multi-dimensional stimuli: P300 component of the event-related brain potential during mental comparison of compound digits. *Biological Psychology, 40,* 17–31.

Ungerleider, L. G. (1995). Functional brain imaging studies of cortical mechanisms for memory. *Science, 270,* 769–775.

Ungerleider, L. G., & Mishkin, M. (1982). Two cortical visual systems. In D. J. Ingle, M. A. Goodale, & R. J. W. Mansfield (Eds.), *Analysis of visual behavior* (pp. 549–586). Cambridge, MA: MIT Press.

Van Oeffelen, M. P., & Vos, P. G. (1982). A probabilistic model for the discrimination of visual number. *Perception & Psychophysics, 32,* 163–170.

Vygotsky, L. S. (1962). *Thought and language (E. Hanfmann & G. Vaker, Trans.).* Cambridge, MA: MIT Press. (Original work published 1934)

Willms, J. D. (1999). Quality and inequality in children's literacy: The effects of families, schools, and communities. In D. Keating & C. Herzman (Eds.), *The developmental health and wealth of nations: social, biological and educational dynamics.* New York: Guilford.

Witelson, S. (1983). Bumps on the brain: Right–left anatomic asymmetry as a key to functional lateralization. In S. J. Segalowitz (Ed.), *Language functions and brain organization.* London: Academic.

Wynn, K. (1992). Addition and subtraction by human infants. *Nature, 358,* 709–750.

Chapter 9

Why Development Does (and Does Not) Occur: Evidence from the Domain of Inductive Reasoning

Deanna Kuhn
Teachers College, Columbia University

It is a particular pleasure to contribute to a volume devoted to mechanisms of developmental change. The article published in 1972 based on my doctoral dissertation was titled "Mechanisms of Change in the Development of Cognitive Structures," and I have been concerned with the problem ever since. The growing interest in the problem has been attributed to the success of the microgenetic method, which I first used in the 1980s (Kuhn & Ho, 1980; Kuhn & Phelps, 1982), as a tool for studying change. The implication is that the study of change mechanisms has been difficult because of the methodological challenge it poses. Although I agree with this assumption and believe the microgenetic method offers the opportunity to make new inroads on the problem, I want to suggest today another source of difficulty in the effort to understand change mechanisms, and that is an impoverished conception of what is developing.

Part of our conceptual legacy in developmental psychology has been the classical model of developmental change in Fig. 9.1. One monolithic underlying structure, responsible for behaviors A, B, and C across diverse domains is transformed into an entirely different, similarly monolithic structure that produces modified behaviors (A´, B´, C´) in each domain. Thus, a huge job needs to be accomplished by the little arrow connecting the two structures, and it has always been less than clear just how this can happen. Characteristics such as disorganization, disequilibrium, and variability have been attributed to the presumably brief period of disruption symbolized by the arrow, in contrast to the stability that characterizes the

221

Classical Model of Developmental Change

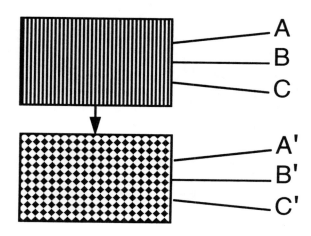

FIG. 9.1.   The classical model of developmental change.

old and new structures. Yet, these characteristics by themselves, even if accurate, do not explain the change.

Most problematic for the classical model has been the repeated finding from microgenetic research that variability is not in fact confined to some brief transition state. My comparative studies of adults and children (Kuhn, Garcia-Mila, Zohar, & Andersen, 1995; Kuhn, Ho, & Adams, 1979; Kuhn, Shaw, & Felton, 1997) have shown that neither is variability confined to critical periods of the life span when development is most likely to occur. Rather, variability is characteristic of the organism, at all times and in all settings. In any situation that elicits action, the individual has a variety of things it can do, with varying probabilities of occurrence (Fig. 9.2). Moreover, these strategies are of differing degrees of effectiveness, ranging from the counterproductive to the optimal effective and efficient strategy. In the case in which a counterproductive strategy is dominant, what we would like to see, and what we would call development, is a shift in the distribution of usage to one in which the optimal strategy is dominant and others have limited probability of occurrence (Fig. 9.3).

Already, then, we have a portrayal of what we mean by development quite different from the classical model. We are still a long way, however, from explaining the mechanism that gets one from the current state to the goal state. Note that traditional learning theories are of no help here, in

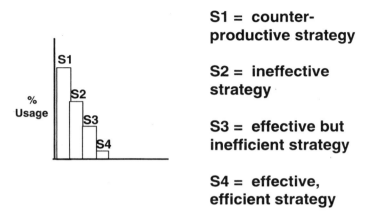

S1 = counter-productive strategy

S2 = ineffective strategy

S3 = effective but inefficient strategy

S4 = effective, efficient strategy

FIG. 9.2.   Typical frequency distribution of strategies applied to a task by an individual at a relatively low level of competence.

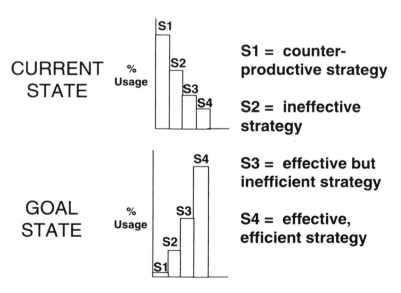

S1 = counter-productive strategy

S2 = ineffective strategy

S3 = effective but inefficient strategy

S4 = effective, efficient strategy

FIG. 9.3.   Developmental change as a shift in the frequency distribution of strategies applied to a task.

particular ones that stipulate a strengthening of behavior through repetition. What we need, instead, is a theory that can account for the finding that some behaviors are weakened, rather than strengthened, in the course of their application. In the absence of any obvious mechanism at the level of performance, it must be assumed that something is missing from the diagram in Fig. 9.3. In other words, more is going on, more is developing, than is observable at the level of performance.

I therefore want to add several entities to this diagram, the first being a representation of the task in relation to which the individual acts. Strategies do not exist in a void, exercised for their own sake. Rather, they function in relation to a goal, defined by a task that the individual has accepted and engaged. The label *strategy* only has meaning and ought properly only to be used in relation to a goal.

To connect the two (task and strategies), and to explain change, something else is needed—another level of operation distinct from strategies, which I will call a *metalevel* of operation. Operators at this level *select* strategies to apply, in relation to task goals, and *manage* and *monitor* their application. I choose the *metalevel* label to emphasize it as a level of operation distinct from and superordinate to the performance level, and I propose that it includes two components, the *metatask* and the *metastrategic* (Fig. 9.4). The role of one is to represent task goals, and the role of the other is to coordinate them with the strategies known to be available. In this process, a subset of strategies are recognized as applicable in relation to task goals. These strategies are selected and applied, with probabilities reflecting each strategy's strength, and the application is monitored by the metalevel.

We now have a considerably more complex, multipart entity (compared to the classical model) that we need to explain the development of; however, paradoxically, the task gets easier. The metalevel directs the application of strategies, and feedback from this application is directed back to the metalevel (Fig. 9.5). This feedback leads to enhanced awareness of the goal and the extent to which it is being met by different strategies, as well as en-

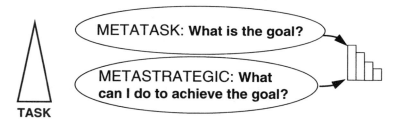

FIG. 9.4.   The metalevel of operation as the conduit between task and strategy application.

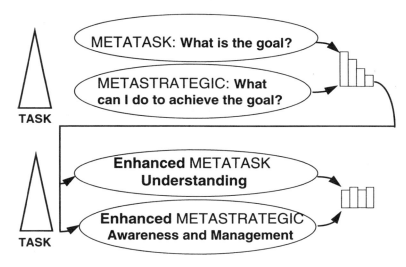

FIG. 9.5.   Performance feedback enhances metalevel understanding.

hanced awareness of the strategies themselves, in particular increased recognition of the power and the limitations associated with each. These enhancements at the metalevel lead to revised strategy selection and hence changes in the distribution of usage observed at the performance level (Fig. 9.5). This modified usage in turn feeds back to enhanced understanding at the metalevel (Fig. 9.6), eventually getting the individual to the perform-

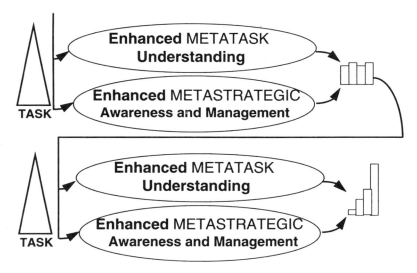

FIG. 9.6.   Metalevel both guides and is modified by performance level.

ance goal state depicted at the bottom of Fig. 9.6. Although Fig. 9.6 portrays a sequence of discrete steps, what is implied is a continuous process, one in which the metaknowing level both guides and is modified by the performance level. (See Siegler & Shipley, 1995; Sophian, 1997, for similar conceptions that have appeared in recent literature.)

In this model, then, changes at the performance level are all channeled, or mediated, through the metalevel. This feature of the model thus privileges the metalevel as the locus of developmental change. This is a strong claim, and, as such, warrants some qualification. I do not deny the importance of phenomena that may be occurring at other levels—the neurological level, certainly, and also the performance level, which is the level at which behaviors are practiced and perfected. My claim, however, is that it is the metalevel (given that it is sufficiently developed and functioning effectively) that is at all times in charge. It remains in charge even if the choice it makes is to shift to "automatic pilot" because nothing much is going on at the moment that requires higher-level management. In other words, it has the final say in what the individual will be doing at any particular moment. In practice, however, this may not always be the case because the metalevel has not reached a point in its development at which it is able to maintain this high level of control, resulting in behaviors that are inconsistent across occasions and highly vulnerable to situational influence. Indeed, the major developmental claim I make here is that increasing metalevel awareness and control constitute the most important dimension in terms of which we see developmental change—important, again, in the sense of determining how the individual is going to behave.

One other implication of the model worth noting is that it readily accounts for the common finding that efforts to induce change directly at the performance level have only limited success, reflected in failure to transfer outside the specific context. Using as an example one of the major strategies examined in this chapter, we can readily tell someone, "If you are trying to find out what makes a difference, change just one thing and leave everything else the same, and see what happens." Even young children can comply with this instruction. If we have not done anything to influence the metalevel, however, the new behavior will disappear quickly once the instructional context is withdrawn and individuals resume metalevel management of their own behavior.

Knowledge Acquisition as a Strategic Task

The discussion until now has been situated at an entirely abstract level. We could be talking about cognitive strategies of almost any sort, applicable to many kinds of tasks (the only exclusion, possibly, being tasks that can be successfully performed through rote learned responses, bypassing strategy selection or application). Here, we narrow the focus and get more specific, but only slightly, because the task and strategies I now turn to are regarded as broad and generic. Indeed, I have studied them for many years precisely because I see them as representative of much of what humans do as cognitive beings. The activity I refer to, again very generically, is *knowledge acquisition*. In these studies, we observe children and adults microgenetically over a period of weeks and months as they engage in the acquisition of new knowledge within a domain.

Knowledge Acquisition as Belief Change. What processes are involved in knowledge acquisition? We know now that knowledge acquisition is *not* the incremental accumulation of facts or associations featured in psychological theories of an earlier era. Instead, if we had to characterize in a single phrase the process by which people's knowledge of the world progresses, *belief change* would be more appropriate. It is now widely accepted that children, from a very early age, construct theories as a means of understanding the world and that these theories undergo revision as children interact in the world and encounter evidence bearing on their theories. It has been my contention that metalevel control over this process is the major dimension in terms of which we see developmental change (Kuhn, 1989, in press).

My research has focused not on theories that affect the formation of concepts (Keil, 1991; Wisniewski & Medin, 1994), but on the second-order theories of how concepts or categories are related to one another, in particular in causal relationships. Understanding the mechanisms that govern formation and revision of causal theories is a scholarly objective, but also one of enormous practical import. Consider, for example, the data in Fig. 9.7, showing that between 1977 and 1998, the percentage of Americans who believe the causes of homosexuality to be innate increased from 13% to 31%, reflecting a belief change on the part of at least several million American adults (who were represented at both times). At least some new empirical evidence bearing on this question has become available between 1977 and 1998, but arguably not of sufficient amount or exposure to account for belief change of this magnitude.

What led, then, to these revisions in causal theories on the part of millions of individuals? Do cultural shifts in attitudes or values, such as tolerance for diversity, dictate shifts toward causal theories that will support these attitudes (rather than the reverse direction of influence we might expect, from beliefs to attitudes)? If so, is evidence sought and attended to selectively to help support new causal theories? These questions point to the importance of studying the process of change in causal theories in the face of new evidence as it occurs within an individual over time. Changes of this sort are likely to have wide-ranging implications, both formal and informal, at many levels of society.

In your view , is homosexuality something a person is
born with, or is homosexuality due to other factors such
as upbringing or environment?

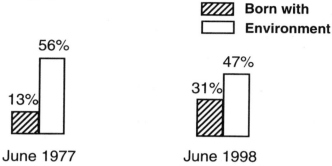

Source: Gallup

FIG. 9.7.   Historical shifts in causal theories about homosexuality.

The Central Role of Causal Belief in Knowledge Acquisition.   Interesting new work has appeared in recent years in the area of multivariable inductive causal inference (Ahn & Bailenson, 1996; Ahn, Kalish, Medin, & Gelman, 1995; Cheng, 1997; Cheng & Novick, 1992; Spellman, 1996). Across several investigators and methods, this work has yielded general agreement that causal inference involves the coordination of theories and evidence—theories of causal mechanism and evidence (primarily covariation) supporting a claim that two factors are causally related. Neither theories of mechanism nor data on patterns of occurrence can alone do the job of accounting for how inferences of causality get made. Cheng (1997) puts it well in claiming that people "postulate unobservable theoretical entities . . . to interpret and explain their observable models" [of the empirical world] (p. 369).

Yet, there has also come to be wide agreement that theories of mechanism are the more influential component of the two. Several kinds of data from independent investigators point to this conclusion. Mechanism information is more influential in causal attribution than covariation information (Ahn & Bailenson, 1996). Mechanism information induces more belief change than covariation information (Slusher & Anderson, 1996). Mechanism information augments belief change produced by covariation information (Koslowski, 1996). People seek mechanism information rather than covariation data to test causal theories (Ahn, Kalish, Medin, & Gelman, 1995). People offer mechanism explanations rather than covariation evidence when asked to justify their causal theories (Kuhn,

1991). People rate mechanism arguments as more persuasive than co-variation arguments (Kuhn & Felton, 2000). People are more likely to acknowledge and interpret covariation evidence if they have a mechanism theory in place (Kuhn, Amsel, & O'Loughlin, 1988).

As compelling as these findings are in pointing to the power of theories of mechanism in causal inference, theories of mechanism have a fatal weakness: They lack the power to falsify a causal belief. In our research, we have abundant evidence of people's attention to empirical data solely for the purpose of illustrating their causal theories, with the data that are attended to chosen selectively to serve this purpose. A goal of our work has been to get people to attend to and process data in a way that would enable them to test their theories, rather than merely illustrate them.

Microgenetic Study of Theory-Evidence Coordination. Conceiving of knowledge acquisition about causal relations as a process of theory–evidence coordination, we have studied people engaged in the process over multiple sessions in a variety of multivariable content domains and formats involving either real objects or computer simulations. The accumulating data that they access must be coordinated with their evolving theories and inferences made about the causal structure of the domain. A constant feature is that the activity is goal-based, the goal being to accurately predict outcomes based on knowledge of the particular constellation of levels of antecedent variables.

In a current version, for example, a computer simulation asks the participant to work building cabins on sites along a lakefront that is susceptible to flooding. The cabins must be elevated on supports to avoid flood damage, but the supports should be no higher than necessary, to minimize cost. The task, therefore, is to predict the varying degrees of flooding that occur at different sites, which can be done only by analyzing the effects of varying features, such as soil type and depth.

In other versions, participants examine variables enabling them to predict how fast a model boat travels down an improvised canal, or, in the social realm, the variables that influence the popularity of children's TV programs. The multivariable context, we find, has the desirable feature of affording high degrees of freedom in causal attribution. If an outcome conflicts with theory-driven expectations with respect to one variable, the implications can be avoided by shifting to another variable to do the explanatory work. The task is thus a rich one for observing the theory–evidence coordination process.

The microgenetic design has the desirable feature of allowing us to simultaneously track changes in knowledge—the set of beliefs about the causal structure of the system—and in the investigative and inference strategies used to acquire that knowledge. These knowledge acquisition strategies, we have found in several different research designs (Kuhn, Garcia-Mila, Zohar, & Andersen, 1995; Kuhn, Schauble, & Garcia-Mila, 1992), show significant generality across content, justifying the assumption that we are observing some broad "modes of knowing" significant enough to warrant study. The fact that strategies often undergo developmental change in the course of their use allows us to closely observe the process in microgenetic study. Fine-grained microgenetic analysis enables us to observe the emergence of new strategies, and we find that they typically appear at about the same time across domains in studies in which we have participants work in two different task domains during the same period of time (Kuhn et al., 1992; Kuhn et al., 1995).

Comparative study of adults and children allows us to identify differences in the knowledge acquisition process at the two different age levels. We could predict that the conflict between the store of feature-related knowledge and new evidence would be greater for adults as they could be presumed to have more entrenched feature-related beliefs. Thus, we would expect more resistance to belief change in adults than children. In fact, however, the difference is just the opposite. Children show greater resistance to abandoning a causal belief in the face of disconfirming evidence than do adults (Kuhn et al., 1995). The implication is that we are observing development of enhanced cognitive flexibility, rather than entrenchment, with the passage from childhood to adulthood (except when development, for one reason or another, is arrested at an early stage). This flexibility takes the form of (a) increased differentiation of theory and evidence as a source of support for one's beliefs, and, hence (b) enhanced ability to represent relations between theories and evidence, that is, the implications of one with respect to the other.

Phases of Knowledge Acquisition Activity

Because participants direct their own investigations of a data base, the work that has been described also falls within the realm of studies of scientific reasoning, and as Klahr (2000) notes, few studies of scientific reasoning examine all phases of scientific activity. In the model presented here, several major phases, or tasks, are distinguished along with a set of strategies associated with each. As we would anticipate based on earlier discussion, in each of these phases individuals show a range of strategies of varying adequacy and a distribution of usage that evolves over time. In addition, each phase is hypothesized to have its associated metalevel of operation (Fig. 9.8).

The task in the first phase of knowledge acquisition (top row of Fig. 9.8) is recognizing that there is a question that can be asked and identifying that question. This is the *inquiry* phase. Examined close-up (Fig. 9.9), the first challenge of this phase is to recognize that the data base I can access yields information that bears on the theories I hold—a recognition that eludes many young theorists. Once the relevance of the data in this respect is rec-

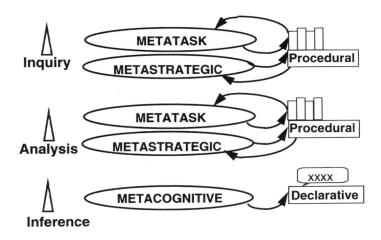

FIG. 9.8.   Phases of knowledge acquisition activity.

ognized, questions can be formulated in a manner conducive to connecting data and theory (left side of Fig. 9.9). The variability and evolution we see in strategies in response to this task are seen on the right side of the figure Here (in contrast to the left side of Fig. 9.9, where objectives are compatible) we have a set of competing strategies that overlap in their usage and are of varying degrees of adequacy (with more adequate strategies appearing

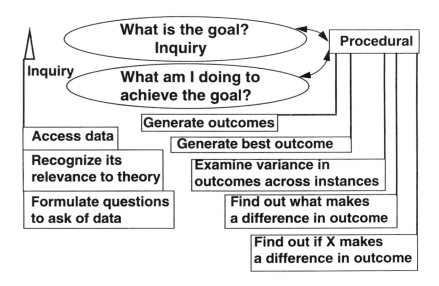

FIG. 9.9.   The inquiry phase.

further down in the figure). At the lowest level, a strategy for some individuals (or for a particular individual some of the time) may be the simple one of activity, that is, generating outcomes or producing the phenomenon—a boat's travel down the canal or a flood of the building site. Later, after the phenomenon has been observed many times, the dominant strategy may become one of producing the most desirable outcome (e.g., the fastest boat). The major developmental shift is one from strategies of activity to genuine inquiry, which in its most rudimentary appearance takes the form of "What is making a difference?" or "What will enable me to predict outcomes?" In more advanced forms, inquiry becomes focused on the specific features in terms of which there is variability, and, ultimately, focused on the effect of a specific feature, "Does X make a difference?"

The next phase is *analysis*, a process that leads to the product phase of *inference*. These are depicted in the second and third rows of Fig. 9.8. To engage in productive analysis (left side of Fig. 9.10), some segment of the data base must be accessed, attended to, processed, and represented as such—that is, as evidence to which one's theory can be related, and these data must be operated on (through comparison and pattern detection), in order to reach the third phase, which yields the product of these operations—inference. The strategies we have observed being applied to this task reflect the struggle to coordinate theories and evidence. As seen on the right side of Fig. 9.10, theory predominates in the lower-level strategies, and only with the gradually more advanced strategies does evidence acquire the power to influence theory.

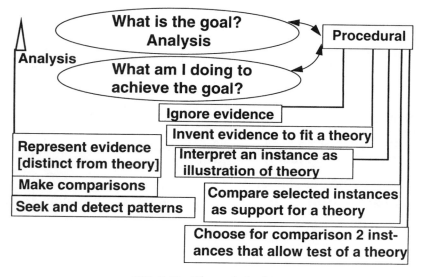

FIG. 9.10. The analysis phase.

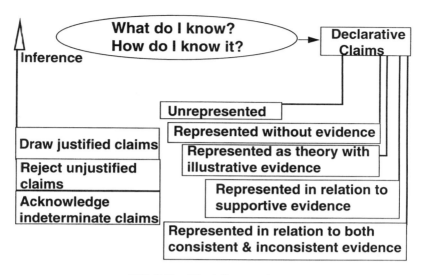

FIG. 9.11.   The inference phase.

The task of *analysis* requires procedural strategies, or a procedural form of knowing. *Inference*, in contrast, requires one to make knowledge claims that are a declarative form of knowing. Hence, the form of knowing at the metalevel is also different (Fig. 9.11). This distinction is noted by referring to procedural metaknowing as *metastrategic* (and *metatask*) and declarative metaknowing as *metacognitive*. As shown in the left side of Fig. 9.11, the inference phase involves inhibiting claims that are not justified, as well as making those that are. The inferential processes we observe being applied to this task (right side of Fig. 9.11) range in adequacy from no processing of the evidence and no conscious awareness of one's theories (so-called "theories in action") to the skilled coordination of theory and evidence, which entails understanding the implications of evidence as supporting or disconfirming one's theories.

Illustrations of Ineffective Knowledge Acquisition

The distinction just noted—between making a justified (high-level) inference and inhibiting an unjustified (low-level) one—becomes important. In individual case studies of participants in repeated sessions over several months struggling to coordinate the initial beliefs they bring to the situation with the accumulating evidence they access, one observation stands out as most striking. Whether the individual is at a level at which counterproductive and ineffective strategies predominate or a more advanced level at which effective and efficient strategies are more frequent, the challenge to further progress appears similar. It is also counterintuitive to the way we typically think about development. Rather than a struggle to find better strategies (these, recall, are typically already present in the repertory, even if used infrequently), the challenge appears to be one of "letting go" of inferior strategies, which remain as a continuing temptation even after they have become infrequent. Interestingly, the concept of multiple players competing for influence (with inferior players needing to be inhibited and stronger ones supported) is one that other chapters in this volume describe as applying at the neural level as well.

Data from the first two sessions of investigative activity by Geoff, one of the students from our community college sample (Kuhn et al., 1995), working on the TV problem, illustrate the continuing threat that inferior strategies pose. These excerpts also illustrate the role that theoretical belief (which continually must be coordinated with new data as they are accessed) plays in this competition among strategies. The task in the TV problem is to determine how various features of children's TV programs influence their popularity among a sample group of children for whom data are available. The participant selects instances (types of TV programs) to examine, makes predictions (regarding their popularity ratings), observes outcomes, and draws inferences. Table 9.1 portrays the sequence of data Geoff selected to examine in his first two sessions. Each of the nine instances consists of a particular combination of program features and an associated outcome (popularity rating by the children), selected from a file box of records of TV programs, representing all the possible feature combinations. Thus, the first instance Geoff chose (first row of Table 9.1) was of a program with commercials but without music or humor, 2 hours long and on Tuesday, with an outcome of *fair*.

Geoff immediately engages and interprets the data he chooses to examine, thus performing at a level above that observed in many younger participants, where the primary strategies are to ignore the data or to invent

TABLE 9.1
Initial Evidence Generated by Geoff for TV Problem

1. -C2ts	fair	
2. MC0wf	excellent	
3. M-1wf	good	**Causal**
4. MC0wf	excellent	Music (M or -)
5. M-2wf	good	Commercials (C or -)
6. ---2ws	poor	Length (0, 1, 2)
7. MC2ts	good	**Noncausal**
8. MC2tf	good	Day (t or w)
9. MC2wf	good	Humor (f or s)

data to fill a theoretical need. Many, for example, justify their inferences entirely on theoretical grounds and simply respond *no* when we ask our evidence-focus probe ("Does any of the information from the records here in the file box tell you about whether it makes a difference?"). Geoff, in contrast, accurately represents the first instance of evidence he generates and interprets it in relation to his theoretical beliefs. His causal attributions were as follows: "You see, this shows you that the factors I was saying about . . . that you have to be funny to make it good or excellent and the day doesn't really matter and it's too long." Geoff's is a classic false inclusion inference, in which one or more features are causally implicated in an outcome based on a single co-occurrence of a particular level of the feature and the outcome. On the basis of this single instance, Geoff implicates both humor and length, while excluding day of the week with no evidence-based justification.

In the second instance he examined, Geoff added humor and music and changed the length to a half-hour (represented as '0' in Table 9.1) and the day to Wednesday. Implicitly comparing this instance to the preceding one, his attribution was as follows: "It does make a difference when you put music and have commercials and the length of time and the humor. Basically the day is the only thing that doesn't really matter." Note Geoff's weak differentiation between theory and evidence as the basis for his claims. He attributes causality to three of the factors that covaried with outcome over these two instances (music, humor, and length). He includes the commercials factor as causal, despite the fact it did not vary, while he nonetheless excluded day, which did vary, as noncausal. This example illustrates how individuals are able to engage the evidence consistently, drawing on it as a basis for their inferences, without its having any influence on their theories. Although illustrated here in a task-constrained context, it is this kind of reasoning I would claim is ubiquitous in everyday life.

Another common tactic is to protect theories by selective application of strategies (thus contributing to the strategy variability we find routinely). In generating Instance 8, for example, Geoff predicted, "I know that if we make it funny it will be even better" (thus showing a high-level inquiry strategy, focused on the effect of a single feature). The outcome in fact remained the same (good), affording Geoff the opportunity for a valid exclusion inference for humor. Yet, he shied away from this interpretation and instead turned to another feature to carry the explanatory burden: "It was less than I expected. This brings me back to what I thought . . . it's rated less because it's too long." In contrast, the very next instance he generated (Instance 9) enabled him to achieve his stated intent of "finding out that the day doesn't make a difference." This time he concluded, "The day doesn't make a difference because the previous one was a different day and it still was good." Thus, instead of excluding day by theoretical fiat (as

he had done earlier), he shows progress by applying the high-level strategy of controlled-comparison analysis with respect to this feature, the
same strategy he had just failed to apply to the humor feature. We thus see
how theoretical belief can contribute to variability in strategy usage (although it is not the sole cause of this variability).

Rather than simply ignore the implications of evidence for the causal
status of one feature and shift to another feature to do the explanatory
work, a more subtle tactic that frequently appears is to "piggyback" a favored feature on to one known to be causal, thus protecting it from the
possibility of disconfirming evidence. We see this tactic for Instances 5 and
6 in Geoff's record. He counts on the music feature, which he has taken to
be causal and which to this point has covaried consistently with outcome,
to produce an expected outcome, and piggybacks the presence or absence
of humor to the presence or absence of music in Instances 5 and 6, thus
setting up the conditions for the causal power of music to diffuse, illegiti-
mately, to humor, which is never subjected to independent test. Another
tactic that individuals frequently use is to particularize the theory, claiming for example that "humor and music together make a difference."

In characterizing the individual's task as one of coordinating theory and
evidence, I need to address an important objection before proceeding. Why
should someone give credence to our minimal data base and give it much
weight in comparison to a lifetime's worth of experience that has presumably gone into one's own convictions about the causal status of these variables? In other words, how can we regard exclusively theory-based justifica-
tions as non-normative or ineffective strategies to apply to this task? A brief
answer is that the task is *not* one that asks the individual to set aside his or
her own understanding of the world in favor of some arbitary new information, but rather one that assesses an individual's ability to examine and represent new evidence and to understand the relation it bears to different theoretical claims. As well as metastrategic management of inquiry and
analysis strategies, the task requires metacognitive awareness of knowledge
claims, as entities susceptible to contradiction by evidence. A participant
could say, "This is what this evidence implies for these theories, although
other sources of support I have for some of these theories leads me to maintain belief in them in the face of your contradictory evidence." This individual will do perfectly well in our task. More troubling are the many participants in our studies whose beliefs *are* influenced by our evidence but who
lack metacognitive awareness of how or why.

Empirical Study of Metalevel Knowing

The particular difficulty of achieving metacognitive awareness of one's
own beliefs warrants emphasis and is reflected in Fig. 9.8. Unlike procedural knowing, which generates feedback in the course of its operation

(making individuals aware that they are doing something), declarative knowing is more a state of being than doing, and as such the knowing itself does not generate any automatic feedback (even though verbal expression of an incorrect belief may be challenged): We have little occasion to say to ourselves, "This is what I know." This difference between metastrategic and metacognitive knowing is reflected in the absence of a reciprocal arrow from the performance level to the metalevel in the case of the declarative knowing at the inference phase (Fig. 9.8). Research with preschool children has identified knowing that or how they know something as a particular difficulty at this age, even when the knowing consists of a simple fact they have just been told (Gopnik & Graf, 1988). In recent work (Kuhn & Pearsall, 2000), we have shown preschoolers to be indifferent in identifying as the source of their knowledge that an event occurred either (a) a theory that makes the event plausible, or (b) evidence showing that the event did occur. These weaknesses in metacognitive knowledge regarding a simple event parallel those observed among older children and adults in the research discussed here involving more complex metaknowing about causal relations.

Metacognitive knowing, I have suggested, is implicit and therefore difficult to observe, but neither is metastrategic knowing readily observable. The variability in strategies portrayed in Fig. 9.8, and the gradual evolution that we see in their frequency distribution as they are applied to a task, are empirical facts. The metalevels of operation portrayed in the center column of Fig. 9.8 are not. They are theoretical constructs. Do we have any evidence that they exist, that they develop along with the strategies that constitute the performance level, and that they mediate strategic development? To obtain this kind of evidence, we need to be able to assess the metalevel components independent of the performance components.

It is not obvious how one might do this, but the approach that Pearsall and I (Kuhn & Pearsall, 1998) took was to externalize metalevel knowing of the procedural type by asking children to explain to another child, who had not participated in the activity, what is to be done and how to do it. Like the participants described earlier (Kuhn et al., 1995), fifth-grade children in the main task selected multivariable instances for examination (cars or boats in this case), made predictions, observed outcomes, and drew inferences regarding the causal status of the various features in affecting outcome (speed of the car or boat). We added the metastrategic assessment at two points, first when the participating child had just begun the activity (Session 2 of 7 sessions) and again after the final session. The observing child had been instructed not to ask questions or otherwise intervene, and so the participating child's verbal communication gave us an indication of his or her understanding of the task and the strategies for accomplishing it. (The latter was not as complete as we would have liked, as

it encompasses the child's understanding only of preferred or recom-
mended strategies; it would be worthwhile to know as well what the child
understood about other, nonrecommended strategies and what the child
understood to be wrong with them.) Metatask understanding (the task
goal) was assessed separately from metastrategic understanding (how to
achieve the goal) and levels of understanding were coded for each. (These
two forms of metalevel knowing, note, are independent: Knowing goals
does not confer knowledge of strategies, nor does awareness of strategies
guarantee that they will be connected to an appropriate goal.)

I do not report these results in detail, as I want to cover more recent
work. For most participants, metalevel understanding, like performance,
shows improvement over time, during a period when strategies are exercised (as the model presented earlier of the knowledge acquisition process
would predict). Most important, correspondences are apparent in the improvements that occur at the two levels. The relationship is one of the
metalevel playing a gatekeeper, or necessary-but-not-sufficient, role with
respect to the performance level. Preferred strategies observed at the performance level rarely exceeded the level of understanding exhibited in assessment of the metalevel, although the reverse was not true—metalevel
understanding sometimes exceeded what was realized in performance.
Typically this happened when individuals understood what was to be accomplished (identification of the causal status of each of the features) but
knew no strategies for accomplishing it.

In each of our studies, regardless of the age of the participants, it has
been the case that exercise of strategies has been sufficient to induce
change (at both performance and metaknowing levels, when both have
been assessed) in a majority, but not all, of the participants. In the cases in
which change does not occur, metalevel functioning is perhaps too weak to
utilize the feedback generated by the performance level. Whether or not
this is exactly the right explanation, the recurring minority of cases in
which change fails to occur led us to consider the possibility of doing
something to stimulate the metalevel directly (rather than depend on the
feedback from strategic performance). How might this be done? Although
we have gone on in current work to experiment with other approaches,
the first approach we tried was collaborative activity.

Collaboration as a Vehicle for Strengthening Metalevel Knowing

Collaborative learning has been proposed as the answer to so many different educational problems that it is important to take a close look at the
reasons we thought it might work. The benefits of collaborative cognition
typically are construed at the performance level and center around two

constructs that the sociocultural literature has made popular, *distributed cognition* and *scaffolding* (Fig. 9.12). Distributed cognition divides the cognitive workload, allowing the pair to accomplish jointly what neither could manage alone. It is particularly advantageous when each partner contributes competencies not in the repertory of the other. Scaffolding occurs when the competency of the less able partner is enhanced through the guidance provided by the more able partner.

What if we extend this analysis of potential benefits of collaboration to the metaknowing level, where it has received much less attention? Two effects might be predicted (Fig. 9.12). First, verbal communication of my metalevel understanding externalizes it, making it more explicit. The benefit here is well captured in the orangutan theory. If I have some new ideas and I go into a room with an orangutan to explain them, the orangutan will simply sit there and eat its banana. I will come out of the room, however, knowing more than I did before.

Second, my explicit communication of metaknowing to a partner, or a partner's explicit communication of his or her metaknowing to me, may influence and guide the recipient of this communication, particularly if the communicator is more capable. This is scaffolding at the metaknowing level. When both partners have some competence, we can predict the most powerful effect—one in which both partners benefit, because each guides the metaknowing of the other. In other words, each partner serves as the other's metaknowing operator, helping to monitor and manage their partner's strategic operations, in a way that the partner is not yet able to do as competently for him- or herself.

How does collaboration enhance cognition?

At the performance level

- **Cognitive work is *socially distributed*.**

- **My performance is *scaffolded* by an able partner.**

At the metaknowing level

- **Metaknowing is *externalized*.**

- **My partner's metaknowing *scaffolds* my metaknowing and performance.**

FIG. 9.12. Potential effects of collaboration on performance-level and metalevel cognition.

In a doctoral dissertation by Andersen (1998), each of these processes can be observed, although there are almost as many occasions where they fail to occur as occasions where they do. In other words, conditions have to be just right for them to work. The illustrations presented here are primarily of the less-studied scaffolding of metaknowing functions, where this is particularly true.

The design of Andersen's study, it should be noted first, works against demonstrating the benefits of peer collaboration, the typical goal of researchers who have studied the process. Each of the fifth-grade participants in his study worked on two problems over the course of 7 weeks (the boat and the racecar problems), one of them alone and the other with a partner (who remained the same over the 7 weeks). It would naturally be anticipated that the competencies individuals bring to the task will affect performance in the collaborative setting. This research design, however, leaves open the additional likelihood that whatever a child does in collaboration with a peer will influence his or performance in the solo condition, thus diminishing the difference between the two conditions. If we wished to maximize the difference between the two conditions, we would design the study as a between-subject comparison of groups representing each of the conditions. In this case, however, we were willing to accept the disadvantage of the within-subject design because the objective was not to maximize difference but rather to be able to precisely compare a child's performance over time in the solo condition to that same child's performance in collaboration with a peer.

Despite the likely diffusion of gains from one condition to the other, the superiority of performance in the peer condition over the solo condition did still appear in overall comparisons. Offering the greatest insight, however, are the individual summaries of how a child performed over the 7 weeks when working alone versus with his or her partner. As causal inferences were the product or goal of the activity, the extent to which they were adequately justified (by data available in the data base that had been accessed) was taken as the critical measure of performance.

The records of 4 of the 12 fifth-grade pairs in Andersen's study appear in Fig. 9.13. These individual case summaries show for each pair the percentage of valid inferences during the early, middle, and later portions of the activity (represented by points 1, 2, and 3, respectively in the figures), compared to the percentages when each member of the pair is working alone. An inference was counted as valid if either spontaneously or in response to the evidence-focus probe ("Does the information here tell you anything about whether it makes a difference?"), the evidence the individual cited (a) was an accurate representation of evidence that actually existed in that individual's accumulated data base, and (b) was sufficient to justify the inference.

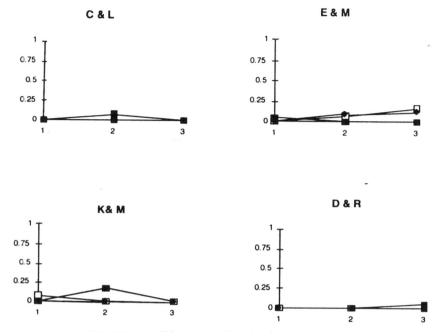

FIG. 9.13.   Collaborating pairs who show no progress.

The first lesson these findings offer is that collaboration is by no means a sufficient condition to enhance the performance that partners show when working alone. The four records in Fig. 9.13 show that if neither member of a pair knows much, the pair working together will not necessarily bootstrap themselves to increased competence. Nor, we see from the next two pairs (Fig. 9.14), is competence on the part of one member a sufficient condition for the enhancement of performance by peer collaboration. The efforts of the more able members of each pair (A and P) to assist their partners in the metaknowing functions of task understanding and strategy selection are evident in both cases (although to a greater extent for A & E). At one point, for example, A says to E, as they construct a new car:

A: The same thing except for the wheels. Change the wheels whatever way you want.
E: Why can't it be a big engine?
A: No, because you're going to find out about the wheels.

Similarly, on another occasion A resists E's attempt to divert attention away from the variable being examined:

A & E

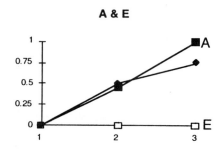

F & P

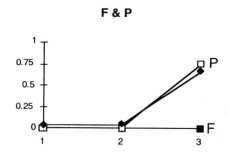

FIG. 9.14.   Illustrations of unsuccessful scaffolding attempts.

A: I'll have the same car except I want to leave off the muffler.

E: Why won't you change the wheel?

A: Cause we're going to find out about the muffler.

And yet again:

A: Same exact car except we want to change it to no fin.

E: We have to go to the muffler now.

A: No, we need to . . . we're still on fin now.

E: I know. . . oh, we are?

A: Yeah, because you got to run it with and without the fin. Then, after this one, we get a chance to find out about the muffler. Not today, but next time.

Note that A is not simply telling E what to do; she is telling her *why* to do it. Despite these repeated efforts on A's part to impart metastrategic knowledge to E, and E's seeming involvement and receptivity, such efforts are unsuccessful, as they are for the other pair in Fig. 9.14 (F & P). The less able partners fail to show any competence working alone, and in the pair condition all of the successful inferences are made by the more capable partner.

In the two pairs, J and O and J and D (Fig. 9.15), in contrast, the scaffolding of metaknowing is successful: The less able partner benefits from the collaboration. In each case, the less able partner begins to show some competence in the solo condition, and in the pair condition the less able partner contributes at least some of the successful inferences (in contrast to the cases in Fig. 9.14). For example, on several occasions, O starts to vary a second variable and J restrains her: "No, you've got to leave it the same, all of these," and O accedes, going on to make a valid inference that most likely would not have occurred had she been working alone.

The pair in Fig. 9.16, P and R, shows a similar pattern: The less able partner (R) benefits from collaboration and contributes to the pair performance. In this case, however, the process appears to be different and describable in terms of the simpler, more familiar mechanism of distributed cognition. Their collaboration evolved into a pattern in which the more able P takes over the most cognitively challenging parts of the task—identifying the question and designing the comparison—allowing the less able R to draw the resulting inference. When working alone, we see in Fig. 9.16, R only rarely makes a successful inference.

Figures 9.17 and 9.18 summarize performance of the remaining three pairs. In comparison to the pairs who show successful collaboration in

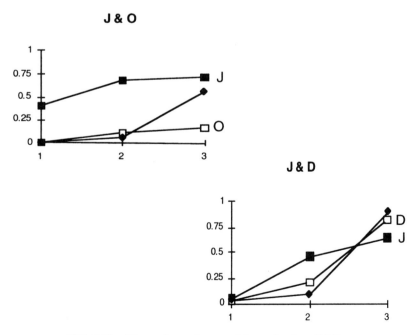

FIG. 9.15.   Illustrations of successful metalevel scaffolding.

P & R

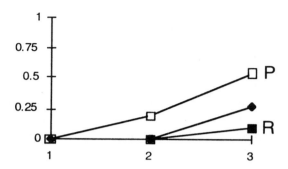

FIG. 9.16.   An illustration of successful performance-level scaffolding.

Figs. 9.15 and 9.16, Fig. 9.17 depicts supersuccessful pairs, in that both members appear to benefit from the collaboration and both contribute to the collaborative performance, with the result that collaborative performance exceeds the performance of either partner alone. In the pair N and S (Fig. 9.17), for example, S proposes:

> S: Why don't we just stick to one subject [i.e., one variable to investigate]? Then we'll be there quicker and get all of them.

At the next session, however, N directs S:

> N: Just change it to the small size.
> S: Change it to the small size. And that's all. And then that way we'll know for sure.

Later in this same session, N monitors and corrects S:

> S: We found out about the weight.
> N: No, about the boat size, that's all.
> S: Oh, the boat size.
> N: Just talk about the boat size.

Similar phenomena occur in the case of Y and M (Fig. 9.17) and show beginning signs in the case of E and J (Fig. 9.18). Our interpretation is that in each of these cases, we are seeing the mutual scaffolding of metaknowing functions. Each partner at different times serves as the other's metaknowing operator, helping to monitor and manage their partner's strategic operations.

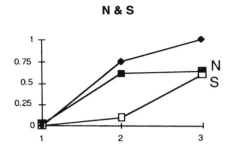

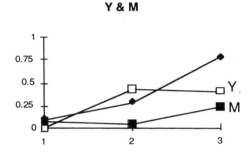

FIG. 9.17.   Illustrations of supersuccessful mutual-scaffolding: Collaboration enhances both partners' performance.

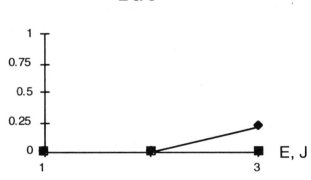

FIG. 9.18.   Mutual-scaffolding at an early stage.

By externalizing metaknowing operations, these case studies of social collaboration make these operations easier to observe and highlight their role as managers of performance. Our data show, however, that this mangerial role is not always actualized. This fact led us to explore whether there may be other, more direct, perhaps more powerful ways to strengthen metaknowing operations. In her doctoral dissertation, Pearsall (1999) investigated one, which is to demand at least the evaluative aspect of metalevel operation by requiring participants to evaluate the performance of others in the kind of task that we have described (an evaluation that includes assigning and justifying a grade for the performance). Whether the effects of this activity will transfer to the participant's own metaknowing operations and strategic performance when they engage the task themselves we have yet to see.

Argument and Values as Dimensions of Knowing

Before concluding it is important to add two further dimensions to the portrayal of the knowledge acquisition process offered in Fig. 9.8. One is a further row (Fig. 9.19), representing an additional phase following analysis and inference, and that is the phase of argument, the other major area of reasoning in which I have done research (Kuhn, 1991). Argument is more complex but, as an intellectual activity, follows directly from the declarative claims that are the product of inference. It entails debate of multiple claims, in a framework of alternatives and evidence. The same structure that I identified in the preceding tasks of analysis and inference, I believe, applies in the case of argument. A range of argumentive strategies can be identified, strategies that an individual draws on with varying degrees of probability. Given sustained exercise, these strategies undergo development (Kuhn,

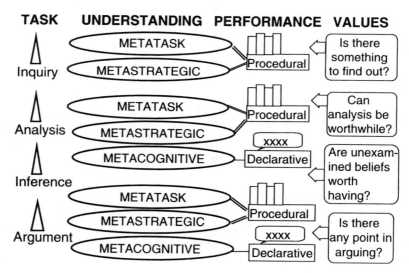

FIG. 9.19.   Additional phases and dimensions of knowledge acquisition activity.

Shaw, & Felton, 1997). The products of the argumentive process are revised and strengthened *claims*, strengthened in the sense of being better supported as an outcome of the argumentive process.

The other dimension to add to this picture is a final column (see Fig. 9.19) representing a different kind of cognitive operation, called *values*. Values—intellectual values, in this case—are slippery phenomena to pin down precisely and psychologists have shied away from studying them. They are anchored, however, by both antecedents and consequences. Feeding into and shaping intellectual values are a person's beliefs about knowing, which have been studied under the heading of *epistemological beliefs* (Hofer & Pintrich, 1997, in press). Intellectual values, in turn, give rise to *dispositions* (a term I prefer to *habits*) to behave in a particular way. Epistemological understanding, however, remains the critical foundation on which intellectual values and dispositions rest. Attainment of a mature level of epistemological understanding is a necessary condition for any of the more specific intellectual values portrayed in Fig. 9.19 to prevail. Unless one believes that analysis and argument are essential, indeed the only route, to knowing (a belief representative of the highest, *evaluative* level of epistemological thought; Kuhn, Cheney, & Weinstock, in press), one is unlikely to hold (and to behave in accord with) the values that analysis is worthwhile, that unexamined beliefs are not worth having, and that there is a point to arguing (Fig. 9.19). In the end, then, we cannot fully understand the kinds of knowing and knowledge acquisition that people engage in without understanding their beliefs about knowing.

CONCLUSION

In conclusion, performance skills or strategies are very often there, in some nascent form at least. When development fails to occur, the failure can often be traced to a failure in the development of the metaknowing components that manage, monitor, and evaluate performance. Metacognitive knowledge of what one knows and how is particularly likely to be fragile, if it is present at all, as we infrequently contemplate what we know. Each time we must decide what to do, in contrast, metastrategic operations are invoked.

More is developing, then, than just the performance we see. Metaknowing, I have argued is a central part of what develops, but so are the intellectual values that we are beginning to see have a critical influence on performance. It remains to add that my message is not entirely new. Educators have long been concerned about the gap between performance and understanding, which is in large part what we have been talking about here under the heading of metaknowing. Empirical research focused on the operations involved in metaknowing, and how they feed both from and to performance, has a potentially valuable contribution to make in helping educators determine how to bridge this critical gap between performance and understanding.

REFERENCES

Andersen, C. (1998). *A microgenetic study of science reasoning in social context.* Unpublished doctoral dissertation. Teachers College, Columbia University.

Ahn, W., & Bailenson, J. (1996). Causal attribution as a search for underlying mechanisms: An explanation of the conjunction fallacy and the discounting principle. *Cognitive Psychology, 31*, 82–123.

Ahn, W., Kalish, C., Medin, D., & Gelman, S. (1995). The role of covariation versus mechanism information in causal attribution. *Cognition, 54*, 299–352.

Cheng, P. (1997). From covariation to causation: A causal power theory. *Psychological Review, 104*(2), 367–405.

Cheng, P., & Novick, L. (1992). Covariation in natural causal induction. *Psychological Review, 99*, 365–382.

Gopnik, A., & Graf, P. (1988). Knowing how you know: Young children's ability to identify and remember the sources of their beliefs. *Child Development, 59*, 1366–1371.

Hofer, B., & Pintrich, P. (1997). The development of epistemological theories: Beliefs about knowledge and knowing and their relation to learning. *Review of Educational Research, 67*(1), 88–140.

Hofer, B., & Pintrich, P. (in press). (Eds.), *Epistemiology: The psychology of beliefs about knowledge and knowing.* Mahwah, NJ: Lawrence Erlbaum Associates.

Keil, F. (1991). The emergence of theoretical beliefs as constraints on concepts. In S. Carey & R. Gelman (Eds.) *The epigenesis of mind: Essays on biology and cognition* (pp. 237–256). Hillsdale, NJ: Lawrence Erlbaum Associates.

Klahr, D. (2000). *Exploring science: The cognition and development of discovery processes.* Cambridge, MA: MIT Press.

Koslowski, B. (1996). *Theory and evidence: The development of scientific reasoning.* Cambridge, MA: MIT Press.

Kuhn, D. (1972). Mechanisms of change in the development of cognitive structures. *Child Development, 43*, 833–844.

Kuhn, D. (1989). Children and adults as intuitive scientists. *Psychological Review, 96*, 674–689.

Kuhn, D. (1991). *The skills of argument.* New York: Cambridge University Press.

Kuhn, D. (1995). Microgenetic study of change: What has it told us? *Psychological Science, 6*, 133–139.

Kuhn, D. (1999). Metacognitive development. In L. Balter & C. Tamis-LeMonda (Eds.), *Child psychology: A handbook of contemporary issues.* Philadelphia: Psychology Press.

Kuhn, D., Amsel, E., & O'Loughlin, M. (1988). *The development of scientific thinking skills.* Orlando, FL: Academic Press.

Kuhn, D., Cheney, R., & Weinstock, M. (in press). The development of epistemological understanding. *Cognitive Development.*

Kuhn, D., & Felton, M. (2000). Developing appreciation of the role of evidence in argument. Paper presented at the Winter Text Conference, Jackson Hole, WY.

Kuhn, D., Garcia-Mila, M., Zohar, A., & Andersen, C. (1995). Strategies of knowledge acquisition. *Society for Research in Child Development Monographs, 60*(4, Serial no. 245).

Kuhn, D., & Ho, V. (1980). Self-directed activity and cognitive development. *Journal of Applied Developmental Psychology, 1*, 119–133.

Kuhn, D., Ho, V., & Adams, C. (1979). Formal reasoning among pre- and late adolescents. *Child Development, 50*, 1128–1135.

Kuhn, D., & Pearsall, S. (1998). Relations between metastrategic knowledge and strategic performance. *Cognitive Development, 13*, 227–247.

Kuhn, D., & Pearsall, S. (2000). Developmental origins of scientific thinking. *Journal of Cognition and Development, 1*, 113–129.

Kuhn, D., & Phelps, E. (1982). The development of problem-solving strategies. In H. Reese (Ed.), *Advances in child development and behavior,* Vol. 17 (pp. 1–43). New York: Academic Press.

Kuhn, D., Schauble, L., & Garcia-Mila, M. (1992). Cross-domain development of scientific reasoning. *Cognition and Instruction, 9*, 285–327.

Kuhn, D., Shaw, V., & Felton, M. (1997). Effects of dyadic interaction on argumentive reasoning. *Cognition and Instruction, 15*, 287–315.

Pearsall, S. (1999). Effects of metacognitive exercise on the development of scientific reasoning. Unpublished doctoral dissertation, Teachers College, Columbia University.

Siegler, R. (1996). *Emerging minds: The process of change in children's thinking.* New York: Oxford University Press.

Siegler, R., & Shipley, C. (1995). Variation, selection, and cognitive change. In T. Simon & G. Halford (Eds.), *Developing cognitive competence: New approaches in process modeling.* Hillsdale, NJ: Lawrence Erlbaum Associates.

Slusher, M., & Anderson, C. (1996). Using causal persuasive arguments to change beliefs and teach new information: The mediating role of explanation availability and evaluation bias in the acceptance of knowledge. *Journal of Educational Psychology, 88*, 110–122.

Sophian, C. (1997). Beyond competence: The significance of performance for conceptual development. *Cognitive Development, 12*, 281–303.

Spellman, B. (1996). Acting as intuitive scientists: Contingency judgments are made while controlling for alternative potential causes. *Psychological Science, 7*, 337–342.

Wisniewski, E., & Medin, D. (1994). On the interaction of theory and data in concept learning. *Cognitive Science, 18*, 221–281.

NEUROSCIENCE APPROACHES

Developing Cortical Specialization for Visual–Cognitive Function: The Case of Face Recognition

Mark H. Johnson
Center for Brain and Cognitive Development,
School of Psychology
Birkbeck College
London, UK

Michelle de Haan
Center for Brain and Cognitive Development,
School of Psychology
Birkbeck College
London, UK

and

Institute of Child Health
University College London
London, UK

Over a century of neuroanatomy and neuropsychology has taught us that the adult cerebral cortex is composed of a number of distinct cytoarchitectonic areas (e.g., Brodman, 1905). It is widely assumed that these structural differences result in areas with different computational properties, and correspondingly there is now a vast cognitive neuroscience literature attempting to localize cognitive functions to particular cortical areas (Posner, Petersen, Fox, & Raichle, 1988). Yet, the more fundamental question of how these structural and functional specializations arise within an individual remains unresolved. Over the past decade, developmental neuroscientists have debated whether cortical specificity arises through molecular and genetic interactions, or through activity-dependent processes (O'Leary, 1989; Rakic, 1988). A parallel debate has raged among cognitive scientists between those who believe that the newborn infant has "innate modules" for cognitive functions such as language acquisition (Pinker, 1994, 1997) and object knowledge (Spelke, 1994), and those who regard such specialization as an emergent product of prenatal and post-natal activity-dependent processes (e.g., Elman, Bates, Johnson, Karmi-loff-Smith, Parisi, & Plunkett, 1996; Munakata, McClelland, Johnson, & Siegler, 1997). In fact, much recent evidence from developmental neuro-science appears inconsistent with the assumption of extensive prespeci-fication, and instead supports the view that there is likely to be some de-gree of postnatal activity-dependent specialization in humans, especially for more cognitive functions (see Johnson, 1997a, 1997b for review).

In this chapter, we start with a brief overview of postnatal brain devel-opment in humans and its relation to the differentiation of the cerebral cortex. We then take one prominent example of cortical specialization, that for face processing, and report recent evidence from our laboratory suggesting that there is a gradual process of increasing specialization. Finally, we summarize some work with artificial neural network models that illustrate the possible mechanisms of specialization. We suggest that similar mechanisms may underlie postnatal cortical specialization in the human child.

POSTNATAL BRAIN DEVELOPMENT
AND THE DIFFERENTIATION
OF THE CEREBRAL CORTEX

A number of lines of evidence indicate that there are substantive changes in the human brain during its postnatal development. At the most gross level of analysis, the volume of the brain quadruples between birth and adulthood. This increase comes from a number of sources such as exten-sion of fiber bundles, and the myelination of nerve fibers in the fatty sheath, which helps conduct electrical signals (Conel 1939–1963; Schade & van Groenigen, 1961). Perhaps the most obvious change as viewed through a standard microscope is in the dendritic tree: For many neurons, its extent and reach may increase dramatically while at the same time be-coming more specific. With the greater power of the electron microscope, a corresponding increase can be seen in the density of synapses in the ce-rebral cortex. For example, in parts of the human visual cortex, the gener-ation of synapses (synaptogenesis) begins around the time of birth and reaches a peak around 150% of adult levels toward the end of the first year (Huttenlocher, 1990). In the frontal cortex the peak of synaptic density occurs later, at around 12 to 18 months of age (Huttenlocher, 1990; Huttenlocher & Dabholkar, 1997). Although there is variation in the time-table, in all regions of cortex studied so far, synaptogenesis begins prena-tally and increases until postnatally peaking at a level well above that ob-served in adults (Huttenlocher & Dabholkar, 1997; Zecevic, 1998).

Somewhat surprisingly, regressive events are also common during the development of brain cells and their connections. Although further research is needed, in most cortical areas studied to date the mean density of synapses per neuron starts to decrease after puberty. In humans, most cortical regions and pathways appear to undergo this "rise and fall" in synaptic density. The postnatal rise and fall developmental sequence can also be seen in other measures of brain physiology and anatomy. For example, Positron Emission Tomography (PET) can be used to measure the glucose uptake of regions of the brain. Glucose uptake is necessary in regions of the brain that are active, and, as glucose is transported by the blood, measuring the glucose uptake also provides a measure of blood flow. Using this method, Chugani, Phelps, and Mazziotta (1987) observed a qualitatively adult-like distribution of resting brain activity within and across brain regions by the end of the first year. However, quantitatively, the overall level of glucose uptake reaches a peak during early childhood when it is much higher than that observed in adults. The rates return to adult levels after puberty for some cortical regions. The extent to which these changes relate to those in synaptic density is currently the topic of investigation.

Is this differentiation of the cortex influenced more by neural activity or by intrinsic molecular and genetic specification? Supporting the importance of the latter processes, Rakic (1988) proposed that the prenatal differentiation of the cortex into areas is due to a protomap. The hypothesized protomap either involves prespecification of the proliferative zone, or intrinsic molecular markers that guide the division of cortex into particular areas. One mechanism for this would be through the radial glial fiber path from the proliferative zone to the cortex. By this view, the differentiation of cortex is mainly due to the unfolding of a genetic plan implemented by molecular markers. The alternative viewpoint, advanced by O'Leary (1989) and Killackey (1990) among others, is that genetic and molecular factors build an initially undifferentiated protocortex, and that the cortex subsequently becomes divided into specialized areas as a result of activity within neural circuits. This activity within neural circuits need not necessarily be the result of input from the external world, but may result from intrinsic, spontaneous, patterns of firing within sensory organs or subcortical structures that feed into cortex, or from activity within the cortex itself (Katz & Shatz, 1996).

Although the neurobiological evidence is complex, overall it currently tends to support the importance of neural activity-dependent processes (see Elman et al., 1996; Johnson 1997a, 1997b for reviews). With several notable exceptions (e.g., in certain primary sensory regions), it seems likely that activity-dependent processes contribute to the differentiation of functional areas of the cortex, especially those involved in higher cognitive functions. During prenatal life this activity may be largely a spontaneous intrinsic process, yet in postnatal life this neural activity is likely also to be influenced by sensory and motor experience (Katz & Shatz, 1996). Factors such as sensorimotor experience may be important in determining cortical organization even into adulthood. For example, the sensorimotor cortex of adults with experience playing a stringed instrument shows a larger representation of fingers in the left hand (which requires greater fine motor learning) compared to fingers the right hand (which requires less fine motor learning as it bows) or fingers of the left-hand area of adults without experience playing a stringed instrument (Elbert, Pantev, Wienbruch, Rockstroh, & Taub, 1995).

It is unlikely that the transition from spontaneous intrinsic activity to that influenced by sensory experience is a sudden occurrence at birth, for in the womb the infant can process sounds and generate spontaneous movement, while in postnatal life the brain maintains spontaneously generated intrinsic electrical rhythms (EEG). Nevertheless, in humans there are two related reasons to believe that there is a substantial effect of postnatal experience. Finlay and Darlington (1995) have proposed an elegant theory of the relation between the ontogeny and phylogeny of the volumes of brain structures, in which it is argued compellingly that there is a general genetic factor relating to overall speed of brain development during ontogeny that can account for many species differences in brain volume. Further, the later developing a brain structure is, the relatively larger it will be with slowed development. Thus, the greatly slowed sequence of brain development observed in humans has a twofold effect: First, it extends the period of brain development well into postnatal life; second, it means the cerebral cortex is relatively larger than other brain structures. Although the former effect offers more opportunity for postnatal experience to shape the later stages of brain development, the latter means that there is less scope for thalamic afferents guided by molecular markers to contribute to cortical specialization. Thus, cortical specialization in humans may be more susceptible to interaction with the postnatal environment than in most mammals, including other primates.

EVIDENCE FOR AND AGAINST AN INNATE CORTICAL MODULE

At the cognitive neuroscience level, some have argued that a cortical "face module" is present early in infancy, and that this module contains a prototype used for face processing (Farah et al., 2000). This conception is somewhat analogous to arguments used about a module for language acquisition (Pinker, 1994, 1997). Three lines of evidence used to support this claim are (a) the specific activation of a ventral temporal lobe area in response to faces—"the fusiform face area," (b) the existence of face-responsive cells in infant monkeys as young as 6 weeks of age, and (c) newborn infants responding preferentially to faces.

Fusiform Face Area. Neuroimaging techniques such as PET and fMRI have been used to demonstrate that specific areas of the human adult's cortex are selectively activated during face processing. One such area studied is the fusiform face area (FFA), a region of ventral temporal cortex that lies anterior to V4 (Halgren, Dale, Sereno, Tootell, Marinkovic, & Rosen, 1999). In a series of studies, it has been shown that this cortical area is more active while adults are viewing faces than while they are viewing objects (Kanwisher, McDermott, & Chun, 1997; Sergent, Ohta, & MacDonald, 1992), hands (Kanwisher et al., 1997) scrambled faces (Puce, Allison, Gore, & McCarthy, 1995), or textures (Malach, Reppas, Benson, Kwong, Jiang, Kennedy, Ledden, Brady, Rosen, & Tootell, 1995; Puce, Allison, Asgari, Gore, & McCarthy, 1996), and while matching faces than while matching locations (Courtney, Ungerleider, Keil, & Haxby, 1997; Grady, Maisog, Horwitz, Ungerleider, Mentis, Salerno, Pietrini, Wagner, & Haxby, 1994; Haxby, Horwitz, Maisog, Ungerleider, Mishkin, Schapiro, Rapoport, & Grady, 1993; Haxby, Horwitz, Maisog, Ungerleider, Pietrini, & Grady, 1994; Horwitz, Grady, Haxby, Schapiro, Rapoport, Ungerleider, & Moscovitch, 1992) houses (Wojciulik, Kanwisher, & Driver, 1998) or hands (Kanwisher et al., 1997). These results support the view that the activation of fusiform face area in response to faces cannot be accounted for by visual complexity, stimulus meaningfulness or simple perceptual properties of faces. If adults are also required to remember facial identity, activation in these ventral occipito-temporal areas is negatively correlated with delay, suggesting that these regions are more involved in the initial encoding of faces and less in retaining memories of them (Courtney, Maisog, Ungerleider & Haxby, 1996; Courtney et al., 1997; Haxby, Ungerleider, Horwitz, Rapoport, & Grady, 1995).

Face-Selective Cells in Monkey Cortex. Studies with nonhuman primates have showed that the activity of even single cells within the cortex can be face-selective. To identify face-selective cells, investigators usually record a cell's firing rate (i.e., number of action potentials over time) while showing the monkey a variety of face and nonface stimuli. The activity of an individual cell can be measured by lowering small microelectrodes into the cortex and measuring the small voltage changes that occur at its tip. These voltage changes emanate from the brain cell closest to the tip (Perrett & Mistlin, 1990). In general, if the magnitude of the cell's response to faces is at least two times greater than the magnitude of its largest response to any nonface object, then the cell is considered to be face-selective (e.g, Rolls & Baylis, 1986).

Face-selective cells have been found in several areas of the adult monkey cortex but are especially common in areas TEa and TEm of the inferior temporal cortex and area TPO of the superior temporal polysensory cortex, where they constitute between 10% to 20% of all visually responsive cells (Baylis, Rolls & Leonard, 1987; Rolls, & Baylis, 1986). In general, activity in single cells in TE is remarkably adultlike when measured in alert infant monkeys. The percentage of cells responsive to non-face visual stimuli and the selectivity of responses were qualitatively similar to responses of adults in monkeys as young as 5 weeks, although the magnitude of responses was lower and the latencies longer and more variable (Rodman, Scalaidhe, & Gross, 1993; Rodman, Skelly, & Gross, 1991).

In studying these cells, a small number of cells responsive to faces have been found. However, it is not clear that these cells show all of the same properties as those identified in adults (e.g., that they are not responding more generally to visual complexity), as their responses have not been studied as extensively. Moreover, in adults, objects other than faces can also show selective response at the single-cell level. For example, in the temporal cortex, investigators have found cells selective for hands (Gross, Rocha-Miranda, & Bender, 1972), for movements of the body (e.g., walking; Oram & Perrett, 1994), and for arbitrary patterns of learned significance (Miyashita, 1988, 1990).

Newborns' Preference for Faces.   During the first hours of life, newborns show preferential orienting to faces. They will move their eyes, and sometimes their heads, further to follow a moving schematic face than several other patterns including faces with features scrambled and nonface patterns (Goren, Sarty, & Wu, 1975; Johnson, Dziurawiec, Ellis, & Morton, 1991; Maurer & Young, 1983). Moreover, newborns will even direct their attention to a facelike pattern when simultaneously presented with a nonface stimulus that has the optimal spatial frequency for the newborn visual system (Valenza, Simion, Cassia, & Umilta, 1996).

Some investigators have used these three lines of evidence to argue that there is an "innate cortical module" for face processing (Farah et al., 2000) similar to that posited for aspects of language acquisition (Pinker, 1994, 1997). An alternative, 'ontogenetic' view of the development of face processing streams in the cortex was advanced independently by deSchonen and Mathivet (1989) and by Johnson and Morton (1991; Morton & Johnson, 1991). The latter authors argued for a two-process model of the development of face processing, which accounts for all of the previous observations without assuming an innate cortical module dedicated for faces. The first process, termed *Conspec*, mediates the tendency for newborns to orient to faces early in life and was argued to be largely mediated by subcortical circuits. This proposal has been supported by subsequent evidence. For example, one study showed that differential orientation to faces occurs when the faces are presented in the temporal, but not the nasal, visual hemifield, indicating that the response is likely mediated by a retinotectal rather than geniculostriate pathway (Simion, Valenza, Umilta, & Barba, 1998). At about 6 weeks, this tracking response declines and disappears (Johnson et al., 1991), and a preference for longer visual fixation of faces compared to other patterns presented in the central visual field emerges (Johnson & Morton, 1991; Maurer & Barrera, 1981). These and other changes in infants response to face-like patterns that occur at this time (e.g., Dannemiller & Stephens, 1988; Maurer & Salapatek, 1976) are thought to reflect emergence of cortical systems in face processing. The second process, which Johnson and Morton termed *Conlern*, develops through activity-dependent specialization of cortical circuits in response to face inputs. This stands in contrast to the notion of prespecified cortical circuits specifically dedicated to face processing.

What type of evidence could we use to discriminate between an innate modular view and an emergent specialization view of the development of face processing within cortex? One type of data more consistent with the latter view is evidence for dynamic changes in the cortical processing of faces during infancy or childhood. When considering this type of evidence we believe it may be useful to distinguish between specialization and localization (Johnson, 2000). *Localization* we take to refer to changes in the extent of cortical area activated by a class of stimulus or task, for example, faces may activate more or less cortical tissue following experience during infancy. *Specialization* we take to refer to increases in the selectivity of response patterns of cortical tissue in response to a particular input for example, a given piece of cortical tissue may originally respond to wide range of objects, but with experience decrease the range of stimuli it responds just to one class of objects, such as faces. Another putative feature of cortical specialization could be that specialization could occur earlier in the cortical processing stream with infants than with adults. This could be indicative of an information processing pathway becoming more specialized throughout its extent as a result of experience. We have attempted to address some of these issues in our laboratory through the use of high-density event-related potentials (ERPs) in infants and adults.

Localization of Function

In order to compare the spatiotemporal pattern of cortical activation during face processing by infants and adults, we recorded their ERPs while they passively viewed upright and inverted human female faces (de Haan, Johnson, & Pascalis, under review). We chose to use inverted faces as "nonface" control stimulus because: (a) several studies have shown that inverted faces do not appear to be processed by the same neural and cognitive mechanisms as upright faces (e.g., Moscovitch, Winocur, & Behrmann, 1997), and (b) upright and inverted faces are matched on low level visual characteristics, so these do not confound the comparison.

In adults, an ERP component called the N170 is observed during passive viewing of faces (Bentin, Allison, Puce, Perez, & McCarthy, 1996). The N170 is a prominent negative deflection peaking between 120 ms and 200 ms after stimulus onset, and is sensitive to faces in that it shows a shorter peak latency to upright faces than to a variety of other stimuli such as scrambled or inverted faces, cars, and butterflies (Bentin et al., 1996; Botzel & Grusser, 1989; Botzel, Grusser, Haussler, & Naumann, 1989; George, Evans, Fiori, Davidoff, & Renault, 1996). The N170 also tends to have a larger amplitude for faces than nonface complex objects, with the exception that inverted faces, scrambled faces, or eyes alone elicited larger responses than objects or upright faces (Bentin et al., 1996; George et al., 1996).

Color Plate 1, top panel, illustrates the N170 component observed in adults in our experiment (de Haan et al., under review). As illustrated in this grand average (the average of the average ERPs for the eleven subjects), we replicated the results of previous experiments showing that the N170 is larger in amplitude and longer in peak latency for inverted compared to upright faces. This can also be seen in the lower panels of Color Plate 1 which shows spline maps of the voltages across the scalp at the peak of the N170: The area of negativity (represented by the blue shades) is larger for inverted than upright faces. In addition, the influence of inversion on the N170 was most prominent over right, lateral recording sites.

The ERPs measured from infants under the same testing conditions differ in a number of respects. We did observe a negative deflection peaking between 260–336 ms after stimulus onset whose general morphology appeared similar to the N170 (see Color Plate 2, top panel). However, this "infant N170" was not influenced by the orientation of the face (see spline maps in Color Plate 2, middle panel). Instead, orientation affected a longer-latency component, the P400. This component is a positive deflection over occipito-temporal electrodes and was larger for upright than inverted faces. When we examined the patterns of lateralization of this effect we found that it occurred over both left and right recording sites. In this sense, processing of faces appears to be less localized in infants than in adults.

Specialization of Function

The next question of interest was to determine whether for infants, as for adults, this pattern of activation showed specificity to human faces. To study this question, we conducted a condition in the experiment just described in which infant and adult subjects viewed upright and inverted

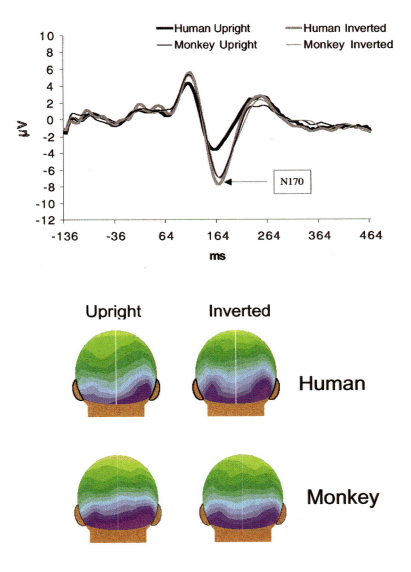

Color Plate 1. The top panel shows the grand average ERPs for adults elicited by upright human faces (thick black), inverted human faces (thick gray) upright monkey faces (thin black) and inverted monkey faces (thin gray) for a representative recording location from posterior cortex. The lower panels show voltage maps created from the ERP data using spherical spline interpolation, illustrating the distribution of voltages at the peak of the adult 'N170.' Blue shades represent negative voltages, red shades positive voltages and yellow/greens surround zero.

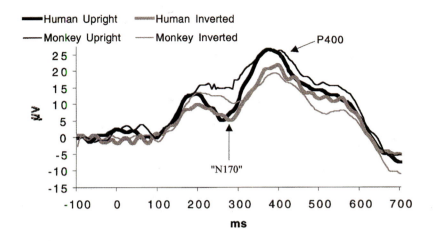

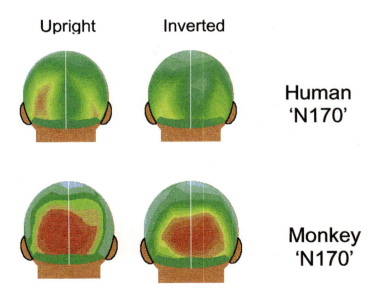

Upright   Inverted

Human 'N170'

Monkey 'N170'

Color Plate 2.   The top panel shows the grand average ERPs for 6-month-olds elicited by upright human faces (thick black), inverted human faces (thick gray) upright monkey faces (thin black) and inverted monkey faces (thin gray) for a representative recording location from posterior cortex. The lower panels show voltage maps created from the ERP data using spherical spline interpolation, illustrating the distribution of voltages at the peak of the infant 'N170.' Blue shades represent negative voltages, red shades positive voltages and yellow/greens surround zero.

monkey faces. We used monkey faces because they are different from, yet closely related to, human faces, and thus likely to be good stimuli for assessing the extent to which cortical processing of faces becomes specialized for our own species.

Color Plate 1, top panel, shows the ERP responses at posterior temporal sensors to the upright and inverted monkey faces in the group of adults. Adults showed no inversion effect for monkey faces. In fact, contrast analyses computed for the four types of face (upright and inverted human and monkey faces) showed that the N170 elicited by upright human faces was unique in that it was of smaller amplitude for upright human faces than for all other stimuli. In contrast, the amplitudes for inverted faces, upright monkey faces, and inverted monkey faces did not differ significantly from one another (see Color Plate 1, lower panels).

At no point in time did the infant ERP show this same degree of specificity to upright human faces. This was not because infants were unable to process the species of the face. Comparison of the results with human and monkey faces showed that the "infant N170" was significantly larger in amplitude to human than monkey faces, regardless of orientation (see Color Plate 2). This suggests that the generators underlying the putative infant N170 are activated more by human faces than nonhuman faces. However, infants further processing of faces appears less specialized than adults: the orientation of the face affected the P400 elicited by monkey faces in just the way it did for human faces, with a larger amplitude for upright than inverted faces.

In summary, processing of faces by 6-month-old infants appears less localized and less specialized than in adults. While infants are able to detect the relevant information specifying the upright human face, processing of this information appears protracted in time and is not integrated into a common, early point in procesing reflected in the adult N170. Thus, it appears as if 6-month-olds are in a transitional phase of partial specialization and localization of function.

RECOGNITION OF FACIAL IDENTITY

In the two-process theory of the development of face recognition, one major role of the cortical system is to encode information about facial identity. However, recognition of at least some individual faces, such as the mother's face, is possible from the first few days of life (Bushnell, Sai, & Mullin, 1989; Field, Cohen, Garcia, & Greenberg, 1984; Pascalis, de Schonen, Morton, Deruelle, & Fabre-Grent, 1995). This early memory for faces seems to rely mainly on the external contours of the face as recognition of the mother's face disappears if only the internal features of her face are visible (Pascalis et al., 1995). In order to account for the neonatal recognition of the mother's face, the two-process theory advanced by Johnson and Morton (1991) must be modified to include a mechanism for encoding facial identity in the first weeks of life.

One possibility is that this early ability to recognize faces is a hippocampal-based preexplicit memory, which is thought to mediate memory measured in visual paired-comparison test during early infancy (Nelson, 1995). In primates, there is evidence that the hippocampus is functioning from around the time of birth (Bachevalier, Brickson, & Hagger, 1983), and that its interactions with the cortex result in postnatal changes in functioning (see Johnson 1997a, 1997b for review). If such a memory system exists, it may represent a pure form of hippocampal encoding relatively unclouded by preprocessing from the still underdeveloped cortex.

This system may allow infants to learn about and recognize visual stimuli, including faces, from birth. However, due to the underspecialized input from cortex at this stage, we suggest that it is unlikely to be specific to faces, or to resemble adult face recognition. It is possible that as the cortical structures that interact with the hippocampus start to become more specialized around 6 to 8 weeks, the nature of face processing and encoding changes. This is reflected in several developmental changes that occur after this age such as an increased use of the internal features of a face for recognition (Morton, 1993; Pascalis et al., 1995), and the ability to recognize a familiar face from novel viewpoints (Pascalis, de Haan, de Schonen, & Nelson, 1998).

In order to examine more clearly the nature of the changes that take place around 6 to 8 weeks, we (de Haan, Johnson, & Maurer, 1998) decided to investigate whether or not infants at different ages used prototype-based encoding like adults, or whether each face is recognized as an isolated individual record. There is now gathering evidence that adults encode individual faces relative to their difference from a prototype. For example, adults recognize atypical faces better than they do those closer to the norm, and caricatures (which exaggerate how features deviate from the norm) are easier to recognize than undistorted faces (see Rhodes et al., 1998 for a review).

To assess infants' abilities to extract a prototype from a series of individual faces, we exposed 1- and 3-month-old infants to a series of four faces with exposure times tailored to the age group of the subjects. Afterwards, we gave them two preference tests with paired stimuli. In one of these tests, infants were allowed to look at one of the individual training faces paired with a novel individual face. Infants at both ages demonstrated that they had encoded each of the training faces by showing a novelty preference for the novel individual face. In the second test, we gave them a choice between one of training faces and a computed average of the four training faces (using a specific method thought to be most suitable for reconstructing a realistic face; Rowland & Perrett, 1995).

The logic behind this comparison was that, if infants were able to construct a prototype, or average, from the input set of faces, this stimulus might be more familiar (less novel) to them than any of the individual training faces. This was what we observed in a group of 3-month-olds, but not in the 1-month-olds. The 1-month-olds' abilities to recognize a familiar individual face show that their failure to recognize the prototype face cannot be attributed to inattention during familiarization or inability to discriminate among the faces. These results suggest that the ability to extract invariances from a set of input faces develops between 1 and 3 months of age.

McClelland, McNaughton, and O'Reilly (1995) argued for a complementary systems view of neocortical and hippocampal memory systems in which the former learns slowly to discover the structure in ensembles of items, whereas the latter permits rapid learning of new items and does not in itself encode relations between items. Extending this general view to infant face processing, we might expect that the hippocampal capacity for rapid item-specific learning would exist in the absence of much cortical learning in the newborn. Thus, the information being encoded about faces by hippocampus would be in the absence of any processing of the relations between faces and their commonalities. With gradual experience over the first few months of life, neocortical circuits on the ventral visual pathway may gradually extract invariances from the array of faces to which the infants has been exposed and discover the structure of this class of stimuli. This process, presumably involving both specialization and localization, would change the nature of the input to the hippocampus and perhaps cause a temporary decline in individual face recognition as has been observed around 2 months (de Schonen & Mancini, 1995).

What are the consequences of this view of the development of neocortical specialization for the effects of early brain damage, or other perturbations of early development? First, the complementary systems view when applied to the development of face processing suggests the possibility that there may be multiple routes to successful individual face recognition following developmental perturbation. Namely, some types of diffuse or specific damage to cortex could result in a form of processing more heavily dependent on the hippocampus. Even though such subjects might score within, or even above, the normal range in certain standardized face tests, in tests of the extent to which their brains have extracted the invariances between faces (such as face prototype effects) they may be deviant (for a variant of this argument see Karmiloff-Smith, 1998).

Turning to the consequences of focal brain damage, in the adult brain that has already specialized and localized for face processing focal damage to structures on the ventral visual pathway can give rise to prosopagnosia. However, we note that natural variation in patterns of localization can be apparent here. In most, but by no means all, normal adults there is so-called "right-hemisphere specialization" for face processing. More specifically, what this means is that face processing goes on in both the right and left ventral pathways, but that it is specialized (in the sense expounded in this chapter) for faces moreso on the right side. On the left side, faces and objects are processed together and in a more similar way (see Wojciulik et al., 1998; although, there may be special face area on the left that is simply weaker than on the right—I. Gauthier, personal communication, June, 1999). However, this pattern of specialization and localization is by no means universal with some healthy normal subjects also showing bilateral specialization and left-sided specialization. These observations fit with claims from the adult neuropsychological literature that bilateral damage is required before prosopagnosia can be observed (Damasio, Damasio, & van Hoesen, 1982).

The results of a recent study of face processing in children who suffered early (perinatal or during the first year) focal cortical lesions is consistent with the emerging specialization view. Subtle deficits in complex face processing tasks were observed in children (ages 5 to 17 years) who suffered right or left posterior, but not anterior, lesions (Mancini, Casse-Perrot, Giusiano, Girard, Camps, Deruelle, & de Schonen, 1998). However, the deficits were relatively mild and always co-occurred with object processing difficulties. One way to interpret these findings is that early damage to the right or left ventral visual pathway results in a compression of face and object specializations resulting in subtle deficits in both. Thus, regions of the right ventral pathway are not prewired for face processing, but, in the majority of individuals, become specialized for this purpose.

Modeling Processes of Localization and Specialization

In order to understand in more detail the neurocomputational mechanisms that may underlie the processes of specialization and localization in neocortex, it is useful to construct neurally plausible computational models. Although we are currently far from a complete "whole-brain" model of the development of face processing, we (Johnson, Oliver, & Shrager, 1998; Oliver, Johnson, & Shrager, 1996; Shrager & Johnson, 1995) have constructed a simplified "cortical sheet" model that incorporates some known aspects of neocortical development such as synaptic attrition. This model displays properties observed in a variety of other neural network models designed to simulate aspects of cortical development (e.g., Miller, Keller, & Stryker, 1989).

Our model neocortex consisted of a "cortical sheet" composed of excitatory and inhibitory nodes, in much the same ratio as inhibitory and excitatory cells in the cortex. Input to the cortical sheet was provided by a sensory sheet that simulated sources of input to the neocortex from subcortical structures. As is standard in such models, the activity of a node depended on a weighted sum of the activity of the other nodes that form connections with it, and these weights are stored in the connections. A general property of the model that resembled postnatal neocortical development is that the nodes are initially overconnected and that pruning of links occurs during development (training).

Connections between nodes (which may correspond to synapses) are initially labile, but during the course of training they can either stabilize or die off. The chance of a connection becoming stable depends on the amount of trophic factor it has received, and a general function associated with the age of the network. Connections that are useful are strengthened according to the association rule and are more likely to become stable, while other connections which do not have their weights increased are more likely to be pruned out by the regressive process.

This model, like many others of its kind, shows forms of both specialization and localization, and illustrates the interdependence between them. Specifically, nodes within the sheet, like those in many other similar models, refine their response properties from responding to most of the possible input patterns, to one, or to some specific combinations of inputs. For example, Shrager and Johnson (1995) classified the response properties of nodes according to Boolean functions. At the same time, nodes sharing common response properties often cluster together, forming blobs with common functionality. Under some conditions, more specific spatial structuring can be obtained, such as a graded topographic map (Oliver et al., 1996). In Fig. 10.1, we illustrate the end state of the cortical sheet in a simulation in which clusters with particular response properties formed. It is possible to analyze the factors critical for the emergence of clusters with unique response properties by manipulating some of the variables in the model.

In Fig. 10.1 we show the result of an identical simulation with the exception that we changed one aspect of the initial state wiring of the network such that both the excitatory and the inhibitory links had the same average length (Oliver, Johnson, Karmiloff-Smith, & Pennington, 2000). We have manipulated a number of different variables such as varying the node firing thresholds, varying the projection patterns of afferent links, increasing the overall rate of link loss, and changing the statistical structure of the inputs, to analyze their importance for the localization and specialization observed in the network.

CONCLUSIONS

The observation that the majority of normal adults tend to have similar functions within approximately the same areas of cortex could equally well be accounted for by a common intrinsic prespecification between shared by individuals (prewiring) as by the very similar pre- and postnatal environments shared by individuals. In this chapter, we have used the exam-

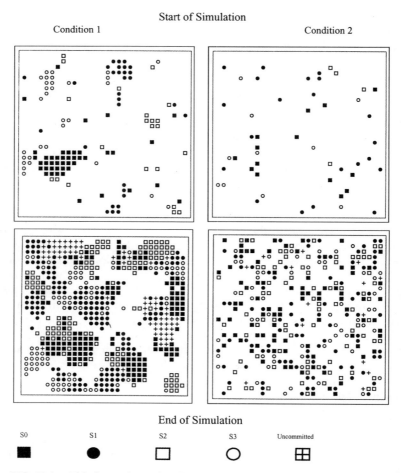

FIG. 10.1. This figure shows the effect of varying endogenous constraints on the development of representations within the cortical sheet. In Condition 1 (1a), lateral links were short-range excitatory, long-range inhibitory. In Condition 2, (1b) both excitatory and inhibitory links were of medium length. The top two panels show the state of the representation at the start of the simulation, while the bottom two panels show the state at the end. The key indicates to which stimuli the cortical nodes were responsive. In condition 1 at the end of the simulation clusters have formed that reflect the similarity structure between the stimuli. In condition 2, no such clustering is seen. Figure taken from Johnson et al. (1998).

ple of face processing as a case study for the plausibility of arguments about innate cortical modules for cognitive functions. Our results, and those of others, support our view that at least some of the patterns of cortical specialization seen in adults are the result of emergent and gradual processes of specialization and localization. These processes can be understood at a lower level through computational modeling, and also have implications for our views on abnormal development.

REFERENCES

Bachevalier, J., Brickson, M., & Hagger, C. (1993). Limbic-dependent recognition memory in monkeys develops early in infancy. *Neuroreport, 4,* 77–80.

Baylis, G. C., Rolls, E. T, & Leonard, C. M. (1987). Functional subdivisions of the neocortical temporal lobe neocortex. *The Journal of Neuroscience, 7,* 294—328.

Bentin, S., Allison, T., Puce, A., Perez, E., & McCarthy, G. (1996). Electrophysiological studies of face perception in humans. *Journal of Cognitive Neuroscience, 8,* 551–565.

Botzel, K., & Grusser, O.-J. (1989). Electric brain potentials evoked by pictures of faces and non-faces: A search for "face-specific" EEG potentials. *Experimental Brain Research, 77,* 349–360.

Botzel, K., Grusser, O.-J., Haussler, B., & Naumann, N. (1989). The search for face-specific evoked potentials. In E. Basar & T. H. Bullock (Eds.), *Brain Dynamics* (pp. 449–466). Berlin: Springer.

Brodman, K. (1905). Beitrage zur histologischen Lockalisation der Grosshirnrinde. Dritte Mitteilung: Die Rindenfelder der niederen Affen. *J. Psychol. Neurol. Lpz., 4,* 177–226.

Bushnell, I. W. R., Sai, F., & Mullin, J. T. (1989). Neonatal recognition of the mother's face. *British Journal of Developmental Psychology, 7,* 3–15.

Chugani, H. T., Phelps, M. E., & Mazziotta, J. C. Positron emission tomography study of human brain functional development. *Annals of Neurology, 22,* 487–497.

Conel, J. (1939–1963). *The postnatal development of the human cerebral cortex* (6 volumes). Cambridge, MA: Harvard University Press.

Courtney, S. M., Ungerleider, L. G., Keil, K., & Haxby, J. V. (1996). Object and spatial visual working memory activate separate systems in human cortex. *Cerebral Cortex, 6,* 39–49.

Courtney, S. M., Ungerleider, L. G., Keil, K., & Haxby, J. V. (1997). Transient and sustained activity in a distributed neural system for human working memory. *Nature, 386,* 608–611.

Damasio, A. R., Damasio, H., & van Hoesen, G. W. (1982). Prosopagnosia: Anatomical basis and behavioral mechanisms. *Neurology, 32,* 331–341.

Dannemiller, J. L., & Stephens, B. R. (1988). A critical test of infant pattern preference models. *Child Development, 59,* 210–216.

de Haan, M., Johnson, M. H., & Maurer, D. (1998). Recognition of individual faces and average face prototypes by 1- and 3-month-old infants. *Developmental Cognitive Neuroscience Technical Report Series of the Human Frontiers Scientific Foundation.* No. 98.8.

de Haan, M., Johnson, M. H., & Pascalis, O. (in press). Specialization of neural mechanisms underlying face recognition in human infants.

de Schonen, S., & Mancini, J. (1995). About functional brain specialization: The development of face recognition. *Developmental Cognitive Neuroscience Technical Report Series of the Human Frontiers Scientific Foundation.* No. 95.1.

de Schonen, S., & Mathivet, E. (1989). First come, first served: A scenario about the development of hemispheric specialization in face recognition during infancy. *European Bulletin of Cognitive Psychology, 9,* 3–44.

Elbert, T., Pantev, C., Wienbruch, C., Rockstroh, B., & Taub, E. (1995). Increased cortical representation of the fingers of the left hand in string players. *Science, 270,* 305–307.

Elman, J., Bates, E., Johnson, M. H., Karmiloff-Smith, A., Parisi, D., & Plunkett, K. (1996). *Rethinking innateness: A connectionist perspective on development.* Cambridge, MA: MIT Press.

Farah, M. J., Rabinowitz, C., Quinn, G. E., & Liu, G. T. (2000). Early commitment of neural substrates for face recognition. *Cognitive Neuropsychology, 17*(1–3), 117–123.

Field, T. M., Cohen, D., Garcia, R., & Greenberg, R. (1984). Mother–stranger face discrimination by the newborn. *Infant Behavior & Development, 7,* 19–25.

Finlay, B. L., & Darlington, R. B. (1995). Linked regularities in the development and evolution of mammalian brains. *Science, 268,* 1578–1584.

George, N., Evans, J., Fiori, N., Davidoff, J., & Renault, B. (1996). Brain events related to normal and moderately scrambled faces. *Cognitive Brain Research, 4,* 65–76.

Goren, C. C., Sarty, M., & Wu, P. Y. K., (1975). Visual following and pattern discrimination of face-like stimuli by newborn infants. *Pediatrics, 56,* 544–549.

Grady, C. L., Maisog, J. M., Horwitz, B., Ungerleider, L. G., Mentis, M., Salerno, J. A., Pietrini, P., Wagner, E., & Haxby, J. V. (1994). Age related changes in cortical blood flow activation during visual processing of faces and location. *The Journal of Neuroscience, 14,* 1450–1462.

Gross, C. G., Rocha-Miranda, C. E., & Bender, D. B. (1972). Visual properties of neurons in the inferotemporal cortex of the macaque. *Journal of Neurophysiology, 35,* 96–111.

Halgren, E., Dale, A. M., Sereno, M. K., Tootell, R. B. H., Marinkovic, K., & Rosen, B. R. (1999). Location of human face-selective cortex with respect to retinotopic areas. *Human Brain Mapping,* 729–737.

Haxby, J. V., Horwitz, B., Maisog, J. M., Ungerleider, L. G., Mishkin, M., Schapiro, M. B., Rapoport, S. I., & Grady, C. L. (1993). Frontal and temporal participation in long-term recognition memory for faces: A PET-rCBF activation study. *Journal of Cerebral Blood Flow and Metabolism (Suppl. 1), 13,* S499.

Haxby, J. V., Horwitz, B., Maisog, J. M., Ungerleider, L. G., Pietrini, P., & Grady, C. L. (1996). The functional organization of human extrastriate cortex: A PET-rCBF study of selective attention to faces and locations. *The Journal of Neuroscience, 14,* 6336–6353.

Haxby, J. V., Ungerleider, L. G., Horwitz, B., Rapoport, S. I., & Grady, C. L. (1995). Hemispheric differences in neural systems for face working memory: A PET-rCBF study. *Human Brain Mapping, 3,* 68–82.

Horwitz, B., Grady, C. L., Haxby, J. V., Schapiro, M. B., Rapoport, S. I., Ungerleider, L. G., & Moscovitch, M. (1992). Functional associations among human posterior extrastriate brain regions during object and spatial vision. *Journal of Cognitive Neuroscience, 4,* 311–322.

Huttenlocher, P. R. (1990). Morphometric study of human cerebral cortex development. *Neuropsychologia, 28,* 517–527.

Huttenlocher, P. R., & Dabholkar, A. S. (1997). Regional differences in synaptogenesis in human cerebral cortex. *Journal of Comparative Neurology, 387,* 167–178.

Johnson, M. H. (1997a). *Developmental cognitive neuroscience: An introduction.* Oxford: Blackwell.

Johnson, M. H. (1997b). The neural basis of cognitive development. In W. Damon, D. Kuhn, & R. Siegler (Eds.), *Handbook of child psychology: Vol. 2. Cognition, perception and language* (5th ed.; pp. 1–49). New York: Wiley.

Johnson, M. H. (2000). Functional brain development in infants: Elements of an interactive specialization framework. *Child Development, 71*(1), 75–81.

Johnson, M. H., Dziurawiec, S., Ellis, H., & Morton, J. (1991). Newborns' preferential tracking of face-like stimuli and its subsequent decline. *Cognition, 40,* 1–19.

Johnson, M. H. , & Morton, J. (1991). *Biology and cognitive development: The case of face recognition.* Oxford, UK: Blackwell.

Johnson, M. H., Oliver, A., & Shrager, J. (1998). The paradox of plasticity: A constrained plasticity approach to the emergence of representations in the neocortex. *Cognitive Studies: Journal of the Japanese Cognitive Science Society, 5,* 5–24.

Kanwisher, M., McDermott, J., & Chun, M. M. (1997). The fusiform face area: a module in human extrastriate cortex specialized for face perception. *Journal of Neuroscience, 17,* 4302–4311.

Karmiloff-Smith, A. (1998). Development itself is the key to understanding developmental disorders. *Trends in Cognitive Sciences, 2*(10), 389–398.

Katz, L. C., & Shatz, C. J. (1996). Synaptic activity and the construction of the cortical circuits. *Science, 274,* 1133–1138.

Killackey, H. P. (1990). Neocortical expansion: An attempt toward relating phylogeny and ontgeny. *Journal of Cognitive Neuroscience, 2,* 1–17.

Malach, R. Reppas, J. B., Benson, R. R., Kwong, K. K., Jiang, H., Kennedy, W. A., Ledden, P. J., Brady, T. J., Rosen, B. R., & Tootell, R. B. (1995). Object-related activity revealed by functional magnetic resonance imaging in human occipital cortex. *Proceedings of the National Academy of Sciences, USA,* 8135–8139.

Mancini, J., Casse-Perrot, C., Giusiano, B., Girard, N., Camps, R., Deruelle, C., & de Schonen, S. (1998). *Face processing development after a perinatal unilateral brain lesion.* Manuscript submitted for publication.

Maurer, D., & Barrera, M. (1981). Infants' perception of natural and distorted arrangements of a schematic face. *Child Development, 52,* 196–202.

Maurer, D., & Salapatek, P. (1976). Developmental changes in the scanning of faces by young infants. *Child Development, 47,* 523–537.

Maurer, D., & Young, R. (1983). Newborns' following of natural and distorted arrangements of facial features. *Infant Behavior and Development, 6,* 127–131.

McClelland, J. L., McNaugton, & O'Reilly, R. C. (1995). Why there are complementary learning systems in the hippocampus and neocortex: Insights from successes and failures of connectionist models of learning and memory. *Psychological Review, 102,* 419–457.

Miller, K. D., Keller, K. B., & Stryker, M. P. (1989). Ocular dominance column development: Analysis and simulation. *Science, 245,* 605–615.

Miyashita, Y. (1988). Neuronal correlate of visual associative long term memory in primate temporal cortex. *Nature, 335,* 817–820.

Miyashita, Y. (1990). Associative representation of visual long term memory in the neurons of the primate temporal cortex. In E. Ewai & M. Miskin (Eds.), *Vision, memory and the temporal lobe* (pp. 75–87). New York: Elsevier.

Morton, J. (1993). Mechanisms in infant face processing. In B. de Boysson-Bardies, S. de Schonen, P. Juszyk, P. McNeilage, & J. Morton (Eds.), *Developmental neurocognition: Speech and face processing in the first year of life* (pp. 93–102). London: Kluwer.

Morton, J., & Johnson, M. H. (1991). CONSPEC and CONLERN: A two-process theory of infant face recognition. *Psychological Review, 2,* 164–181.

Moscovitch, M., Winocur, G., & Behrmann, M. (1997). What is special about face recognition? Nineteen experiments on a person with visual object agnosia and dyslexia but normal face recognition. *Journal of Cognitive Neuroscience, 9,* 555–604.

Munakata, Y., McClelland, J. C., Johnson, M. H., & Siegler, R. S. (1997). Rethinking infant knowledge: Toward an adaptive process account of successes and failures in object permanence tasks. *Psychological Review, 104,* 686–713.

Nelson, C. A. (1995). The ontogeny of human memory: A cognitive neuroscience perspective. *Developmental Psychology, 31,* 723–738.

O'Leary, D. D. M. (1989). Do cortical areas emerge from a protocortex? *Trends in Neurosciences, 12,* 400–406.

Oliver, A., Johnson, M. H., Karmiloff-Smith, A., & Pennington, B. (2000). Deviations in the emergence of representations: A neuro-constructivist framework for analysing developmental disorders. *Developmental Science, 3,* 1–23.

Oliver, A., Johnson, M. H., & Shrager, J. (1996). The emergence of heirarchical clustered representations in a Hebbian neural network model that simulates development in the neocortex. *Network: Computation in Neural Systems, 7,* 291–299.

Oram, M. W., & Perrett, D. I. (1994). Responses of anterior suprior temporal polysensory (STPa) neurons to "biological motion" stimuli. *Journal of Cognitive Neuroscience, 6,* 99–116.

Pascalis, O., de Haan, M., de Schonen, S., & Nelson, C. A. (1998). Long-term recognition memory for faces assessed by visual paired comparison in 3- and 6-month-old infants. *Journal of Experimental Psychology: Learning, Memory and Cognition, 24,* 249–260.

Pascalis, O., de Schonen, S., Morton, J., Deruelle, C., & Fabre-Grent, M. (1995). Mother's face recognition by neonates: A replication and an extension. *Infant Behavior and Development, 18,* 79–95.

Perrett, D. I., & Mistlin, A. J. (1990). Perception of facial characteristics by monkeys. In W. C. Stebbins & M. A. Berkley (Eds.), *Comparative perception: Vol. 2, Complex signals* (pp. 187–215). New York: Wiley.

Pinker, S. (1994). *The language instinct.* New York: William Morrow.

Pinker, S. (1997). *How the mind works.* New York: Norton.

Posner, M. I., Petersen, S. E., Fox, P. T., & Raichle, M. E. (1988). Localization of cognitive function in the human brain. *Science, 240,* 1627–1631.

Puce, A, Allison, T, Asgari, M, Gore, J. C., & McCarthy, G. (1996). Differential sensitivity of human visual cortex to faces, letterstrings, and textures: A functional magnetic. *Journal of Neuroscience, 16,* 5205–5215.

Puce, A., Allison, T., Gore, J. C., & McCarthy, G. (1995). Face-sensitive regions in human extrastriate cortex studied by functional MRI. *Journal of Neurophysiology, 74,* 1192–1199.

Rakic, P. (1988). Specialization of cortical areas. *Science, 241,* 170–176.

Rhodes, G., Carey, S., Byatt, G., & Proffitt, F. (1998). Coding spatial variations in faces and simple shapes: A test of two models. *Vision Research, 38,* 2307–2321.

Rodman, H. R., Scalaidhe, S. P., & Gross, C. G. (1993). Response properties of neurons in temporal cortical visual areas of infant monkeys. *Journal of Neurophysiology, 70,* 1115–1136.

Rodman, H. R., Skelly, J. P., & Gross, C. G. (1991). Stimulus selectivity and state dependence of activity in inferior temporal cortex in infant monkeys. *Proceedings of the National Academy of Science USA, 88,* 7572–7575.

Rolls, E. T., & Baylis, G. C. (1986). Size and contrast have only small effects on the response to faces of neurons in the cortex of the superior temporal sulcus of the monkey. *Experimental Brain Research, 65,* 38–48.

Rowland, D. A., & Perrett, D. I. (1995). Manipulating facial appearance through shape and color. *IEEE Computer Graphics and Applications, 15,* 70–76.

Schade, J. P., & van Groenigen, D. B. (1961). Structural organization of the human cerebral cortex: I. Maturation of the middle frontal gyrus. *Acta Anatomica, 47,* 74–111.

Scherg, M., & Berg, P. (1996). *BESA-Brain Electromagnetic Source Analysis User Manual Version 2.2,* MEGIS Software GMBBH, Scheinkelstr. 37, D-80805 Munchen, Germany.

Sergent, J., Ohta, S., & MacDonald, B. (1992). Functional neuroanatomy of face and object processing. *Brain, 115,* 15–36.

Shrager, J., & Johnson, M. H. (1995). Waves of growth in the development of cortical function: A computational model. In I. Kovacs & B. Julesz (Eds.), *Maturational windows and adult cortical plasticity* (pp. 31–44). Reading, MA: Addison-Wesley.

Simion, F., Valenza, E., Umilta, C., Barba, D. B. (1998). Preferential orienting to faces in newborns: A temporal-nasal asymmetry. *Journal of Experimental Psychology: Human Perception & Performance, 24,* 1399–1405.

Spelke, E. (1994). Initial knowledge: Six suggestions. *Cognition, 50,* 431–445.

Valenza, E., Simion, F., Cassia, V. M., & Umilta, C. (1996). Face preference at birth. *Journal of Experimental Psychology: Human Perception & Performance, 22,* 892–903.

Wojciulik, E., Kanwisher, N., & Driver, J. (1998). Covert visual attention modulates face-specific activity in the human fusiform gyrus: fMRI study. *Journal of Neurophysiology, 79,* 1574–1578.

Zecevic, N. (1998). Synaptogenesis in layer I of the human cerebral cortex in the first half of gestation. *Cerebral Cortex, 8,* 245–252.

Chapter 11

Variability of Developmental Plasticity

Helen J. Neville
University of Oregon

Daphne Bavelier
University of Rochester

Students of development, including parents, educators, cognitive scientists, and neuroscientists agree that knowledge of the neural substrate that permits and implements cognitive growth will promote a basic understanding of the time course of cognitive development, of individual differences in cognitive development and, in addition, will contribute to the design of programs that optimize cognitive development. Everyone acknowledges that there is an important role for the genotype in human development and that experience is essential in the expression of that genotype. A key issue within neuroscience and within psychological science is to specify the degree to which the genetic constraints limit the scaffolding and to know the degree to which these constraints can be modified and the times in human development when they can be modified.

In this chapter, we summarize research in which we have studied the effects of early sensory experience and early language experience on the development of the specialized neural systems that mediate different aspects of cognition. The main point we wish to emphasize, based on results from several different studies, is that there is considerable variability and specificity in the degree to which different neurocognitive systems are modified by early experience. The mechanisms responsible for this variability in experience-dependent changes in the human brain are not yet understood. They may be mediated by differences in maturation rates, in the initial degree of overlap of connectivity between and within different brain regions, in the degree and timing of inhibitory mechanisms in the different brain regions or perhaps by differences in the overproduction and pruning of synapses and receptors in different brain regions. The variability in experience-dependent changes in human development can be important in helping to identify different subsystems and in specifying the timing when different kinds of subsystems can be responsive to and modified by input from the environment.

We present research in which we have compared different and specific aspects of cerebral organization in normal adults and in adults who have had specific alterations in early experience. Normal hearing adults and congenitally deaf adults are compared to assess the impact of the absence of auditory input on visual development. Parallel studies of blind individuals are reported. We also assess the effects of different types of language experience (e.g., late acquisition; acquisition of signed vs. spoken language) on the development of the language systems of the brain. We observe that distinct functional subsystems within vision, audition, and language display different degrees of modifiability by experience during development. These differences may depend on several factors including the rate of maturation of the different neural systems that mediate these functions. More generally, it may be that different neural systems employ fundamentally different learning/change mechanisms that permit different patterns of developmental plasticity. We discuss this hypothesis after summarizing results from several experiments.

In our studies, the physiology of sensory and language processing in humans is studied using two brain imaging techniques: event-related brain potentials (ERPs), and functional magnetic resonance imaging (fMRI). Both of these techniques involve recording indices of brain activity while subjects perform different visual or language tasks. ERPs are voltage fluctuations in the electroencephalogram extracted by signal averaging techniques. The latency of different positive and negative components in an ERP reveal the time course of the activation (within microseconds) of the underlying neural populations. The distribution of ERP activity between and within the hemispheres is determined by the anatomical position and geometry of the contributing neurons. The spatial resolution of the ERP can be enhanced by transforming voltage maps of electrical activity to current source density maps that provide a reference-free estimate of the instantaneous electrical currents flowing from the brain perpendicular to the scalp at each location at the specified time point. Thus, ERP recordings can provide exquisite information about the timing of sensory and language processes as well as constraints on their location. The fMRI technique permits the monitoring of the local increase in oxygen delivery that occurs in neurally active cerebral tissue. As the ratio of oxygenated to deoxygenated hemoglobin increases within the microvasculature of metabolically active areas, an increased magnetic resonance (MR) signal relative to the resting state is observed from these areas. This noninvasive technique has spatial resolution of about a millimeter.

VISUAL PROCESSING

Although there is considerable anecdotal evidence that deprivation of input in one sensory modality leads to compensatory increases in the functioning of the remaining modalities, the pertinent human behavioral literature presents a confusing array of results. Some studies report increases, some report decreases, and others no differences in the abilities of remaining modalities after deafness or blindness compared with those abilities in intact control subjects (see Neville & Lawson, 1987b, for review; Reynolds, 1978). This variability may be due in large part to differences in the age of onset, completeness, and etiology of the sensory deprivation (deafness or blindness). For these reasons, our studies have focused on deaf individuals who were born profoundly deaf due to a genetic etiology in which the central nervous system is not directly affected. These subjects learned ASL from their deaf parents at the same age that normal hearing children acquire spoken language.

In an early study, we observed that ERPs to peripheral and foveal (i.e., center of the visual field) visual stimuli differed in morphology and distribution over the scalp in normal adults in a manner that is consistent with the hypothesis that they were generated by different cortical systems. Results from congenitally deaf adults in the same paradigm showed that whereas ERPs to foveal stimuli were similar in the two groups, ERPs to peripheral stimuli were two to three times larger in deaf (than in hearing) subjects over superior temporal cortical areas (Neville, Schmidt, & Kutas, 1983). We hypothesized that the transient visual system, proposed to mediate the processing of peripheral, spatial, and motion information may, through a process of competitive interactions, take over what would normally be auditory cortical fields in primary or secondary auditory areas or within multimodal temporal areas.

To test the generality of these findings, we extended these studies to conditions under which visual attention was required. We tested the hypothesis that attention to central and peripheral visual space is mediated by different neural systems in normal-hearing adults and, further, that the systems important in processing peripheral motion information are more altered by auditory deprivation than are those important in processing central visual information (Neville & Lawson, 1987a, 1987b, 1987c). ERPs were monitored while subjects focused their eyes straight ahead and attended to a white square presented either in the periphery or in the center of the visual field. Subjects' task was to detect the direction of motion of the square.

ERPs elicited by peripheral visual stimuli displayed attention-related increases that were several times larger in deaf subjects than in normally hearing subjects. Additionally, the attention effects were distributed differently in the two groups. In particular, while for the hearing subjects, the principal effects of attention occurred over the parietal region contralateral to the attended visual field, in deaf subjects the effect was also observed over the occipital regions of both hemispheres. Moreover, deaf subjects also displayed considerably larger attention-related increases over the left temporal and parietal regions than did the hearing subjects.

This specific pattern of group difference can be considered with respect to anatomical and physiological evidence from the animal literature, which shows two major types of change can occur following unimodal sensory deprivation from birth. First, there is evidence of increased growth and activity of the remaining sensory systems (Ryugo, Ryugo, Globus, & Killackey, 1975). The bilateral increase of attention-related changes in occipital regions of the deaf may represent this type of change. Second, there is evidence that the brain systems that would normally subserve functions that are lost—audition and auditory language skills in the case of deaf—may become organized to process other information from remaining modalities (Rebillard, Rebillard, & Pujol, 1980). The larger activation within the left temporal cortex may represent this type of change. In addition, performance results show that these changes carry functional significance. Deaf individuals were faster and more accurate than hearing subjects in detecting the direction of motion of the peripheral stimuli.

Studies of hearing subjects have shown that visual analysis of a scene relies on at least two main processing streams that are anatomically quite separate. The "what" pathway projects from primary visual cortex to the temporal lobe and is specialized for the identification of objects and the processing of color and of fine visual details, while the "where" pathway projects to the parietal cortex and is specialized for the perception of motion and for the localization of objects. In view of our earlier results, we recently investigated the hypothesis that auditory deprivation alters primarily processing along the "where" visual pathway. In order to investigate this issue, Armstrong, a graduate student in our lab, recorded ERPs to stimuli designed to selectively activate either the ventral visual pathway (high spatial frequency colored gratings that occasionally changed color), or the dorsal pathway (low spatial frequency, grey gratings that occasionally displayed apparent motion) (Armstrong, Mitchell, Hillyard, & Neville, 1885; Mitchell, Armstrong, Hillyard, & Neville, 1997). In response to the color changes ERP responses from hearing and deaf subjects were similar (Fig. 11.1a). By contrast, motion elicited ERPs were significantly enhanced in the deaf as compared to hearing subjects (Fig. 11.1b). These results suggest that there is considerable specificity in the effects of auditory deprivation on the processing of visual information. Current source density analyses of these data suggested that there were additional visual gen-

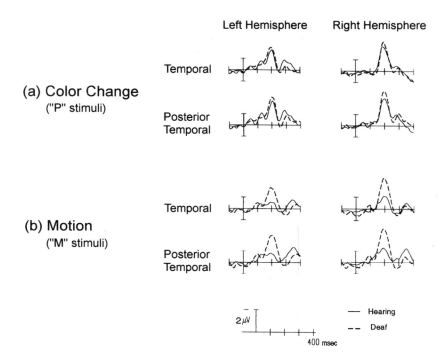

FIG. 11.1.   ERPs elicited by (a) color change and (b) motion in normally hearing and congenitally deaf adults. Recordings from temporal and posterior temporal regions of the left and right hemispheres. (Fig. 2 from "*The Cognitive Neurosciences, 2nd ed.*," edited by M. S. Gazzaniga, 1999. Copyright © by MIT Press. Reprinted with permission.)

erators within temporal cortex of the deaf subjects that were not apparent in the normal hearing subjects.

We have further pursued the hypothesis that deafness alters the functional organization of the dorsal visual stream, by employing fMRI (Bavelier et al., 1998). Specifically, we assessed whether early auditory deprivation alters cerebral activation during motion processing and, in addition, we hypothesized that these changes would be most marked when visual attention was required, in view of the central role of dorsal parietal regions in spatial attention. Motion processing was compared between congenitally deaf (native signers/born to deaf parents) and hearing individuals as visual attention was manipulated. Subjects fixated centrally and viewed an alternation of radial flow fields (converging and diverging) and static dots. Although the first run required only passive viewing, visual attention was manipulated in all other runs by asking subjects to detect velocity and/or luminance changes.

Under conditions of active attention, deaf individuals showed a greater number of voxels on the MRI activated and a larger percent signal change than did hearing subjects in temporal cortex including areas MT-MST (Color Plate 3). Further studies are needed to establish this pattern and to determine whether activations also occur within primary, other secondary and tertiary auditory areas. Thus, congenital deafness alters the cortical organization of motion processing, especially when attention is required. Interestingly, the recruitment of the intraparietal sulcus was also significantly larger in deaf than in hearing subjects. This result, like our earlier ERP study of spatial attention (Neville & Lawson, 1987b; Neville, 1995), suggests that early auditory deprivation may alter the cortical organization of visual attention. Ongoing studies will determine the precise location and the specificity of these effects.

Together, these data suggest that there is considerable specificity in the aspects of visual processing that are altered following auditory deprivation, specifically that the dorsal visual processing stream may be more modifiable in response to alterations in afferent input than is the ventral processing pathway. This hypothesis is in broad agreement with the proposal put forward by Chalupa and Dreher (1991) that components of the visual pathway that are specialized for high acuity vision exhibit fewer developmental redundancies (errors), decreased modifiability and more specificity than do those displaying less acuity and precision. It may also be that the dorsal visual pathway has a more prolonged maturational time course than the ventral pathway, permitting extrinsic influences to exert an effect over a longer time. Although little evidence bears directly on this hypothesis, anatomical data suggest that in humans, neurons in the parvocellular layers of the LGN mature earlier than those in the magnocellular laminae (Hickey, 1977) and in nonhuman primates, the peripheral retina is slower to mature (Lachica & Casagrande, 1988; Packer, Hendrickson, & Curcio, 1990; Van Driel, Provis, & Billson, 1990). Additionally, data suggest that the development of the Y-cell pathway (which is strongest in the periphery of the retina) is more affected by visual deprivation than is development of the W-and X-cell pathways (Garraghty, 1993; Sherman & Spear, 1982).

Investigators have also reported that the effects of congenital visual deprivation (due to cataracts) has more pronounced effects on peripheral than foveal vision (and by implication on the dorsal pathway) (Bowering, Maurer, Lewis, & Brent, 1997; Mioche & Perenin, 1986). In addition, in developmental disabilities including dyslexia, specific language impairment, and Williams syndrome, visual deficits are more pronounced for dorsal than ventral visual pathway functions (Atkinson et al., 1997; Eden et al., 1996; Lovegrove, Garzia, & Nicholson, 1990). An additional hypothesis that may account for the greater effects on peripheral vision is

MOTION RELATED ACTIVATIONS
MONITOR BRIGHTNESS CHANGES

HEARING

DEAF

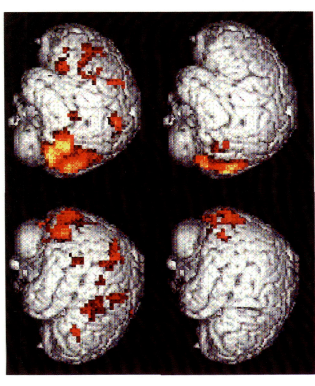

Color Plate 3. Activation (fMRI, 1.5T) in normally hearing and congenitally deaf adults in response to visual motion. (Fig. 4, from *"The Cognitive Neurosciences, 2nd ed.,"* edited by M. S. Gazzaniga 1999. Copyright © by MIT Press. Reprinted with permission.)

Written English -- Native Speakers

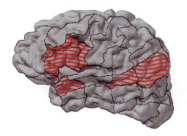 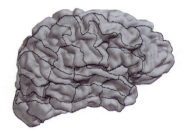

American Sign Language -- Native Signers

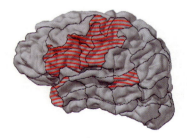 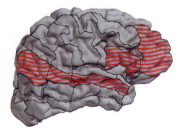

Color Plate 4.   Red stripes indicate areas that display significant increases in MR signal intensity when subjects process sentences. Results show that when native speakers read sentences in English, strongly left-sided activation is seen. Results from native signers processing sentences in ASL indicate strong left- and right-hemisphere activation.

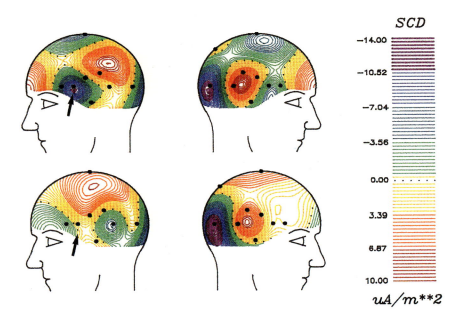

Color Plate 5.   Distribution of current flow for the N280 peak elicited by closed class words in English. Maps in the top row show the prominent response in the left hemisphere of hearing subjects (blue, marked by arrow). Bottom row maps display results from congenitally deaf individuals who lack the response (arrow). (Fig. 9, "Fractioning language: Different neural subsystems with different sensitive periods" by H. J. Neville, D. Mills, & D. Lawson, 1992, *Cerebral Cortex, 2*, p. 244–253. Copyright © 1992 by Oxford University Press. Reprinted with permission.)

that in development the effects of deprivation and enhancement are equivalent within all cortical regions. Those areas with less extent to begin with (e.g., MT, peripheral visual representations) would display the largest proportional effects of both enhancement and vulnerability. A similar hypothesis has been proposed to account for the larger effects of visual deprivation on ocular dominance formation within the periphery in monkeys (Horton & Hocking, 1997).

It is likely that these potential structural substrates for enhanced peripheral processing interact with the many altered processing demands of deafness, including the acquisition of American Sign Language (ASL), to produce the changes we have observed. One way we have assessed the impact of acquiring ASL is to study hearing children born to deaf parents who also have acquired ASL as a first language, but who have no auditory deprivation. In previous behavioral and ERP studies, these subjects do not display the same enhancement of peripheral processing observed in deaf native signers, suggesting auditory deprivation per se is an important factor. Because deaf individuals cannot rely on audition to detect potentially important events in the periphery, it is likely that an interaction between increased attention to the visual periphery and the structural constraints described previously produces the observed effects in visual processing in the deaf.

We have observed that individuals who became deaf after the age of 4 years (due to delayed expression of the gene that leads to cochlear degeneration) typically do not display the increased visual ERPs that we attributed to auditory deprivation (Neville, Schmidt, & Kutas, 1983; Neville & Lawson, 1987c). We considered several mechanisms that might mediate the effects themselves and the developmental time limits on them. One possibility is that they are mediated by an early, normally transient, redundancy of connections between the auditory and visual systems (as has been observed in cats and hamsters, Dehay, Bullier, & Kennedy, 1984; Frost, 1984; Innocenti & Clarke, 1984). In the absence of competition from auditory input, visual afferents may be maintained on what would normally be auditory neurons.

Our results from studies of later deafened individuals suggest that in humans this redundancy may diminish by the fourth year of life. One way we tested this hypothesis was to study the differentiation of visual and auditory sensory responses in normal development (Neville, 1995). In normal adults, auditory stimuli elicit ERP responses that are large over temporal brain regions but small or absent over occipital regions. By contrast, in 6-month-old children we observed that auditory ERPs are equally large over temporal and visual brain regions, consistent with the idea that there is less specificity and more redundancy of connections between auditory and visual cortex at this time. Between 6 and 36 months, however, we observed a gradual decrease in the amplitude of the auditory ERP over visual areas, while the amplitude over the temporal areas was unchanged. These results suggest that early in human development there exists a redundancy of connections between auditory and visual areas and that this overlap gradually decreases between birth and 3 years of age. This loss of redundancy may be the boundary condition that determines when auditory deprivation can result in alterations in the organization of the visual system. Ongoing studies of infants and children employing the parvo and magno stimuli described before will test for the specificity of these effects (Mitchell, Wolter, & Neville, 1999).

Developmental plasticity has also been documented in humans within other sensory modalities. There are several reports that early blindness due to abnormalities within the retina leads to changes in the visually deprived cortex. Measures of glucose utilization have shown an increased metabolism in the visual cortex of early blind humans as compared to subjects who became blind after the completion of visual development. These studies report that metabolic activity within the occipital cortex of early blind individuals is higher than that found in blindfolded sighted subjects and equivalent to that of sighted subjects with their eyes open (Uhl et al., 1994; Veraart et al., 1990; Wanet-Defalque et al., 1988). Additionally, ERP studies indicate a larger slow negative DC potential shift over the occipital lobe in early blind than sighted persons during tactile and auditory tasks (Röder, Rösler, Hennighausen, & Näcker, 1996; Röder, Teder-Salejarvi, et al., 1999; Uhl, Franzen, Lindinger, Lang, & Deecke, 1991; Uhl, Kretschmer, et al., 1994). Recently, a number of studies have confirmed the functional participation of visual areas during somatosensory tasks in early blind individuals. Using PET, Sadato et al. (1996) compared tactile discrimination in early blind braille readers and control subjects. Blind subjects revealed activation of visual cortical areas, whereas these regions were deactivated in controls. The functional relevance of visual areas in tactile discrimination was further established in a transcranial magnetic stimulation experiment (Cohen et al., 1997). Transient stimulation of the occipital cortex induced errors on a tactile task in early blind subjects but had no effect on the sighted controls. It is worth noting that not all aspects of somatosensory processing recruit visual areas in blind subjects. For example, simple tactile stimuli that did not require discrimination produced little activation in visual areas of blind subjects (Sadato et al., 1996). This finding is in agreement with the hypothesis that there is considerable specificity in those neurocognitive systems and subsystems that are modified by altered experience.

This point is further supported by the work of Röder et al. (1999) who have studied auditory localization abilities in blind humans. ERPs were recorded as congenitally blind adults and sighted controls attended either to central or peripheral sound sources in order to detect a rare frequency shift among noise bursts either at the 0- or the 90-degree loudspeaker (on different blocks). Behavioral data revealed a higher spatial resolution in the blind, particularly when they were attending to the periphery. Gradients of ERP amplitudes suggested a sharper auditory spatial attention focus in the blind compared to the sighted. The results suggest that across auditory and visual modalities the representation of peripheral space is more altered by early sensory experience than is the representation of central space. As discussed previously, this effect is likely due to an interaction between various structural constraints and the need to attend to peripheral locations to detect potentially important events. More generally, results from studies of the effects of sensory deprivation demonstrate marked variability in the degree to which, and time periods during which, different functional subsystems are modified by altered sensory experience.

LANGUAGE PROCESSING

It is reasonable to assume that the rules and principles governing development of the sensory systems also guide the development of cognitive and language relevant brain systems. This approach hypothesizes that, as for vision and audition, there exist biological constraints on the forms of natural language and the organization of the brain systems that mediate natural language and that, in addition, different subsystems within language will display different degrees of modification by environmental input. The study of deaf individuals whose first language is American Sign Language (ASL, i.e., a visual–manual language) and their comparison with native English speakers offers a unique opportunity to assess the idea that there exist certain biological constraints on the organization of natural languages that are independent of the modality through which the language is acquired or the structural characteristics of the language, and to observe the effects of language modality on the differentiation of the language systems of the brain.

Consideration of the formal linguistic structure of ASL and psycholinguistic data from adults and children all support the claim that ASL is a fully developed natural language just as are English and French (Klima & Bellugi, 1979). Thus, ASL exhibits grammatical structure at all linguistic levels including phonology, morphology, and syntax. Grammatical properties that hold for spoken languages are also found in sign languages. Children acquiring ASL as a native language from their deaf parents go through the same stages at the same ages as hearing children acquiring a spoken language. These results suggest that there are constraints on the organization of all natural languages that operate independently of the modality through which language is acquired. At the same time, the modality of transmission clearly impacts other aspects of language acquisition and processing. For example, the modality of the language affects the nature of the grammatical devices that the language exploits. In particular, within signed languages, morphological and lexical information is most often conveyed concurrently, reflecting the capacity of the visual system to process information simultaneously.

Grammatical distinctions are often conveyed by planes of signing space and spatial loci within these planes (see Fig. 11.2). The pattern of similarities and differences imply that there are likely to be parallels in the identity and operation of the neural systems that mediate spoken and signed languages, but that there are likely to be some differences as well. In particular, while aural–oral language processing has been consistently associated with the peri-sylvian cortex within the left hemisphere, studies of visuospatial functions in hearing subjects indicate that this type of processing is mediated by parietal cortex primarily within the right hemisphere. The observation that ASL relies on visual spatial processing at most of the different stages of linguistic analysis raises the hypothesis that the right hemisphere and/or parietal structures participate to a greater extent when subjects process ASL.

One approach to this issue has been to compare the effects of damage to specific brain regions on particular aspects of oral and sign language production and comprehension. The initial studies along these lines reported similar patterns of hemispheric asymmetries and similar patterns of anterior–posterior dependencies in language breakdown following cortical damage in speaking and signing patients (Poizner, Klima, & Bellugi, 1987). More recently, additional studies have reported discrepancies between the findings of aural–oral and sign language breakdown (Corina, 1997).

FIG. 11.2. The ASL sign "APPOINTMENT" (arrow indicates motion).

To further assess this issue, we have been comparing cerebral organization for language processing in normal hearing, monolingual, native English speakers and congenitally deaf, native ASL signers. ERPs were recorded while subjects either read English sentences or viewed ASL sentences. It is commonly accepted that language processing can be decomposed into separate subsystems including semantic processing (e.g., as occurs in processing the meaning of nouns and verbs that make reference to specific objects and events) and grammatical processing (as occurs in processing the structural or relational information) provided in English primarily by words such as articles, conjunctions, and auxiliaries. In normal hearing adults processing their native language (English), open-class words elicited ERPs that were characterized by a negative component that became maximal approximately 350 ms after the word onset (N350) and was largest over the posterior regions of both hemispheres.

In contrast, ERPs to closed-class words displayed a negative potential largest at 280 ms (N280) that was localized to anterior temporal regions of the left hemisphere (Neville, Mills, & Lawson, 1992) (Fig. 11.3). When deaf individuals processed ASL, they, just like hearing people processing English, displayed more negative ERP responses for open-class signs over posterior regions, while for closed-class signs the ERPs responses were more negative over anterior regions (Fig. 11.3). This similarity of anterior/posterior ERP pattern as a function of word/sign class membership is consistent with the proposal that there is significant overlap in the identity and organization of the neural systems within a hemisphere that mediate the processing of all formal languages, independently of the modality through which they are acquired. However, while in hearing subjects the ERPs to closed-class words were strongly lateralized to the left hemisphere, in deaf subjects they were symmetrical, suggesting a greater contribution of the right hemisphere during ASL processing (Neville et al., 1997).

Recently we employed the fMRI technique to characterize more precisely the brain structures that mediate language processing in these populations. We compared hearing and deaf individuals as they read English sentences and viewed ASL sentences (see Color Plate 4). Consistent with our ERP results, when native signers viewed ASL sentences, robust activation was observed within classical language areas of the left hemisphere, as observed when native speakers of English read English sentences. These results imply that there is a strong bias for these regions to process language independently of the modality of the language. At the same time however, native signers, but not native speakers, displayed robust and extensive activation within the right hemisphere (Neville et al., 1998). These results are consistent with the idea that the early acquisition of ASL leads to an increased role for right-hemisphere structures in language process-

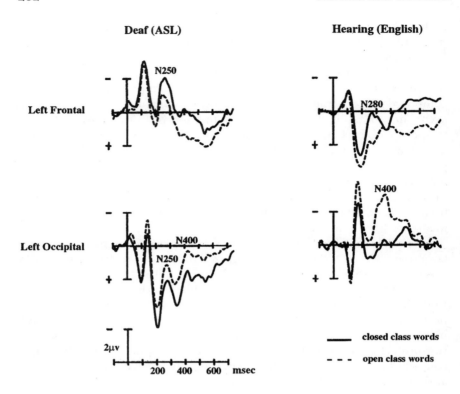

FIG. 11.3.   ERPs to open- and closed-class signs in ASL sentences averaged across 10 deaf subjects from left frontal and occipital areas. ERPs to
open- and closed-class words in English sentences from 17 normally hearing subjects. (Fig. 2, from "Neural systems mediating American Sign Language: Effects of sensory experience and age of acquisition" by H. J. Neville
et al., 1997, *Brain and Language, 57*, pp. 258–303. Copyright © 1997 by Academic Press.)

ing. This may occur in response to the important role of visual spatial information in processing ASL.

The comparison of the processing of written English in deaf and hearing subjects provides an important opportunity to study sensitive periods
during development. Although hearing subjects learn English from birth,
deaf individuals are introduced to English much later (in the group we
studied American Sign Language was the first language, English the second language, and it was learned late). ERPs recorded while deaf and
hearing subjects read English sentences suggest that semantic and grammatical processing are differentially vulnerable to altered early language
experience (Neville, Mills, & Lawson, 1992). The deaf subjects displayed

ERPs to open-class words and other semantic information in English that were similar to those observed when normal-hearing subjects process English. These results suggest that aspects of semantic processing are robust following deaf subjects' altered early language experience.

In contrast, deaf subjects' ERPs to closed-class words that carry grammatical information were markedly different from those of hearing subjects reading the same sentences. The ERPs of deaf subjects lacked the negative (N280) potential over anterior regions of the left hemisphere and did not display any evidence of left hemisphere advantage (see Color Plate 5). These results are in accord with the idea that language experience has different effects on the development of the several different brain systems that mediate language. Brain systems that mediate grammatical aspects of language processing appear to be more sensitive to altered language experience. This idea is supported by our observation that deaf subjects whose grammar skills were excellent displayed an N280 response that was prominent and asymmetrical just as in normal-hearing subjects (Neville, 1991).

Similarly, in an fMRI study comparing hearing and congenitally deaf individuals processing English, we observed marked differences in the brain systems activated in the two groups. Whereas hearing subjects displayed the characteristic left temporal asymmetry, deaf individuals had very little left temporal activity but robust temporo-parietal activation within the right hemisphere (Neville et al., 1998). Further evidence along these lines comes from the study of bilinguals who acquired English at different ages. Early learners of English showed a large asymmetrical response to grammatical information in English, while in late learners this response is altered both in latency and morphology (Weber-Fox & Neville, 1996, 1999).

The results from the language studies taken as a whole support the hypothesis that there are constraints on the organization of the neural systems that mediate formal language independently of the modality through which language is acquired. The biases include different specializations of anterior and posterior cortical regions in aspects of grammatical and semantic processing. Additionally, however, it is clear that the nature and timing of sensory and language experience significantly affect the development of the language systems of the brain. Deaf subjects showed an increased role for the right hemisphere when processing ASL, and a lack of the left-hemisphere specialization for grammatical processing of English. These results suggest that functions such as the acquisition of grammar that are dependent on the rapid, online computation of relationships between items are maturationally constrained and display distinct time periods in development when they require specific types of environmental inputs.

SUMMARY AND CONCLUSIONS

The results from the language studies taken as a whole point to different developmental time courses and developmental vulnerabilities of aspects of grammatical and semantic/lexical processing. They thus provide support for conceptions of language that distinguish these subprocesses within language. Similarly, following auditory deprivation, processes associated with the dorsal visual pathway were more altered than were functions associated with the ventral pathway, providing support for conceptions of visual system organization that distinguish functions along these lines. A general hypothesis that may account for the different patterns of plasticity within both vision and language is that systems employing fundamentally different learning mechanisms display different patterns of developmental plasticity.

It may be that systems that display experience-dependent change throughout life, including the topography of sensory maps (Gilbert, 1995; Kaas, 1995; Merzenich, Recanzone, Jenkins, Allard, & Nudo, 1988), lexical acquisition (i.e., object–word associations), and the establishment of form, face and object representations (i.e., ventral pathway functions) rely upon general, associative learning mechanisms that permit learning and adaptation throughout life. By contrast, systems that are important for computing, online, dynamically shifting relations between locations, objects, and events (including the dorsal visual pathway and the systems of the brain that mediate grammatical and phonological processing) appear dependent on and modifiable by experience primarily during more limited periods in development.

This could account for both the greater developmental deficits and enhancements of dorsal pathway function following various developmental anomalies and for the greater effects of altered language experience on grammatical and phonological functions. Further research is necessary to characterize systems that become constrained in this way and those that can be modified throughout life. In addition, as described in McClelland (this volume), intensive training may improve language functions, even within non-native, adult speakers. It will be important to determine the neural substrate and mechanisms underlying these effects and to compare them with those occurring in normal development. This type of developmental evidence can contribute to fundamental descriptions of the architecture of different cognitive systems. Additionally, in the long run, they may contribute to the design of educational and habilitative programs for both normally and abnormally developing children.

ACKNOWLEDGMENTS

This research has been supported by grants from National Institutes of Health, DC00128 and DC00481 (to Helen Neville); and from the DANA

Foundation and the McDonnell-Pew Foundation (to Daphne Bavelier). We are grateful to our many collaborators on the several studies summarized here, and to Linda Heidenreich for manuscript preparation.

REFERENCES

Armstrong, B., Mitchell, T., Hillyard, S. A., & Neville, H. J. *Effects of auditory deprivation on color and motion processing: An ERP study*, UCSD, La Jolla, CA.

Atkinson, J., King, J., Braddick, O., Nokes, L., Anker, S., & Braddick, F. (1997). A specific deficit of dorsal stream function in Williams' syndrome. *NeuroReport, 8*(8), 1919–1922.

Bavelier, D., Tomann, A., Mitchell, T., Corina, D., Pouget, A., Hutton, C., Liu, G., & Neville, H. (1998). Cortical re-organization for visual functions in congenitally deaf subjects: Part I. Motion processing. *Society for Neuroscience Abstracts, 24*(2), 2094.

Bowering, E. R., Maurer, D., Lewis, T. L., & Brent, H. P. (1997). Constriction of the visual field of children after early visual deprivation. *Journal of Pediatric Ophthalmology and Strabismus, 34*(6), 347–356.

Chalupa, L., & Dreher, B. (1991). High precision systems require high precision "blueprints": A new view regarding the formation of connections in the mammalian visual system. *Journal of Cognitive Science, 3*, 209–219.

Cohen, L. G., Celnik, P., Pascual-Leone, A., Corwell, B., Faiz, L., Dambrosia, J., Honda, M., Sadato, N., Gerloff, C., Catala, M. D., & Hallett, M. (1997). Functional relevance of crossmodal plasticity in blind humans. *Nature, 389*, 180–183.

Corina, D. (1997). Aphasia in users of signed languages. In P. Coppens (Ed.), *Language in atypical population* (pp. 261–310). Mahwah, NJ: Lawrence Erlbaum Associates.

Dehay, C., Bullier, J., & Kennedy, H. (1984). Transient projections from the fronto-parietal and temporal cortex to areas 17, 18, and 19 in the kitten. *Experimental Brain Research, 57*, 208–212.

Eden, G. F., VanMeter, J. W., Rumsey, J. M., Maisog, J. M., Woods, R. P., & Zeffiro, T. A. (1996). Abnormal processing of visual motion in dyslexia revealed by functional brain imaging. *Nature, 382*, 66–69.

Frost, D. O. (1984). Axonal growth and target selection during development: Retinal projections to the ventrobasal complex and other "nonvisual" structures in neonatal Syrian hamsters. *Journal of Comparative Neurology, 230*, 576–592.

Garraghty, P. E. (1993). Competitive interactions influencing the development of retinal axonal arbors in cat lateral geniculate nucleus. *Proceedings of the National Academy of Sciences, 73*, 529–545.

Gilbert, C. D. (1995). Dynamic properties of adult visual cortex. In M. S. Gazzaniga (Ed.), *The cognitive neurosciences* (pp. 73–89). Cambridge, MA: MIT Press.

Hickey, T. L. (1977). Postnatal development of the human lateral geniculate nucleus: Relationship to a critical period for the visual system. *Science, 198*, 836–838.

Horton, J. C., & Hocking, D. R. (1997). Timing of the critical period for plasticity of ocular dominance columns in macaque straite cortex. *Journal of Neuroscience, 17*(10), 3684–3709.

Innocenti, G., & Clarke, S. (1984). Bilateral transitory projections to visual areas from auditory cortex in kittens. *Developmental Brain Research, 14*, 143–148.

Kaas, J. H. (1995). The reorganization of sensory and motor maps in adult mammals. In M. S. Gazzaniga (Ed.), *The cognitive neurosciences* (pp. 51–71). Cambridge, MA: MIT Press.

Klima, E. S., & Bellugi, U. (1979). *The signs of language*. Cambridge, MA: Harvard University Press.

Lachica, E. A., & Casagrande, V. A. (1988). Development of primate retinogeniculate axon arbors. *Visual Neuroscience, 1,* 103–123.

Lovegrove, W., Garzia, R., & Nicholson, S. (1990). Experimental evidence for a transient system deficit in specific reading disability. *Journal of the American Optometric Association, 61,* 137–146.

Merzenich, M., Recanzone, G., Jenkins, W., Allard, T., & Nudo, R. (1988). Cortical representational plasticity. In P. Rakic & W. Singer (Eds.), *Neurobiology of neocortex* (pp. 41–67). New York: John Wiley & Sons Ltd.

Mioche, L., & Perenin, M. (1986). Central and peripheral residual vision in humans with bilateral deprivation amblyopia. *Experimental Brain Research, 62*(2), 259–272.

Mitchell, T. V., Armstrong, B. A., Hillyard, S. A., & Neville, H. J. (1997). Effects of Auditory Deprivation on the Processing of Motion and Color. *Society for Neuroscience, 23*(2), 1585.

Mitchell, T. V., Wolter, A., & Neville, H. J. (1999). The electrophysiology of motion and color perception from early school years to adulthood. *Cognitive Neuroscience Society Abstract, 6,* 75.

Neville, H. (1991). Neurobiology of cognitive and language processing: Effects of early experience. In K. Gibson & A. Petersen (Eds.), *Brain maturation and cognitive development: Comparative and cross-cultural perspective* (pp. 355–380). Hawthorne, NY: Aldine de Gruyter.

Neville, H. J. (1995). Developmental specificity in neurocognitive development in humans. In M. Gazzaniga (Ed.), *The cognitive neurosciences* (pp. 219–231). Cambridge, MA: MIT Press.

Neville, H. J., & Bavelier, D. (1999). Specificity and plasticity in neurocognitive development in humans. In M. S. Gazzaniga (Ed.), *The cognitive neurosciences; 2nd ed.* (pp. 83–98). Cambridge, MA: MIT Press.

Neville, H. J., Bavelier, D., Corina, D., Rauschecker, J., Karni, A., Lalwani, A., Braun, A., Clark, V., Jezzard, P., & Turner, R. (1998). Cerebral organization for language in deaf and hearing subjects: Biological constraints and effects of experience. *Proceedings of the National Academy of Science, USA, 95*(3), 922–929.

Neville, H. J., Coffey, S. A., Lawson, D. S., Fischer, A., Emmorey, K., & Bellugi, U. (1997). Neural systems mediating American Sign Language: Effects of sensory experience and age of acquisition. *Brain and Language, 57,* 285–308.

Neville, H. J., & Lawson, D. (1987a). Attention to central and peripheral visual space in a movement detection task: An event-related potential and behavioral study. I. Normal hearing adults. *Brain Research, 405,* 253–267.

Neville, H. J., & Lawson, D. (1987b). Attention to central and peripheral visual space in a movement detection task: An event-related and behavioral study. II. Congenitally deaf adults. *Brain Research, 405,* 268–283.

Neville, H. J., & Lawson, D. (1987c). Attention to central and peripheral visual space in a movement detection task. III. Separate effects of auditory deprivation and acquisition of a visual language. *Brain Research, 405,* 284–294.

Neville, H. J., Mills, D., & Lawson, D. (1992). Fractionating language: Different neural subsystems with different sensitive periods. *Cerebral Cortex, 2,* 244–258.

Neville, H. J., Schmidt, A., & Kutas, M. (1983). Altered visual-evoked potentials in congenitally deaf adults. *Brain Research, 266,* 127–132.

Packer, O., Hendrickson, A., & Curcio, A. (1990). Developmental redistribution of photoreceptors across the Macaca nemestrina (Pigtail Macaque) retina. *Journal of Comparative Neurology, 298,* 472–493.

Poizner, H., Klima, E. S., & Bellugi, U. (1987). *What the hands reveal about the brain.* Cambridge, MA: MIT Press.

Rebillard, G., Rebillard, M., & Pujol, R. (1980). Factors affecting the recording of visual-evoked potentials from the deaf cat primary auditory cortex (AI). *Brain Research, 188,* 252–254.

287

Reynolds, H. (1978). Perceptual effects of deafness. In R. D. Walk & H. L. Pick (Eds.), *Perception and experience* (pp. 241–259). New York: Plenum.

Röder, B., Rösler, F., Hennighausen, E., & Näcker, F. (1996). Event-related potentials during auditory and somatosensory discrimination in sighted and blind human subjects. *Cognitive Brain Research, 4*, 77–93.

Röder, B., Teder-Salejarvi, W., Sterr, A., Rösler, F., Hillyard, S. A., & Neville, H. J. (1999). Improved auditory spatial tuning in blind humans. *Nature, 400*, 162–166.

Ryugo, D. K., Ryugo, R., Globus, A., & Killackey, H. P. (1975). Increased spine density in auditory cortex following visual or somatic deafferentation. *Brain Research, 90*, 143–146.

Sadato, N., Pascual-Leone, A., Grafman, J., Ibanez, V., Deiber, M.-P., Dold, G., & Hallet, M. (1996). Activation of the primary visual cortex by Braille reading in blind subjects. *Nature, 380*, 526–528.

Sherman, S., & Spear, P. (1982). Organization of visual pathways in normal and visually deprived cats. *Psychological Reviews, 62*, 738–855.

Uhl, F., Franzen, P., Lindinger, G., Lang, W., & Deecke, L. (1991). On the functionality of the visually deprived occipital cortex in early blind persons. *Neuroscience Letters, 124*, 256–259.

Uhl, F., Kretschmer, T., Lindinger, G., Goldenberg, G., Lang, W., Oder, W., & Deecke, L. (1994). Tactile mental imagery in sighted persons and in patients suffering from peripheral blindness early in life. *Electroencephalography and Clinical Neurophysiology, 91*, 249–255.

Van Driel, D., Provis, J. M., & Billson, F. A. (1990). Early differentiation of ganglion, amacrine, bipolar, and Muller cells in the developing fovea of human retina. *Journal of Comparative Neurology, 291*, 203–219.

Veraart, C., DeVolder, A., Wanet-Defalque, M., Bol, A., Michel, C., & Goffinet, A. (1990). Glucose utilization in human visual cortex is abnormally elevated in blindness of early onset but decreased in blindness of late onset. *Brain Research, 510*, 115–121.

Wanet-Defalque, M., Veraart, C., DeVolder, A., Metz, R., Michel, C., Dooms, G., & Goffinet, A. (1988). High metabolic activity in the visual cortex of early blind human subjects. *Brain Research, 446*, 369–373.

Weber-Fox, C., & Neville, H. J. (1996). Maturational constraints on functional specializations for language processing: ERP and behavioral evidence in bilingual speakers. *Journal of Cognitive Neuroscience, 8*(3), 231–256.

Weber-Fox, C., & Neville, H. J. (1999). Functional neural subsystems are differentially affected by delays in second-language immersions: ERP and behavioral evidence in bilingual speakers. In D. Birdsong (Ed.), *New perspectives on the critical period for second language acquisition* (pp. 23–38). Hillsdale, NJ: Lawrence Erlbaum Associates.

COMMENTARY

Chapter 12

Time Matters in Cognitive Development

David Klahr
Carnegie Mellon University

"Time has no divisions to mark its passage."
Thomas Mann, *The Magic Mountain*

The Holy Grail for researchers in cognitive development is a theory that explains the emergence of intelligent behavior in humans from birth through early childhood through adolescence. Regardless of the specific focus of the research, be it face perception, language, reaching, counting, or scientific reasoning, the goal is to propose a set of processes that can not only account for the vast body of data at hand but also make novel predictions. This is a daunting task and the organizers and contributors to this symposium are to be commended for rising to the challenge implicit in the title of this volume: "Mechanisms of Cognitive Development." But what does this focus on *mechanism* mean? What constitutes a mechanism of cognitive development, and what do the chapters reviewed here tell us about such mechanisms?

WHAT IS A MECHANISM?

At the most abstract level, we can characterize a mechanistic theory as one that has (a) some elementary parts, (b) some organization to those parts, (c) some input and output processes, and (d) some dynamic rules and procedures that determine how the parts relate to one another and to their environment over time. Each of the chapters in this section of the book proposes a mechanism to account for a particular aspect of cognitive development, so in a very general sense, one could argue that there is what discussants like to proclaim as an encouraging or remarkable convergence in a potentially disparate set of papers. However, closer examination of the particular mechanisms that are either explicitly or implicitly invoked in this set of chapters does not support such a claim, revealing instead substantial differences among the chapters with respect to each of the four components of mechanism: elementary parts, their organization, their input/output process, and their operating principles and processes. Nevertheless, the fact that each of the projects described here is struggling with mechanism represents an important advance in our field.

A Brief History of Mechanistic Accounts of Cognitive Development

Because claims about advances can only be evaluated in an historical context, I will briefly review the history of mechanistic accounts of cognitive development before turning to the specific chapters. The first serious effort to provide an account of developmental mechanisms can be attributed to Piaget. He characterized the cognitive system in terms of the formalisms available to him at the time. From logic and mathematics, he constructed a representational system. From biology and Baldwin, he borrowed the mechanisms of assimilation and accommodation (cf. Case, 1997).

However, Piaget's initial characterization of these processes was highly abstract, and he continually reconceptualized and refined his formulation of the equilibration process. Thus, as late as 1975, he was using representations like the one shown in Fig. 12.1. Although the notation gives the appearance of a more precisely conceptualized account of equilibration, the accompanying text renders the mechanism obscure. Consider the following:

> Before sufficiently precise models are achieved, therefore, one witnesses a succession of states indicating progressive equilibration. The initial states of this progression achieve unstable forms of equilibrium only because of lacunae, because of perturbations, and above all, because of real or potential contradictions. (Piaget, 1985/1975, p. 47)

Unfortunately for all of us, no one has yet figured out how to translate those ideas into an unambiguous operational system. This lack of a convincing and coherent account of developmental mechanism has vexed many psychologists in recent decades. For example, Flavell (1984) observed that

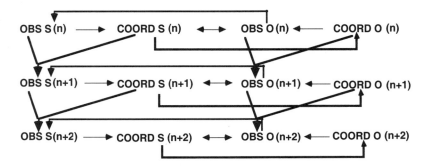

FIG 12.1.   A depiction of part of Piaget's model of assimilation and accommodation. OBS S – "observables relative to the subject's action"; OBS O – "observables relative to objects"; COORD S – "inferential coordinations of the subject's actions or operations"; COORD O – "inferential coordinations among objects" (Piaget, 1985/1975, p. 44).

. . . serious theorizing about basic mechanisms of cognitive growth has actually never been a popular pastime, . . . It is rare indeed to encounter a substantive treatment of the problem in the annual flood of articles, chapters, and books on cognitive development. The reason is not hard to find: Good theorizing about mechanisms is very, very hard to do (p. 189).

A few years later, Allen Newell (1990) echoed Flavell's lament:

I have asked some of my developmental friends where the issue stands on transitional mechanisms. Mostly, they say that developmental psychologists don't have good answers. Moreover, they haven't had the answer for so long now that they don't very often ask the question anymore—not daily, in terms of their research (p. 462).

But even as these gloomy assessments were being made, many researchers were beginning to devote serious attention to the issue of mechanism. For example, the previous decade produced Sternberg's (1984) edited volume *Mechanisms of Cognitive Development*, MacWhinney's (1987) edited volume *Mechanisms of Language Acquisition*, Siegler's (1989) *Annual Review* chapter devoted to transition mechanisms, and an entire volume devoted to production-system models of learning and development (Klahr, Langley, & Neches, 1987).

So, the questions are being asked. What about the answers? What do they look like? Increasingly, they are coming in the form of computational models. Only a few of the chapters in the Sternberg (1984) volume specify mechanisms any more precisely than at the flowchart level, and most of the proposed mechanisms are at the soft end of the "hard-core to soft-core" information-processing spectrum (Klahr, 1992). However, only 5 years later, Siegler (1989)—in characterizing several general categories for transition mechanisms (neural mechanisms, associative competition, encoding, analogy, and strategy choice)—was able to point to computationally-based exemplars for all but the neural mechanisms (Bakker & Halford, 1988; Falkenhainer et al., 1989; Holland, 1986; MacWhinney, 1987; Rumelhart & McClelland, 1986; Siegler, 1988). The recent Halford and Simon (1995) book, consisting entirely of computational models of developmental processes, provides a clear indication of this trend toward "hardening the core" (Klahr, 1992).

However, even my advocacy for computational approaches does not blind me to the fact that, when situated in the vast literature on cognitive development, such models still represent the minority vote, even among developmental theorists. Indeed, of the five chapters in this section of the book—all of which deal with multiple level mechanisms and with highly complex sets of phenomena, only the Johnson and de Haan chapter goes so far as to posit and evaluate a computational model for their domain (cortical specialization for face recognition). Nevertheless, the evolution of theoretical stances taken by noncomputationalists parallels an important trend among computation modelers, and that is the attempt to seamlessly integrate performance with change.

Consider the trend in the computational sphere: The earliest proposals for computational models of cognitive development (e.g., Baylor & Gascon, 1974; Klahr & Wallace, 1976; Simon, 1962a; Young, 1976) suggested that theorists should follow a two-stage approach: first formulate a performance model and then seek an independent set of *transition mechanisms* that operate on that performance model. However, over the years, the sharp distinction between performance models and learning models has become blurred. Today, the most promising approaches—both computational and noncomputational—are those that formulate models that are always undergoing self-modification, even as they perform at a given level or stage.

It is not easy to construct models, computational or otherwise, that achieve an appropriate balance between performance and adaptation. Although the issue is endemic to our area, it is most clearly articulated in the sphere of computational modeling, as two relatively distinct—and at times adversarial—approaches to computational modeling of developmental phenomena have emerged: production systems and connectionist systems. They approach issues of performance and adaptation from different points of departure. In general, production systems emphasize performance over adaptation, while connectionist systems emphasize adaptation over performance. However, the distinctions between the two approaches are diminishing as both fields devote more effort to addressing developmental issues (Klahr & MacWhinney, 1997). Moreover, the distinction between performance and adaptation hinges to a large extent on the time scale of the phenomenon being studied. In the next section, I address that issue.

TIME AND RATE IN MECHANISTIC THEORIES

Perhaps the most novel and important feature of this volume and the symposium on which it is based is the way in which the chapters are organized. One might have expected the conventional dichotomy between cognitive level analyses, such as the topics of mathematics and scientific reasoning, on one hand, and neurocognitive issues, such as motor behavior and early language development, on the other. However, although that is the minor organization within each section—here labeled as either behavioral or neuroscience approaches—the main headings divide the topics according to the *duration* of the processes under investigation. Thus, chapters in this section of the book have been classified by the editors as focusing on "change over long time scales" in contrast to the first set of chapters classified as addressing microgenetic studies focusing on change over short time scales. The classification is of interest because of its inherent theoretical assumption that time scale matters. More specifically, it suggests that both the developmental phenomena being investigated and the mechanisms that underlie them might be qualitatively different for different temporal durations.

Interestingly, neither production system architectures nor PDP models have any principled temporal constraints built into them. Certainly, for any given model, one can attempt to map various aspects of the time course of the phenomenon being investigated to specific temporal parameters of the architecture. For example, one can make assumptions about the basic recognition-act cycle time for a production system, or a learning epoch for a connectionist model. Presently we have no consistent way to do this. If time scale matters, then this is an important theoretical weakness.

The fundamental importance of different temporal scales is elegantly presented by Newell (1990, chapter 3). The first basic assumption is that complex systems, including the human cognitive architecture, involve multiple levels, and that "in a system with multiple levels, the *components* at one level are realized by *systems* at the next level below, and so on for each successive level" (p. 117, emphasis added). Kuhn's depiction of different meta levels, as well as Piaget's diagram (Fig. 12.1) represent attempts to capture this notion that each level of the system comprises a set of lower level components that are themselves composed of subparts. The second basic assumption, derived from Simon's (1962b) "Architecture of Com-

TABLE 12.1
Time Scale of Human Action
(Adapted from Newell, 1990, Figure 3.3)

Approximate scale range (secs)	Conventional time units	Exact secs/unit	Band
$10^7 \sim 10^8$	years	31,536,000	
$10^6 \sim 10^7$	months	2,592,000	
$10^5 \sim 10^6$	weeks	604,800	Social
$10^4 \sim 10^5$	days	86,400	
$10^3 \sim 10^4$	hours	3600	
$10^2 \sim 10^3$	10 min	600	Rational
$10^1 \sim 10^2$	minutes	60	
10^1	10 s	10	
10^0	1 s	1	Cognitive
10^{-1}	100 ms	0.1	
10^{-2}	10 ms	0.01	
10^{-3}	1 ms	0.001	Biological
10^{-4}	100 μs	0.0001	

plexity," is that "stability dictates that systems have to be hierarchical. Attempts to build complicated systems without first building stable subassemblies will ultimately fail—the entire structure will disintegrate before it all gets put together" (Newell, 1990, p. 117).

Newell then turns to the issue raised earlier in this commentary: the time scale of human action. He organizes these time scales into a set of bands as illustrated in Table 12.1. Any mechanistic theory will have elementary components that operate at the basic cycle time, and the system based upon such components will be constrained by the time cycle of its components. In some areas of psychology, such temporal analyses have been highly productive at explanation and prediction (e.g., Card, Moran, & Newell, 1983). However, in other areas, it may be extraordinarily difficult, and perhaps unproductive, to attempt to cross too many temporal levels in a single sweep. Next, I examine the chapters in this section through this lens of temporal scale.

Scale Crossings

This analysis of scale crossings can be applied to these chapters by estimating the temporal durations for three aspects of each of the phenomena under investigation (see Table 12.2). The first aspect is the duration of the underlying basic component of the system (component band). The second is the focal activity, task, or phenomenon being explained (focal band). These first two bands represent system performance at any point in the child's development in terms of the duration of the focal performance and

TABLE 12.2

Temporal relations between the basic elements (component band) and the phenomena being investigated (focal band) in the chapters reviewed here

Chapter	Phenomenon	Performance		Development	Time Factor (×10)
		Component Band	Focal Band		
Kuhn	Scientific Reasoning	Cognitive: Decisions about causal factors in experimental outcomes 10^0 (secs)	Metacognitive: Decisions about how to make such decisions $10^0 - 10^2$ (secs–mins)	Changes in how such decisions are reached $10^6 - 10^8$ (weeks–years)	6–8
Case & Mueller	Quantitative Comparison	Neural: Activation in central parietal region $10^{-1} - 10^0$ (100 msecs–secs)	Cognitive: Deciding on relative magnitude of numbers 10^0 (secs)	Acquisition of improved representations for number and magnitude 10^8 secs (years)	8–9
Thelen	Goal-directed Reaching	Cognitive: Basic motor skills $10^{-1} - 10^0$ (100 msecs–secs)	Cognitive: Reaching for desired object 10^0 (secs)	Dynamic cascade of motivation, control of movement speed, reduction of degrees of freedom, use of proximal muscle group 10^7 (4–8 months)	7–8
Johnson & de Haan	Face Recognition	Neural: Cortical activation 10^{-1} (1/10 sec)	Distinguishing familiar from unfamiliar human faces 10^0 (secs)	Neocortical specialization and localization 10^5 (6–8 weeks)	6
Neville & Bavelier	Speech and Vision Processing	Neural: ERPs during cortical activation $10^{-1} - 10^0$ (1/2 sec)	Word recognition 10^0 (secs)	Neural: Synaptogenesis $10^5 - 10^8$ (days to years)	5–9

Time factor shows the approximate temporal ratio (in powers of 10) between component band and developmental process of interest.

the duration of the component parts of the processes underlying that performance. The third aspect is the developmental process that produced the elements in the component band. The extent of scale crossing can be estimated by comparing the time scale of the component band to the time scale of the developmental process. In the final column in Table 12.2, I have estimated the number of factors of 10 that separate the time cycle of the basic component from the time cycle of the putative developmental period. The main point to be made here is that the larger this number, the less likely we are to construct a satisfactory mechanistic account of the process.

For example, consider Case and Mueller's intellectual tour de force. They summarize and integrate three venerable and foundational epistemological traditions, apply them to a well-defined and focused problem of early quantitative development—connected to Case's own elaborate task-analysis of the rudiments of number sense—and then they go on to explain some of the phenemonena surrounding a single component of this knowledge: the ability to compare single digit names. The empirical support for this account is based on the neural activity engendered by such tasks. For all of the broad and scholarly sweep of their paper, I remain unconvinced that the analysis of neural firing rates tells us much about the emergence of even a simple cognitive task such as the comparison of two numerical magnitudes. As indicated in Table 12.2, the neural activation in the central parietal region that Case and Meuller take as evidence for the basic component in their model has a temporal parameter on the order of hundreds of msec, while the acquisition process that leads to numerical representations takes place over years—10^8 secs. Although we may indeed find that there is some gross activity in one part of the brain rather than another during number comparisons, the vast time scale here makes it difficult to pin down the responsible developmental mechanisms. The difficulty is a consequence, I suggest, of attempting to cross too many temporal levels at once, without positing a sufficiently detailed set of intermediate level processes.

Neville and Bavelier proposed several mechanisms to capture the complementary phenomena of specificity and plasticity in neurocognitive development: maturation of neural systems, bursts of synaptogenesis, synapse proliferation and elimination, rises in cerebral metabolism, and decrease in brain glucose metabolism. The mechanisms in this account are not really complete systems of the type described earlier. One mechanism might be "an early, normally transient, redundancy of connections between the auditory and visual systems" (chap. 11, this volume). The explanation for the intriguing phenomena reported by Neville and Bavelier, however, remains highly nonspecific: "systems employing fundamentally different learning mechanisms display different patterns of developmental plasticity." Clearly, much remains to be specified here about the precise nature of such mechanisms. Here the prognosis for a sucessful theoretical account is improved—at least over the shorter time spans—because of the smaller time factor between basic components and developmental period.

Note that my view of the importance of better understanding these time scale crossings is not unique. Thelen makes a similar point in her chapter. ". . . the major challenge—at whatever level—is to understand how activity on the short time scales, say on the order of milliseconds or seconds, becomes transduced to the behavioral changes we are most interested in, which happen in days, weeks, and months." Yet, it is not clear how the kind of dynamic systems that she proposes to account for the development of reaching can make the kind of successful distinctions between levels that have served the physical sciences so well. Given this vast range of time scales between the biological and the rational (see Table 12.1), one can ask about the likely success of any single mechanistic theory in which the time scale of the basic elements is far distant from the time scale of the phenomena being explained.

These problems of temporal scale are exacerbated when we attempt to compare the mutual contribution and constraint from one type of chapter to the other in this volume—that is when we attempt to make the transition from brain to cognition or—as the editors of this volume have cast it—from neuroscience approaches to behavioral approaches. Consider, as an extreme case, the extent to which advances in temporal and spatial localization of neuroimaging techniques can contribute to our understanding of the metastrategic issues addressed by Kuhn in this volume. Of course, Kuhn makes no such claim, but as discussant, I did make an attempt—unsuccessful, alas—to find some way to connect the two types of approaches.

On the other hand, Johnson and de Haan's work on face recognition exemplifies the kind of tight temporal linking that I believe can be most productive. The temporal scale for their basic ERP data is on the order of tenths of seconds, and the overall duration of the process they are investigating is only about one order of magnitude larger than that. More importantly, the focal phenomenon—the ability to extract invariances from a set of faces—develops over only a few months. Although much of the credit here goes to the ingenuity of the investigators, part of their success, even to the extent of constructing some computational models of the process, may derive from the few temporal boundaries that they attempt to cross in their research.

CONCLUDING COMMENTS

Rather than continue in this vein about the dangers of scale leaping, I prefer to end on a constructive note. I think there is a way for all of these different temporal levels to contribute to an overall theory of the mechanisms of cognitive development, but the emphasis must be on the plural nature of mechanisms. One alone will not do the job. The physical and biological sciences have constructed some remarkable theoretical edifices, by adhering to Simon's stable substructure principle in the very components of their theories. The laws governing subnuclear components are of little use in explaining the behavior of biological processes or planetary dynamics.

I think that cognitive developmental theory can evidence similar advances if we eschew grand leaps across temporal boundaries, and instead attempt to use well-grounded mechanisms at one level as the building blocks for theory at the next level. Interestingly, the early days of the connectionist movement suggested such an orientation to theory building. The subtitle of the seminal PDP volume (Rumelhart & McClelland, 1986) was "the microstructure of cognition." Implicit in that phrase are the notions that (a) higher order cognition is symbolic, and (b) it is time to figure out what symbols are made of. Somewhere along the way, perhaps as a consequence of the phenomenal and captivating advances in neuroimaging technology, the incremental approach appears to have been replaced by efforts to explain high-level, long-duration processes of cognition and cognitive development in terms of low-level components of very short duration. This strikes me as analogous to attempting to describe a web page or a word processor in terms of binary vectors. But it may not be a productive approach.

As we all know, at the very bottom level, all computing processing consists of the manipulation of binary vectors, or even more basically, of the presence or absence of microvoltages. But no one would attempt to directly explain the workings of the "undo" function on my word processor in terms of these basic components. Instead, one would offer a progression of explanatory models, working from binary vectors up through an account of how they are processed by elementary gates and registers, and then how the registers are organized into functional units for symbol manipulation, and then how those units provide the basis for machine code, which in turn is organized into a vast set of specific functional subroutines, and so on up the programming hierarchy until we get to the basic conceptual models that users of word processors can understand. Interestingly, each level functions over a longer temporal duration than do its component processes. Similarly, I believe that attempts to build theories of cognitive development that span a vast temporal and conceptual spectrum may be transforming an already daunting task into one that is nearly impossible. There are other ways, and I have suggested a few of them here.

REFERENCES

Bakker, P. E., & Halford, G. S. (1988). *A basic computational theory of structure-mapping in analogy and transitive inference* (Tech. Rep.). St. Lucia, Australia: University of Queensland, Centre for Human Information Processing and Problem Solving.

Baylor, G. W., & Gascon, J. (1974). An information processing theory of aspects of the development of weight seriation in children. *Cognitive Psychology, 6*, 1–40.

Card, S., Moran, T. P., & Newell, A. (1983). *The psychology of human-computer interaction*. Hillsdale, NJ: Lawrence Erlbaum Associates.

Case, R. (1997). The development of conceptual structures. In D. Kuhn & R. S. Siegler (Eds.), *Handbook of child psychology (5th ed.) Vol. 2: Cognition, perception, and language* (pp. 745–800). New York: Wiley.

Falkenhainer, B., Forbus, K. D., & Gentner, D. (1989). The structure-mapping engine. In *Proceedings of the American Association for Artificial Intelligence*. Philadelphia: American Association for Artificial Intelligence.

Flavell, J. H. (1984). Discussion. In R. J. Sternberg (Ed.), *Mechanisms of cognitive development* (pp. 187–210). New York: Freeman.

Halford, G., & Simon, T. (Eds.). (1995). *Developing cognitive competence: New approaches to process modeling*. Hillsdale, NJ: Lawrence Erlbaum Associates.

Holland, J. H. (1986). Escaping brittleness: The possibilities of general purpose machine learning algorithms applied to parallel rule-based systems. In R. S. Michalski, J. G. Carbonell, & T. M. Mitchell (Eds.), *Machine learning: An artificial intelligence approach* (pp. 593–624). Los Altos, CA: Morgan-Kaufmann.

Klahr, D. (1992). Information processing approaches to cognitive development. In M. H. Bornstein & M. E. Lamb (Eds.), *Developmental Psychology: An advanced textbook* (3rd ed., pp. 273–335). Hillsdale, NJ: Lawrence Erlbaum Associates.

Klahr, D., & MacWhinney, B. (1997). Information Processing. In D. Kuhn & R. S. Siegler (Eds.), W. Damon (Series Ed.). *Handbook of child psychology (5th ed.): Vol. 2: Cognition, perception, and language* (pp. 631–678). New York: Wiley.

Klahr, D., & Wallace, J. G. (1976). *Cognitive development: An information-processing view*. Hillsdale, NJ: Lawrence Erlbaum Associates.

Klahr, D., Langley, P., & Neches, R. (Eds.). (1987). *Production system models of learning and development*. Cambridge, MA: MIT Press.

MacWhinney, B. J. (Ed.). (1987). *Mechanisms of language acquisition*. Mahwah, NJ: Lawrence Erlbaum Associates.

Newell, A. (1990). *A unified theory of cognition*. Cambridge, MA: Harvard University Press.

Piaget, J. (1985). *The equilibration of cognitive structures*. (Trans. T. Brown & J. Thampy). Chicago: University of Chicago Press. (Original work published 1975)

Rumelhart, D. E., & McClelland, J. L. (1986). *Parallel distributed processing: Explorations in the microstructure of cognition*. Cambridge, MA: MIT Press.

Siegler, R. S. (1988). Strategy choice procedures and the development of multiplication skill. *Journal of Experimental Psychology: General, 117*, 258–275.

Siegler, R. S. (1989). Mechanisms of cognitive development. *Annual Review of Psychology, 40*, 353–379.

Simon, H. A. (1962a). An information processing theory of intellectual development. *Monographs of the Society for Research in Child Development, 27* (2, Serial No. 82).

Simon, H.A. (1962b). The architecture of complexity. *Proceedings of the American Philosophical Society, 106*, 467–482.

Sternberg, R. J. (Ed.). (1984). *Mechanisms of cognitive development*. New York: Freeman.

Young, R. M. (1976). *Seriation by children: An artificial intelligence analysis of a Piagetian task*. Basel: Birkhauser.

Part III

DEVELOPMENTAL DISORDERS

DYSLEXIA

Chapter 13

Neural Plasticity in Dyslexia: A Window to Mechanisms of Learning Disabilities

Albert M. Galaburda
Glenn D. Rosen
Beth Israel Deaconess Medical Center
and
Harvard Medical School, Boston, MA

A main current debate in dyslexia research today considers whether the condition arises from a primary developmental disorder of language processing, either linguistic *per se* or metalinguistic, or from a fundamentally sensory/perceptual failure. If linguistic/cognitive, does the problem affect more than just language? If sensory/perceptual, what modalities might be involved—auditory, visual, others? Furthermore, to what extent is the problem longitudinal as well as parallel, that is, does it involve multiple levels of processing? Does it involve processing in more than one pathway? The purpose of this review is to ask the question from the point of view of neuroanatomy, in the hope that there will be new light shed on the question of the functional deficit in dyslexia.

DEFINING AND DESCRIBING THE PROBLEM

From the time of its description (Hinshelwood, 1917; Morgan, 1896; Orton, 1925), the definition of developmental dyslexia has proved to be quite controversial, and often reflected the dichotomy between the sensory/perceptual and language processing camps. The most recently accepted definition by the National Institutes of Health, for example, keenly reflects this issue:

> Dyslexia is one of several distinct learning disabilities. It is a specific language-based disorder of constitutional origin characterized by difficulties in

307

single word decoding, usually reflecting insufficient phonological processing abilities. These difficulties in single word decoding are often unexpected
in relation to age and other cognitive and academic abilities; they are not
the result of generalized developmental disability or sensory impairment.
Dyslexia is manifested by variable difficulty with different forms of language,
often including, in addition to problems reading, a conspicuous problem
with acquiring proficiency in writing and spelling (Lyon, 1995).

It is useful to note that although this definition does not speak specifically to the question of etiology, it explicitly endorses the notion that
developmental dyslexia is a problem in language processing. That the disorder is not the result of sensory impairment is meant to reflect the exclusion or individuals with primary hearing, vision, or tactile impairment
from the definition, and not necessarily the more subtle sensory/percep-
tual problems to be discussed next. Nevertheless, by describing developmental dyslexia at is does, the definition makes specific assumptions about
the nature of the disorder.

Cognitive/Linguistic Evidence

It is well established that dyslexics, at least the bulk of dyslexics, have difficulties with phonological tasks, for example, pseudoword reading, phoneme segmentation, rhyming, phoneme deletion tasks (Bradley & Bryant,
1981; Bryant, Nunes, & Bindman, 1998; Liberman & Shankweiler, 1985;
Morais, Luytens, & Alegria, 1984). These tasks are thought to tap into cognitive processes that could be described as metaphonological (Brady,
Poggie, & Rapala, 1989; Shankweiler et al., 1995). This is in contrast to
phonological processes involved in speech comprehension and speech
productions, which occur automatically without conscious awareness of the
sound structure of the language. These findings do not imply that developmental dyslexia is simply a disorder of language. There is evidence, for
instance, that dyslexics have problems processing information in other
modalities. For example, dyslexics have an impaired verbal working memory system, and this is true whether the material is presented in visual or
auditory form (Brady, 1991; Crain & Shankweiler, 1990; Liberman, Shankweiler, & Liberman, 1989; Swanson, 1993). Interestingly, among the deaf,
those with better working memories are also better readers (Hanson,
1991; Liberman & Mattingly, 1989). That said, defects in phonological
skills occur more frequently and are more strongly correlated with reading deficits than are verbal working memory deficits (Brady, 1991;
Shankweiler et al., 1995). Common to all these findings is the notion that
the primary problem with dyslexics lies in the cognitive domain, as illus-

trated by the deficits in parsing of words into component phonemes and manipulating words within working memory.

Sensory/Perceptual Evidence

Psychophysical evidence indicates that the visual system is abnormal in dyslexics (Chase & Jenner, 1993; Geiger & Lettvin, 1987; Lovegrove, Garzia, & Nicholson, 1990; Slaghuis, Lovegrove, & Davidson, 1993; Williams & Lecluyse, 1990; Williams, May, Solman, & Zhou, 1995). These researchers suggested that dyslexics had specific difficulties in the processing of fast visual information. Subsequent work using neuroscience-based methods for assessing physiology have supported these findings (e.g., Demb, Boynton, & Heeger, 1998; Eden, Vanmeter, Rumsey, & Zeffiro, 1996b; Galaburda, Menard, & Rosen, 1994; Livingstone, Rosen, Drislane, & Galaburda, 1991). We showed that, as evidenced by evoked potentials, dyslexics had difficulties processing rapidly changing, low-contrast checkerboard patterns presented on a television screen (Livingstone et al., 1991). The abnormalities occurred before and up to 100 ms, which indicated involvement of early visual processors anatomically placed before or up to the primary visual cortex. Others (Demb et al., 1998; Eden et al., 1996a; Eden & Zeffiro, 1998) have showed dysfunction of the motion perception area, which is also a rapid perceptual processor, only one or two synapses beyond the primary cortical visual input zone.

In the auditory system, too, dyslexics and other developmentally language impaired children have shown difficulties in processing rapidly changing sounds (Anderson, Brown, & Tallal, 1993; Merzenich et al., 1996; Tallal, 1977; Tallal & Piercy, 1973; Tallal et al., 1995), which suggested to Paula Tallal and colleagues that this was the primary reason for the phonological problems dyslexics exhibit. The argument was that the during language acquisition dyslexics never hear certain sounds and therefore can not represent a full set of phonemes for the given language. Further, if dyslexics do not have a full set of phonemes, they could not map these representations onto the graphemes of written language, hence the difficulty with acquiring reading.

Points of Contention Between the Two Models

The low-level hypothesis argues that the low-level disturbances in sensory input are, in and of themselves, sufficient to cause the high-level problems associated with the disorder. Thus, deficient low level processing of some types of sounds that would be useful for establishing normal representations of the phonology of the language lead to secondary phonological problems, and third hand metaphonological problems. Those who argue that dyslexia is a linguistic/cognitive matter claim that no sensory/perceptual deficit so far described in dyslexics could account for the difficulties in phonological processing exhibited by the dyslexic patients. These opponents of the sensory/perceptual explanation do not deny that these problems exist; rather, they assert that they are irrelevant to the process of learning to read in dyslexia (Mody, Studdert-Kennedy, & Brady, 1997; Studdert-Kennedy & Mody, 1995).

If the phonological architecture is corrupted, then it should be measurable in all tasks that depend on this phonology, including oral language comprehension, speech production, and reading. Thus, one of the problems with the low-level hypothesis is that it fails to account for how well dyslexics speak and comprehend spoken language. Therefore, those in the cognitive/linguistic camp argue that there could not be anything significantly wrong with the phonological architecture in dyslexic brains, irrespective of the low-level problems. Instead, they argue that the problem is purely metaphonological and consists of a difficulty with *consciously* segmenting words into phonemes, which is a prerequisite for learning to read (Mody et al., 1997).

As we know it, the anatomy of the brain is topographically unambiguous regarding the parts of the cerebral cortex dedicated to sensory/perceptual as opposed to cognitive/associational activities. The sensory/perceptual hypothesis implicates early stations in the sensory pathway, which include thalamic relay nuclei, primary cortices (visual, auditory), and adjacent association areas. The linguistic/cognitive hypothesis, on the other hand, implicates stations further downstream in anterior temporal cortex, inferior parietal lobule, hippocampal formation, and frontal lobe. It would make sense, therefore, to examine brain structures in the dyslexic we know to be involved in sensory/perceptual behaviors and those involved in cognitive, including linguistic and other cognitive behaviors. We could compare these to the equivalent structures in nondyslexic brains and draw some conclusions about which are anomalous and which are not. This is, of course, a first pass, as we realize that sensory/perceptual and cognitive behaviors are subject to bottom–up and top–down influences during learning, and their respective anatomical substrates, too, are subject to trophic or antitrophic influences from connected structures upstream and downstream, especially during early development. Of course, the discovery of anatomic anomalies would not permit us to draw conclusions about causality, but we could at least state whether cognitive areas, or sensory/perceptual areas, or both, are or are not abnormal in the dyslexic brains. We can go further into determining causal links only if we could develop animal models. Recent research on anatomical models of developmental dyslexia have helped to get us closer to the answers to these questions, and have revealed interesting facts about developmental plasticity, including a previously unsuspected sex difference.

BRAIN FINDINGS IN DYSLEXICS

There are two types of findings made on autopsied dyslexic brains, which bear on the debate outlined above, and one of these is particularly relevant. The first has to do with alterations in the pattern of cerebral asymmetry. For instance, the planum temporale (Fig. 13.1), a region on the upper surface of the temporal lobe, which is asymmetric in two thirds of the population, fails to show asymmetry in dyslexics in most published studies (Galaburda, Sherman, Rosen, Aboitiz, & Geschwind, 1985; Hynd, Semrud-Clikeman, Lorys, Novey, & Eliopulos, 1990; Morgan & Hynd, 1998; Steinmetz, Volkmann, Jancke, & Freund, 1991). The asymmetry in question is caused at least in part by an asymmetry of posterior temporal area, Tpt. This cortex is one synapse or two away from the primary auditory cortex, and still unimodal enough to be mainly involved in sensory/perceptual and early cognitive tasks associated with auditory function. Area Tpt is just as far from the primary auditory cortex as area MT, one of the motion perception areas, is from the primary visual cortex. Lack of Tpt asymmetry may have some implications concerning early sound and phonological processing in dyslexics, although this is not as yet specified. It is less likely that area Tpt is solely responsible for phonological awareness tasks, which probably also involve the frontal lobe.

A second finding is the presence of focal cortical malformations consisting of nests of ectopic neurons and glia in the first cortical layer, which alter focally the architecture of the affected cortical areas (Galaburda et al., 1985). The malformations, termed *ectopias*, represent altered neuronal migration to the cortex. They are present mostly in perisylvian location, but do not typically affect parts of the brain associated with reading and writing after acquired brain injury, that is, the inferior parietal lobule and occipital lobe (Fig. 13.2). Instead, the malformations cluster around perisylvian language areas, that is, the superior temporal gyrus (which contains Wernicke's area and area Tpt) and the inferior premotor and prefrontal cortex (which contains Broca's area and Brodmann areas 44 and 45). Their prevalence is greater in the left hemisphere. Clearly, some of these ectopias affect cortex that is involved in linguistic/cognitive functions, which is located many synapses away from sensory input stations, for example, primary auditory cortex. As many as 150 instances of ectopia have been seen in a single dyslexia brain specimen. Similar ectopias have been seen in other developmental disorders in humans, including the fe-

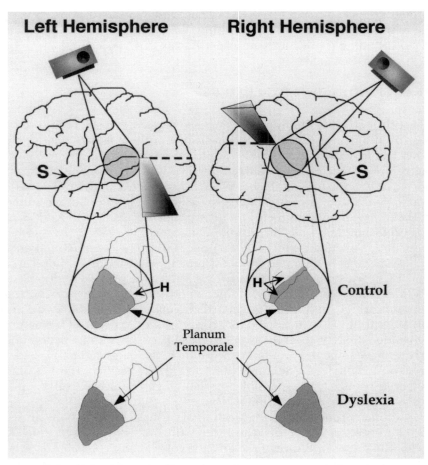

FIG. 13.1. Schematic illustration of measurement of the planum temporale. In both the left and right hemisphere, the Sylvian Fissure (**S**) is followed posteriorly and a knife cut is made from that point through the hemisphere. Note that the Sylvian fissures of the two hemispheres are often differently shaped. The exposed superior surface of the temporal lobe is photographed. The planum temporale is the triangularly shaped region at the posterior portion of the temporal lobe, bounded anteriorly by Heschl's gyrus (**H**). In the situation when there are two Heshl's gyri (most often in the right hemisphere), only the first is included in the measurement. This schematic illustrates typical planum from controls (where ⅔ to ⅘ of the brains are asymmetric in favor of the left hemisphere) and dyslexics where symmetry is the norm. Note that in the case where the plana are symmetric, it is because the right hemisphere is larger than normal.

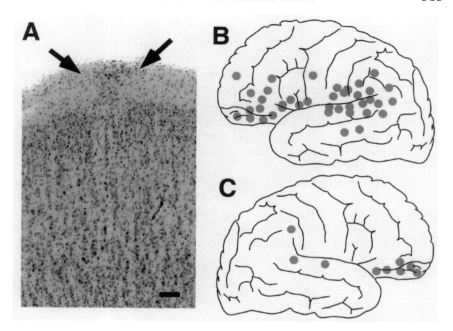

FIG. 13.2. **A.** Photomicrograph of neocortical ectopia in the brain of a dyslexic subject. Arrows denote collections of neurons in layer I of the cortex, a layer that is normally cell sparse. Bar = 100 μm. **B.** Schematic illustrating the typical distribution of ectopias in the Left and Right (**C**) hemispheres of dyslexic subjects. In the left hemisphere, the ectopias are found predominantly in the perisylvian regions and in the prefrontal cortex. No ectopias have been seen in occipital cortex. In the right hemisphere, the distribution is similar to that in the left hemisphere, but there are far fewer ectopias present.

tal alcohol syndrome and many mental retardation syndromes (Friede & Mikolasek, 1978), although we are not aware that the distribution and numbers of the malformations are equivalent.

We have gathered evidence in dyslexic brains to show that some thalamic nuclei exhibit changes in the size of neurons (Galaburda et al., 1994; Livingstone et al., 1991). Changes have been seen in the LGN, a nucleus connected to the retina and to the primary visual cortex, and in the MGN, a nucleus connected to the primary auditory cortex and to auditory brainstem nuclei. These nuclei either have smaller neurons or an increased proportion of small neurons. Because smaller neurons conduct nerve impulses more slowly, smaller neurons in visual and auditory relay nuclei in the thalamus may be interpreted as one possible anatomical substrate underlying slower visual and auditory perceptual processing in dyslexics. The primary visual cortex also shows and alteration in neuronal sizes and a change in the normal pattern of cellular asymmetry (Jenner, Rosen, & Galaburda, 1999).

In summary, there are neuronal changes in thalamic nuclei and primary cortex, which could explain the sensory/perceptual problems exhibited by dyslexics. There are also ectopias and anomalies of asymmetry affecting both sensory/perceptual and linguistic/cognitive areas, which could account for the linguistic/cognitive deficits. The ectopias contain enough connectivity during development to influence development of other areas, including sensory/perceptual areas. If this indeed happens, one could propose that sensory/perceptual deficits are secondary to changes downstream in the pathway, or that linguistic/cognitive changes are secondary to changes upstream in sensory/perceptual areas.

This hypothesis can be tested in experimental animals in which one can produce cortical malformations in high level association cortex and watch for changes in the thalamus, for instance. As the migrational changes, causing ectopias, arise early in development, and changes in cell size can occur at anytime in life, we have thought it more plausible that the ectopias, if at all, cause the changes in the thalamus. A corollary hypothesis is that the earlier presence of ectopias than thalamic changes means that phonological and other linguistic/cognitive changes come first and sensory/perceptual changes result from secondary changes in the thalamus. This is the opposite picture to that suggested by the sensory/perceptual hypothesis and would instead support the linguistic/cognitive hypothesis.

In recent years, we have embarked on a project to discern the effects of malformations of the cortex on brain and behavior of animals. Evidence gathered from this research program, which is discussed next, suggests important linkage between these malformations and subsequent sensory/perceptual defects.

HELP FROM ANIMAL MODELS

Ectopias and other, related, brain malformations have been modeled in rats and mice (Humphreys, Rosen, Sherman, & Galaburda, 1989; Sherman, Rosen, & Galaburda, 1988; Tamagawa, Scheidt, & Friede, 1989). In certain strains of mice, ectopias occur spontaneously (Fig. 13.3A), which has helped us in the search for genes that relate to their production. In rats, the malformations can be experimentally induced (Fig. 13.3B), which can help us with questions regarding causality and for determining exact anatomical and developmental characteristics (Humphreys et al., 1989; Rosen, Jacobs, & Prince, 1998; Rosen, Sherman, Richman, Stone, & Galaburda, 1992).

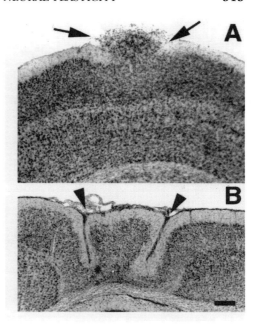

FIG. 13.3. Photomicrographs illustrating two types of malformations seen in the rodent neocortex. **A.** *Spontaneous* neocortical ectopia in the mouse from the BXSB strain. These collections of neurons in layer I (arrows) are similar in appearance to those seen in the dyslexic brain (Figure 1A). **B.** *Induced* malformation in the brain of a rat. In this case, injury to the developing cortex on the first day of life caused a focal malformation that resembles human microgyria. The normally unfolded rat cortex now has two sulci (arrowheads) which form a gyrus between. Bar for panel A = 100 μm. Bar for panel B = 200 μm.

Ectopias arise developmentally before the completion of the period of neuronal migration to the neocortex, which in humans takes place roughly between 16 and 20 weeks of gestation and in mice and rats includes the first 2 days after birth. The time of ectopia formation is long, and therefore the anomalies contain neurons born at different times during histogenesis of the cortex (Sherman et al., 1992). The significance of this finding is that it sets up the possibility for wide-ranging connectivity between neurons in the ectopia and other cortical and subcortical areas. In fact, we have shown in that the ectopias are abnormally connected to the thalamus, ipsilateral cortex, and contralateral cortex (Jenner, Galaburda, & Sherman, 1995; Jenner, Galaburda, & Sherman, 1997; Rosen, 1996; Rosen, Burnstein, & Galaburda, 2000). Thus, by virtue of their connectivity, these anomalies are poised to affect the development of other brain areas with which they are connected, including the thalamus.

We have studied mutant mice with spontaneous ectopias and rats with induced ectopias and related anomalies of cortical development. We have shown that in the mice with ectopias there is an alteration in the pattern of asymmetry in the primary visual cortex (Fig. 13.4). Normally, as the coefficient of asymmetry increases (more asymmetry), the left-plus-right volume decreases. This suggests that asymmetry is a curtailed form of symmetry, rather than one overgrown side (see Galaburda, 1995; Rosen, 1996). In the presence of ectopias, the degree of asymmetry diminishes, but, in addition, the inverse relationship between asymmetry and volume disap-

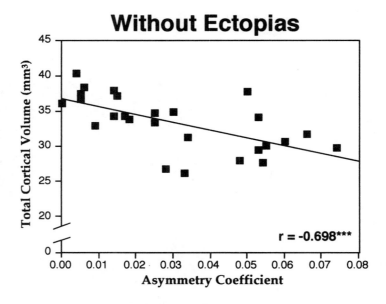

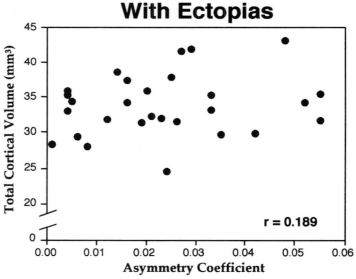

FIG. 13.4. **A**. Scatterplot showing the typical pattern of the relationship between total brain volume and asymmetry. As the total brain volume increases, asymmetry decreases. This relationship has been shown in humans (Galaburda, Corsiglia, Rosen, & Sherman, 1987), rats (Rosen, Sherman, & Galaburda, 1989; Rosen, Sherman, & Galaburda, 1993), and in this particular case, mice (Rosen, Sherman, Mehler, et al., 1989). ***p < 0.001. **B**. Scatterplot showing no correlation between brain volume and asymmetry in mice with ectopias. These mice are the same strain of mice as depicted in Panel **A**. From Rosen, Sherman, Mehler, et al., 1989.

pears (Rosen, Sherman, Mehler, Emsbo, & Galaburda, 1989). This suggests that in dyslexic brains, manifestation of asymmetry may not only be decreased, but is also anomalous.

We used experimental rats to induce migrational anomalies similar to those seen in the dyslexic brains (Fitch, Brown, Tallal, & Rosen, 1997; Herman, Galaburda, Fitch, Carter, & Rosen, 1997). The anomalies were induced bilaterally in the prefrontal, parietal, or occipital cortices. Control animals received sham surgery with no resulting anomalies. We were able to demonstrate that after the induction of cortical malformations, changes in neuronal sizes appeared in thalamic nuclei. Specifically, cortical malformations, produced an excess of small neurons in the MGN of the rat, which mimicked the situation in the human dyslexic. Thus, we demonstrated that the thalamic changes could be a consequence of the cortical anomalies. The complementary finding—to cause shrinkage of thalamic neurons and observe for the presence of ectopias in the cortex—presents experimental difficulties that we have not overcome as yet, although it seems unlikely that such a finding would be made. In a personal communication, Pasko Rakic ( ) informed us that enucleation of an eye in a fetal monkey can lead to migration arrest of some neurons, which then remain in the subcortical white matter. However, no migration beyond the cellular layers of the cortex, that is, layer one, which is the case of the ectopias under present discussion, have been reported.

Animals with cortical anomalies and secondary changes in neuronal size in the MGN performed abnormally on a task that required them to process fast, rapidly changing acoustic stimuli (Fitch, Tallal, Brown, Galaburda, & Rosen, 1994; Fitch et al., 1997; Herman et al., 1997). Specifically, rats were asked to discriminate between two tones, and the interstimulus interval (ISI) between the two tones was progressively shortened. Although both sham and induced ectopic animals were able to perform the discrimination at the long ISIs, only the animals with cortical anomalies had difficulties when the ISIs were shortened (Fig. 13.5). In other words, the animals with the malformations and the thalamic findings (which paralleled the situation in the human dyslexics), behaved similarly on a temporal processing auditory task. Yet, we still could not tell that the temporal processing deficit was caused by the thalamic changes and not by the cortical ectopias. However, an unexpected discovery helped us better to understand the role of the cortical anomaly versus the thalamic changes in the functional deficit.

Sex Differences

The first set of experiments demonstrated that male rats with induced malformations had defects in fast auditory processing as compared with sham-operated rats (Fitch et al., 1994). We subsequently replicated that

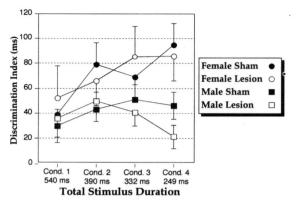

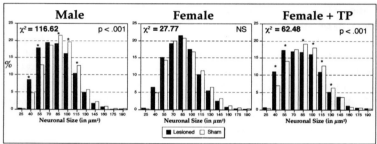

FIG. 13.5. Data illustrating male–female differences in response to in-duced malformations of the cortex. Line graph at the top illustrates the re-sults of an auditory temporal discrimination task over four testing condi-tions. Male (square) and female (circle) lesioned (filled) and sham (open) groups were tested on a two-tone discrimination task, and the numbers un-der each condition are total stimulus time (pre-tone/ISI/post-tone). The higher the discrimination score (Y-axis), the better the performance. These data illustrate (1) The male animals with induced malformations cannot perform the discrimination only at the fastest condition, and (2) that female animals with identical induced malformations have no difficulty perform-ing the task at all stimulus conditions. From (Fitch et al., 1997; Herman, et al., 1997).

The three bar graphs illustrate the frequency distribution of neuronal size in the medial geniculate nucleus of rats with (black bars) or without (white bars) induced malformations of the cortex. In males, there were more small and fewer large neurons among the animals with the induced malformations. In females, there was no difference in neuronal size distri-bution between the two. Perinatal exposure of genetic females to testoster-one propionate (TP) induces a male-like response in the rats with the mal-formations. From (Rosen et al., 1999).

finding, but unexpectedly found that the difference between the sham-operated rats and those with induced malformations did not hold for female rats (Fitch et al., 1997). A closer look at the data showed that, whereas males showed secondary changes in neuronal size in the MGN, females, despite identical induction procedures and the development of identical cortical malformations, did not show the thalamic changes (Herman et al., 1997). This indicated that it was the secondary thalamic changes and not the primary cortical malformations that were likely responsible for the changes in auditory task performance. The cortical changes themselves, however, may be responsible for other behavioral anomalies seen in animals with malformations (Boehm et al.,1996; Boehm et al., 1998; Denenberg, Sherman, Rosen, & Galaburda, 1988; Denenberg, Sherman, Schrott, Rosen, & Galaburda, 1991; Rosen, Waters, Galaburda, & Denenberg, 1995; Schrott, Denenberg, Sherman, Rosen, & Galaburda, 1992; Waters, Sherman, Galaburda, & Denenberg, 1997).

As females did not show the same response in the thalamus to anomaly induction in the cortex, they appeared to be protected from (at least) some of the effects of having a cortical malformation. Sex differences may be the result of hormonal actions; therefore, we exposed pregnant rats to the male hormone testosterone and again examined the male and female offspring after induction of cortical anomalies. Daughters of rats receiving testosterone showed changes in the thalamus (Rosen, Herman, & Galaburda, 1999)—testosterone-exposed rats with induced malformations had more small and fewer large cells in the MGN than their unaltered counterparts (Fig. 13.5). The study of the behavioral effects of this manipulation is ongoing.

SUMMARY AND CONCLUSIONS

The research results presented here indicate that anatomical changes in low-level sensory processors in the brain may be the consequence of earlier anatomical developmental changes taking place in higher-order cortices. Furthermore, based on this anatomical discovery, it is possible that the sensory/perceptual deficits described in dyslexics, both in the visual and auditory modalities, could be the consequence rather than the result of cognitive deficits, which in turn occur first. This scenario would propose that cognitive cortex is defective and therefore does not support development of lower-level centers that would provide it with unusable information. Alternatively, ectopias lead to lower-level problems, which in turn produce defective information for downstream cognitive processors and secondary cognitive changes. The proof of this hypothesis would come from experiments showing that females, which do not have lower level problems, still have cognitive problems.

This proof was made available by our work with Victor Denenberg and colleagues, who showed that both male and female rodents with induced malformations showed deficits in a variety of cognitive tasks, including Lashley Maze, Morris Maze, and discrimination learning (Rosen et al., 1995). The debate is not over, but the anatomical evidence indicates that there is no reason to accept the proposition that problems exist at only one level and in one pathway of information processing and that influences across levels occur in both directions, bottom–up and top–down. A population of subjects that needs to be discovered are dyslexic women with phonological awareness deficits but no temporal processing abnormalities.

REFERENCES

Anderson, K. C., Brown, C. P., & Tallal, P. (1993). Developmental language disorders: Evidence for a basic processing deficit. *Current Opinion in Neurology and Neurosurgery, 6,* 98–106.

Boehm, G. W., Sherman, G. F., Hoplight, B. J., 2nd, Hyde, L. A., Bradway, D. M., Galaburda, A. M., Ahmed, S. A., & Denenberg, V. H. (1998). Learning in year-old female autoimmune BXSB mice. *Physiology and Behavior, 64*(1), 75–82.

Boehm, G. W., Sherman, G. F., Hoplight, B. J., Hyde, L. A., Waters, N. S., Bradway, D. M., Galaburda, A. M., & Denenberg, V. H. (1996). Learning and memory in the autoimmune BXSB mouse: Effects of neocortical ectopias and environmental enrichment. *Brain Research, 726*(1–2), 11–22.

Bradley, L., & Bryant, P. (1981). Visual memory and phonological skills in reading and spelling backwardness. *Psychological Research, 43*(2), 193–199.

Brady, S. A. (1991). The role of working memory in reading disability. In S. A. Brady & D. P. Shankweiler (Eds.), *Phonological Processes in Literacy-A Tribute to Isabelle Y. Liberman* (pp. 125–151). Hillsdale: Lawrence Erlbaum Associates.

Brady, S., Poggie, E., & Rapala, M. M. (1989). Speech repetition abilities in children who differ in reading skill. *Language and Speech, 32*(Pt 2), 109–122.

Bryant, P., Nunes, T., & Bindman, M. (1998). Awareness of language in children who have reading difficulties: historical comparisons in a longitudinal study. *Journal of Child Psychology and Psychiatry, 39*(4), 501–510.

Chase, C., & Jenner, A. (1993). Magnocellular visual deficits affect temporal processing of dyslexics. In P. Tallal, A. M. Galaburda, R. Llinas, & C. von Euler (Eds.), *Temporal Information Processing in the Nervous System, with Special Reference to Dyslexia and Dysphasia* (Vol. 682, pp. 326–329). New York: New York Academy of Sciences.

Crain, S., & Shankweiler, D. (1990). Explaining failures in spoken language comprehension by children with reading disabilities. In D. A. Balota, G. B. Flores d'Arcais, & K. Rayner (Eds.), *Comprehension Processes in Reading* (pp. 539–555). Hillsdale, NJ: Lawrence Erlbaum Associates.

Demb, J. B., Boynton, G. M., & Heeger, D. J. (1998). Functional magnetic resonance imaging of early visual pathways in dyslexia. *The Journal of Neuroscience, 18*(17), 6939–6951.

Denenberg, V. H., Sherman, G. F., Rosen, G. D., & Galaburda, A. M. (1988). Learning and laterality differences in BXSB mice as a function of neocortical anomaly. *Society for Neuroscience Abstracts, 14,* 1260.

Denenberg, V. H., Sherman, G. F., Schrott, L. M., Rosen, G. D., & Galaburda, A. M. (1991). Spatial learning, discrimination learning, paw preference and neocortical ectopias in two autoimmune strains of mice. *Brain Research, 562*(1), 98–104.

Eden, G. F., Vanmeter, J. W., Rumsey, J. M., Maisog, J. M., Woods, R. P., & Zeffiro, T. A. (1996). Abnormal processing of visual motion in dyslexia revealed by functional brain imaging. *Nature, 382*(6586), 66–69.

Eden, G. F., Vanmeter, J. W., Rumsey, J. M., & Zeffiro, T. A. (1996). The visual deficit theory of developmental dyslexia. *Neuroimage, 4*(3 Part 3), S108–S117.

Eden, G. F., & Zeffiro, T. A. (1998). Neural systems affected in developmental dyslexia revealed by functional neuroimaging. *Neuron, 21*(2), 279–282.

Fitch, R. H., Brown, C. P., Tallal, P., & Rosen, G. D. (1997). Effects of sex and MK-801 on auditory-processing deficits associated with developmental microgyric lesions in rats. *Behavioral Neuroscience, 111*(2), 404–412.

Fitch, R. H., Tallal, P., Brown, C., Galaburda, A. M., & Rosen, G. D. (1994). Induced microgyria and auditory temporal processing in rats: A model for language impairment? *Cerebral Cortex, 4*(3), 260–270.

Friede, R. L., & Mikolasek, J. (1978). Postencephalitic porencephaly, hydranencephaly or polymicrogyria. A review. *Acta Neuropathologica (Berlin), 43*(1–2), 161–168.

Galaburda, A. M. (1995). Anatomic basis of cerebral dominance. In R. J. Davidson, & K. Hugdahl (Eds.), *Brain Asymmetry* (pp. 51–73). Cambridge, MA: MIT Press.

Galaburda, A. M., Corsiglia, J., Rosen, G. D., & Sherman, G. F. (1987). Planum temporale asymmetry: Reappraisal since Geschwind and Levitsky. *Neuropsychologia, 25*, 853–868.

Galaburda, A. M., Menard, M. T., & Rosen, G. D. (1994). Evidence for aberrant auditory anatomy in developmental dyslexia. *Proceedings of the National Academy of Sciences (USA), 91*(17), 8010–8013.

Galaburda, A. M., Sherman, G. F., Rosen, G. D., Aboitiz, F., & Geschwind, N. (1985). Developmental dyslexia: Four consecutive cases with cortical anomalies. *Annals of Neurology, 18*, 222–233.

Geiger, G., & Lettvin, J. Y. (1987). Peripheral vision in persons with dyslexia. The *New England Journal of Medicine, 316*, 1238–1243.

Hanson, V. L. (1991). Phonological processing without sound. In S. A. Brady & D. P. Shankweiler (Eds.), *Phonological Processes in Literacy-A Tribute to Isabelle Y. Liberman* (pp. 153–161). Hillsdale, NJ: Lawrence Erlbaum Associates.

Herman, A. E., Galaburda, A. M., Fitch, H. R., Carter, A. R., & Rosen, G. D. (1997). Cerebral microgyria, thalamic cell size and auditory temporal processing in male and female rats. *Cerebral Cortex, 7*, 453–464.

Hinshelwood, J. (1917). *Congenital Word-blindness*. London: Lewis.

Humphreys, P., Rosen, G. D., Sherman, G. F., & Galaburda, A. M. (1989). Freezing lesions of the newborn rat brain: A model for cerebrocortical microdysgenesis. *Society for Neuroscience Abstracts, 15*, 1120.

Hynd, G., Semrud-Clikeman, M., Lorys, A., Novey, E., & Eliopulos, R. (1990). Brain morphology in developmental dyslexia and attention deficit disorder/hyperactivity. *Archives of Neurology, 47*, 919–926.

Jenner, A. R., Galaburda, A. M., & Sherman, G. F. (1995). Connectivity of cortical ectopias in autoimmune mice. *Society for Neuroscience Abstracts, 21*, 1712.

Jenner, A. R., Galaburda, A. M., & Sherman, G. F. (1997). Thalamocortical and corticothalamic connections in New Zealand Black mice. *Society for Neuroscience Abstracts, 27*, 1365.

Jenner, A. R., Rosen, G. D., & Galaburda, A. M. (1999). Neuronal asymmetries in the primary visual cortex of dyslexic and nondyslexic brains. *Annals of Neurology, 46*(2), 189–196.

Liberman, A. M., & Mattingly, I. G. (1989). A specialization for speech perception. *Science, 243*, 489–494.

Liberman, I. Y., & Shankweiler, D. (1985). Phonology and the problems of learning to read and write. *Remedial and Special Education, 6,* 8–17.

Liberman, I. Y., Shankweiler, D., & Liberman, A. M. (1989). The alphabetic principle and learning to read. In D. Shankweiler & I. Y. Liberman (Eds.), *Phonology and Reading Disability* (pp. 1–33). Ann Arbor: Michigan Press.

Livingstone, M., Rosen, G., Drislane, F., & Galaburda, A. (1991). Physiological and anatomical evidence for a magnocellular defect in developmental dyslexia. *Proceedings of the National Academy of Sciences (USA), 88,* 7943–7947.

Lovegrove, W., Garzia, R., & Nicholson, S. (1990). Experimental evidence for a transient system deficit in specific reading disability. *Journal of the American Optometric Association, 2*(2), 137–146.

Lyon, G. R. (1995). Toward a definition of dyslexia. *Annals of Dyslexia, 45,* 3–27.

Merzenich, M. M., Jenkins, W. M., Johnston, P., Schreiner, C., Miller, S. L., & Tallal, P. (1996, January). Temporal processing deficits of language-learning impaired children ameliorated by training. *Science, 271,* 77–80.

Mody, M., Studdert-Kennedy, M., & Brady, S. (1997). Speech perception deficits in poor readers: Auditory processing or phonological coding? *Journal of Experimental Child Psychology, (64),* 199–231.

Morais, J., Luytens, M., & Alegria, J. (1984). Segmentation abilities of dyslexics and normal readers. *Percep Motor Skills, 58,* 221–222.

Morgan, A. E., & Hynd, G. W. (1998). Dyslexia, neurolinguistic ability, and anatomical variation of the planum temporale. *Neuropsychology Review, 8*(2), 79–93.

Morgan, W. P. (1896). A case of congenital word-blindness. *Speech, 23,* 357–377.

Orton, S. T. (1925). "Word-blindness" in school children. *Archives of Neurology and Psychiatry, 14*(Archives of Neurology and Psychiatry), 581–615.

Rosen, G. D. (1996). Cellular, morphometric, ontogenetic and connectional substrates of anatomical asymmetry. *Neuroscience Biobehavioral Review, 20*(4), 607–615.

Rosen, G. D., Burstein, D., & Galaburda, A. M. (2000). Changes in efferent and afferent connectivity in rats with cerebrocortical microgyria. *The Journal of Comparative Neurology, 418*(4), 423–440.

Rosen, G. D., Herman, A. E., & Galaburda, A. M. (1999). Sex differences in the effects of early neocortical injury on neuronal size distribution of the medial geniculate nucleus in the rat are mediated by perinatal gonadal steroid. *Cerebral Cortex, 9*(1), 27–34.

Rosen, G. D., Jacobs, K. M., & Prince, D. A. (1998). Effects of neonatal freeze lesions on expression of parvalbumin in rat neocortex. *Cerebral Cortex, 8*(8), 753–761.

Rosen, G. D., Sherman, G. F., & Galaburda, A. M. (1989). Interhemispheric connections differ between symmetrical and asymmetrical brain regions. *Neuroscience, 33,* 525–533.

Rosen, G. D., Sherman, G. F., & Galaburda, A. M. (1993). Neuronal subtypes and anatomic asymmetry: Changes in neuronal number and cell-packing density. *Neuroscience, 56*(4), 833–839.

Rosen, G. D., Sherman, G. F., Mehler, C., Emsbo, K., & Galaburda, A. M. (1989). The effect of developmental neuropathology on neocortical asymmetry in New Zealand Black mice. *International Journal of Neuroscience, 45,* 247–254.

Rosen, G. D., Sherman, G. F., Richman, J. M., Stone, L. V., & Galaburda, A. M. (1992). Induction of molecular layer ectopias by puncture wounds in newborn rats and mice. *Developmental Brain Research, 67*(2), 285–291.

Rosen, G. D., Waters, N. S., Galaburda, A. M., & Denenberg, V. H. (1995). Behavioral consequences of neonatal injury of the neocortex. *Brain Research, 681*(1–2), 177–189.

Schrott, L. M., Denenberg, V. H., Sherman, G. F., Rosen, G. D., & Galaburda, A. M. (1992). Lashley maze learning deficits in NZB mice. *Physiology & Behavior, 52*(6), 1085–1089.

Shankweiler, D., Crain, S., Katz, L., Fowler, A. E., Liberman, A. M., Brady, S. A., Thornton, R., Lundquist, E., Dreyer, L., Fletcher, J. M., Stuebing, K. K., Shaywitz, S. E., & Shaywitz,

B. A. (1995). Cognitive profiles of reading-disabled children: Comparison of language skills in phonology, morphology, and syntax. *Psychological Science, 6*(3), 149–156.

Sherman, G. F., Rosen, G. D., & Galaburda, A. M. (1988). Neocortical anomalies in autoimmune mice: A model for the developmental neuropathology seen in the dyslexic brain. *Drug Development Research, 15*, 307–314.

Sherman, G. F., Stone, L. V., Walthour, N. R., Boehm, G. W., Denenberg, V. H., Rosen, G. D., & Galaburda, A. M. (1992). Birthdates of neurons in neocortical ectopias of New Zealand Black mice. *Society for Neuroscience Abstracts, 18*, 1446.

Slaghuis, W. L., Lovegrove, W. J., & Davidson, J. A. (1993). Visual and language processing deficits are concurrent in dyslexia. *Cortex, 29*(4), 601–615.

Steinmetz, H., Volkmann, J., Jancke, L., & Freund, H.-J. (1991). Anatomical left-right asymmetry of language-related temporal cortex is different in left- and right-handers. *Annals of Neurology, 29*(3), 315–319.

Studdert-Kennedy, M., & Mody, M. (1995). Auditory temporal perception deficits in the reading-impaired: a critical review of the evidence. *Psychonomic Bulletin and Review, 2*(4), 508–514.

Swanson, H. L. (1993). Working memory in learning disability subgroups. *Journal of Experimental Child Psychology, 56*, 87–114.

Tallal, P. (1977). Auditory perception, phonics and reading disabilities in children. *Journal of the Acoustic Society of America, 62*, S100.

Tallal, P., Miller, S., Fitch, R. H., Stein, J. F., McAnally, K., Richardson, A. J., Fawcett, A. J., Jacobson, C., & Nicholson, R. I. (1995). Dyslexia Update. *The Irish Journal of Psychology, 16*(3), 194–268.

Tallal, P., & Piercy, M. (1973). Defects of non-verbal auditory perception in children with developmental aphasia. *Nature, 241*, 468–469.

Tamagawa, K., Scheidt, P., & Friede, R. L. (1989). Experimental production of leptomeningeal hetertopias from dissociated fetal tissue. *Acta Neuropthologica (Berlin), 78*, 153–158.

Waters, N. S., Sherman, G. F., Galaburda, A. M., & Denenberg, V. H. (1997). Effects of cortical ectopias on spatial delayed-matching-to-sample performance in BXSB mice. *Behavioral Brain Research, 84*(1–2), 23–29.

Williams, M. C., & Lecluyse, K. (1990). Perceptual consequences of a temporal processing deficit in reading disabled children. *Journal of the American Optometric Association, 61*, 111–121.

Williams, M. C., May, J. G., Solman, R., & Zhou, H. (1995). The effects of spatial filtering and contrast reduction on visual search times in good and poor readers. *Vision Research, 35*(2), 285–291.

ATTENTION DEFICIT DISORDER

Disruption of Inhibitory Control in Developmental Disorders: A Mechanistic Model of Implicated Frontostriatal Circuitry

B. J. Casey
Sackler Institute for Developmental Psychobiology
Weill Medical College for Cornell University

A core deficit observed in a number of disorders of childhood is difficulty suppressing inappropriate thoughts and behaviors. For example, children with Attention Deficit/ Hyperactivity Disorder (ADHD) have problems focusing attention and are often characterized as distractible and impulsive (Barkley, 1994; Casey et al., 1997; Trommer et al., 1991). Other childhood disorders have a similar problem suppressing inappropriate behaviors, but the nature of the deficit appears more specific to a particular behavior. For example, children with Tourette Syndrome have difficulty suppressing often quite complex movements in addition to vocalizations that are sometimes emotionally provocative in content (Leckman et al., 1987). Obsessive Compulsive Disorder (OCD) in children and adults alike is characterized by an inability to stop intrusive thoughts and ritualistic behaviors that appear to be specific in content (Insel, 1988). Stereotypes and repeated self-injurious behaviors are also examples of inhibitory problems observed in a wide range of children including those with autism and mental retardation. Even in Childhood Onset Schizophrenia, the child appears unable to stop attending to irrelevant thoughts and information (Asarnow et al., 1995). Clearly, the prevalence of these inhibitory problems in children with developmental disabilities highlights the need for a clearer understanding of these behaviors and their biological bases.

At least two brain regions have been implicated consistently in all of these disorders: the frontal lobes and the basal ganglia. Abnormalities in size, asymmetry, and/or glucose metabolism have been observed. For example, imaging studies at the National Institutes of Mental Health (NIMH) have revealed decreased metabolism in the frontal lobes of patients with ADHD using positron emission tomography (PET) (Zametkin et al., 1990) and abnormalities in the size and symmetry of the prefrontal cortex and basal ganglia using magnetic resonance imaging (MRI) (Castellanos et al., 1994, 1996). Abnormalities in the basal ganglia, specifically the caudate nucleus, in children with Tourette Syndrome have been revealed in PET studies by Wolf and colleagues (1996). Studies of OCD have revealed hypermetabolic activity in the basal ganglia and prefrontal cortex (Baxter et al., 1988; Swedo et al., 1989) and MRI-based decreases in the size of the orbitofrontal cortex and caudate nucleus (Rosenberg et al., 1997). Both PET and MRI studies of adult and adolescent patients with schizophrenia have revealed decreased dorsolateral prefrontal activity (Berman et al., 1988) and decreased volume of the basal ganglia (Frazier et al., 1996). So, common brain regions appear to be involved in a range of disorders that are characterized by inhibitory deficits. This may not be surprising given that the most commonly observed trait of frontal-lobe patients is difficulty regulating behavior (see Kolb & Whishaw, 1990 for a review).

Inhibitory Control: Is There a Single Underlying Mechanism?

The common problem of inhibitory deficits and implicated circuitry across this range of disorders suggests a single underlying biological mechanism. Yet, the observed clinical symptoms and type of abnormality within common brain regions (i.e., hypermetabolic vs. hypometabolic activity) across these disorders are distinct. In an effort to consolidate a model and framework for understanding the underlying mechanisms of inhibitory processes, we turned to the neurobiology of the frontal cortex and basal ganglia. At least five parallel circuits involving the frontal cortex and basal ganglia have been identified by Alexander and colleagues (1991). These basal ganglia thalamocortical circuits involve the same brain regions, but generally speaking, differ in the projection zones within three regions: the basal ganglia, thalamus, and cortex. For example, within the frontal cortex, the projection zones of the motor circuit are in primary motor, supplementary motor, and premotor cortex; the projection zones of the oculomotor circuit are in the supplemental and frontal eye fields; the projection zones for the two prefrontal circuits are in the dorsolateral and lateral orbital frontal cortex; and the projection zones for the limbic circuit are in the medial prefrontal cortex (i.e., anterior cingulate and medial orbitofrontal cortex).

Each of these basal ganglia thalamocortical circuits controls a different set of cortically mediated behaviors that range from skeletal and eye movements to cognitive and emotional actions as illustrated in Fig. 14.1.

Parallel Basal Ganglia Thalamocortical Circuits

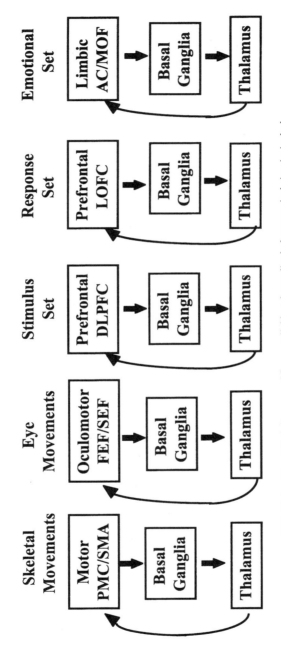

FIG. 14.1. Simplified diagram of five parallel basal ganglia thalamocortical circuits including (1) the motor circuit with primary projection zones in the premotor cortex (PMC), supplementary motor areas (SMA), and primary motor cortex; (2) the oculomotor circuit with primary projection zones in the supplementary eye fields (SEF) and frontal eye fields (FEF); (3) dorsolateral prefrontal (DLPFC) circuit; (4) lateral orbital frontal (LOFC) circuit; and (5) the limbic circuit with primary projection zones in the anterior cingulate (AC) and medial orbitofrontal cortex (MOF).

For example, the sensorimotor circuits (motor and oculomotor) subserve voluntary skeletal and eye movement control. The circuits involving association cortex (dorsolateral and lateral orbital frontal cortex) are involved in representing and maintaining stimulus information (e.g., object, spatial, verbal, etc.) and response information (e.g., different behavioral sets). Finally, the limbic circuit is thought to involve the representation of emotional information such as that important in determining whether to approach or avoid an event.

The projections within the basal ganglia are depicted in Fig. 14.2. The basal ganglia are defined loosely here. For the purposes of simplicity, the substantia nigra and subthalamic nuclei are included in the diagram, although these regions are not always defined as part of the basal ganglia. This cartoon shows how thalamocortical circuits are modulated by the basal ganglia via a direct (excitatory) and an indirect (inhibitory) pathway. The frontal cortex projects to different areas of the striatum (i.e., putamen or

Basal ganglia-Thalamocortical Circuit

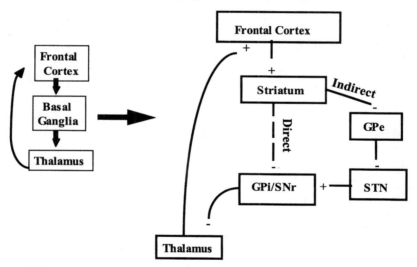

FIG. 14.2.   Basic circuitry of the basal ganglia thalamocortical circuit. The frontal cortex projects to different areas of the striatum (i.e., putamen or caudate nuclei) and then projects to either the direct or indirect pathway. The direct pathway (- - -) involves an inhibitory projection to the internal capsule of the globus pallidus (GPi) and substantia nigra (SNr) resulting in the dampening of an inhibitory projection to the thalamus, which results in disinhibition of the thalamus. The indirect pathway (—) consists of an inhibitory projection to the external capsule of the globus pallidus (GPe), which dampens the inhibitory projection to the subthalamic nuclei (STN) resulting in excitation of the internal capsule of the globus pallidus and substantia nigra. This, in turn, leads to an inhibition of the thalamus.

caudate nuclei) and then projects to the direct or indirect pathway. The direct pathway involves an inhibitory projection to the internal capsule of the globus pallidus (GPi) and substantia nigra (SNr) resulting in the dampening of an inhibitory projection to the thalamus, which results in disinhibition of the thalamus. This direct pathway is facilitative of cortically mediated behavior and results in disinhibition of the thalamus (i.e., an overall increase in thalamic activity). The indirect pathway consists of an inhibitory projection to the external capsule of the globus pallidus (GPe), which dampens the inhibitory projection to the subthalamic nuclei (STN) resulting in excitation of the internal capsule of the globus pallidus and substantia nigra. This in turn leads to an inhibition of the thalamus. This pathway is thought to inhibit cortically mediated or conflicting behaviors.

How does this circuitry relate to inhibitory behavior and, perhaps more importantly, how does this circuitry contribute to the symptoms and behaviors observed in developmental disorders? Assuming that the direct pathway is involved in facilitating cortically mediated behaviors, then its disruption may result in constantly interrupted behaviors such as those observed in ADHD or constantly interrupted thoughts such as those observed in schizophrenia. In contrast, if the indirect pathway is involved in inhibiting cortically mediated behavior, then its disruption may result in irrepressible repetitive behaviors and thoughts similar to those observed in OCD and Tourette syndrome. Alternatively, imbalances (e.g., hypermetabolic activity) in regions of the direct or indirect pathways could lead to similar aberrant behavior with overactivity of the direct pathway leading to irrepressible repetitive behaviors and overactivity of the indirect pathway leading to constantly interrupted behaviors.

Accordingly, we have recently developed a model of inhibitory control whereby the basal ganglia are involved in inhibition of behaviors (Mink, 1996) while the frontal cortex is involved in representing and maintaining information and conditions to which we respond or act (Cohen & Servan-Schrieber, 1992). We hypothesize that the basal ganglia, which consist primarily of inhibitory projections, are involved in the inhibition of inappropriate behaviors and disruption in the development of this brain region results in deficits in inhibitory control. Further, we hypothesize that the frontal cortex, which consists primarily of excitatory projections, is involved in maintenance of information for action; disruption of this brain region results in deficits in the facilitation of these actions. As such, hypofrontality as observed in schizophrenia (e.g., Berman et al., 1988) and ADHD (e.g., Zametkin et al., 1990) would result in constantly interrupted thoughts or actions. In contrast, hyperfrontality as observed in OCD (Swedo et al., 1989) would result in irrepressible behaviors.

The varied repertoire of behaviors supported by basal ganglia thalamocortical circuits is assumed to reflect different types of information being represented and maintained in the frontal lobes. For instance, the dorsolateral prefrontal circuit represents stimulus information (object, spatial, verbal, etc.), while the lateral orbitofrontal circuit is involved in representation and maintenance of a response or behavioral set and the limbic circuit is thought to represent emotionally relevant information for guiding approach and avoidance behaviors. Thus, distinct symptomatology across disorders is reflected in which one of these circuits is disrupted as different types of information are represented in each. Accordingly, we propose that the inhibitory mechanism is the same across circuits (i.e., at the level of the indirect pathway of the basal ganglia), but the type of information represented by each circuit within the frontal cortex is different. More generally, we propose that the basal ganglia thalamocortical circuits underlie inhibitory control and that inhibitory deficits observed in a range of developmental disorders reflect a disruption in the development of these circuits.

How does this circuitry relate to the normal development of inhibitory control? Here, emphasis is placed on the role of the prefrontal cortex in supporting different types of information (e.g., verbal, spatial, motor, emotional) against interference over time or from competing sources (Cohen & Servan-Schreiber, 1992; Goldman-Rakic, 1987). A number of classic developmental studies have demonstrated that these memory-related processes develop throughout childhood and adolescence (Case, 1972; Flavell et al., 1966; Keating & Bobbitt, 1978; Pascual-Leone, 1970). Further, there is converging evidence of prolonged development and reorganization of prefrontal cortex throughout childhood and adolescence (Bourgeois et al., 1994; Chugani et al., 1987; Huttenlocher, 1979; Rakic et al., 1994). The most important difference between the development of the prefrontal cortex relative to other cortical regions (e.g., visual cortex) is in the gradual decrease in synapses into young adulthood (Bourgeois et al., 1994). This decrease in synaptic density does not start until roughly puberty and coincides with the continued development of cognitive capacities. Accordingly, increasing cognitive capacity coincides with a gradual loss rather than formation of new synapses. These cognitive and biological processes may represent the behavioral and ultimate physiological suppression of competing, irrelevant behaviors as appropriate behaviors are reinforced and enhanced (i.e., inhibitory control).

There are at least four lines of converging evidence for the model of inhibitory mechanisms of attention presented. This work comes largely from our laboratory and consists of a collection of clinical or neuropsychological studies, magnetic resonance imaging (MRI)-based morphometry studies, functional MRI studies, and lesion studies. The remainder of the chapter focuses on this work.

Inhibitory Control: Behavioral Disruption in Clinical Populations

Alexander and colleagues (1991) first proposed that various clinical disorders reflected disruptions in the different basal ganglia thalamocortical circuits. These speculations were based largely on existing neuroimaging, clinical, and animal studies implicating specific regions of the frontal cortex in certain disorders and behaviors observed in these disorders. Specifically, they suggested that schizophrenia involved a disruption in the dorsolateral prefrontal cortex circuit; obsessive compulsive disorder involved a disruption in the lateral orbitofrontal cortex circuit; and, Tourette Syndrome involved the limbic circuit. Given the presumed functions associated with each circuit, it is possible to empirically test whether distinct behaviors thought to be subserved by these three circuits are disrupted in schizophrenia, OCD, and Tourette Syndrome with the development of carefully designed behavioral probes.

The design of any cognitive paradigm requires an operational definition of the cognitive process(es) in question. The use of the term *inhibition* throughout this chapter thus far reflects both behavioral phenomenona and also underlying mechanisms. This duality in use of the term may be confusing. For the purposes of clarity, the use of the term inhibition to describe behavior will be replaced with the term *suppression* (see Estes, 1975) except in cases of descriptive phenomena (e.g., inhibition of return) and psychological constructs (e.g., inhibitory control).

Accordingly, the cognitive processes of interest are selective processes that enable one to attend and respond to relevant events in the presence of interfering salient, competing, and compelling, yet otherwise irrelevant events. For instance, how is it that someone is able to concentrate on grocery items needed during a quick trip to the supermarket when there is an abundance of distracting grocery items on the shelves? How do we ignore and select from competing sources of information (stimulus selection), or suppress competing choices (response selection), or for that matter, suppress a behavior or response altogether (response execution)?

One approach to answering these questions is to examine suppression of information at different stages of attentional processing: stimulus selection, response selection, and response execution. These stages of attentional processing map well onto the proposed functions of the association and limbic basal ganglia thalamocortical circuits. For example, the dorsolateral prefrontal circuit is thought to represent stimulus information (object, spatial, verbal, etc.), and the lateral orbitofrontal circuit is involved in representation and maintenance of a response set, and the limbic circuit is thought to represent emotionally relevant information for guiding approach and avoidance behaviors. Accordingly, a battery of tasks was designed (Casey et al., 1993, Casey, Castellanos, et al., 1997; Casey, Vauss, & Swedo, 1994) that require suppression of a stimulus set; a response set; and, an avoidance set. These tasks are described in detail next.

The *stimulus selection task* requires the subject to select which of three objects presented on a computer screen is unique based on the stimulus attributes of color and shape. The unique attribute changes from trial-to-trial such that if the relevant attribute was color on the previous trial, shape is the relevant attribute on the subsequent trial. Performance on these trials is compared to performance during a control condition in which trials are blocked by stimulus attribute. This task requires suppression of a previously attended or salient stimulus attribute. The *response selection task* consists of selecting responses to specific stimuli that are based on compatible (i.e., well learned) and incompatible mappings. In the compatible mapping condition, a subject is presented with one of the digits 1, 2, 3, or 4 centrally on a computer screen. The subject is instructed to press the corresponding 1st, 2nd, 3rd, or 4th button on a response box. In the incompatible mapping condition, the subject has to reverse these responses so that the numbers 1, 2, 3, and 4 correspond to the 4th, 3rd, 2nd, and 1st buttons, respectively. When a 2 is presented, the 3rd button is pressed; when a 1 is presented, the 4th button is pressed, and so on. This task requires suppression of a competing overlearned response set (i.e., incompatible mappings). The *response execution task* is a version of the classic go–no-go task. The subject is instructed to respond to any letter but "X." Trials are programmed so that 75% of the trials are targets to which the subject should respond. The control condition consists of blocks of trials with only 25% targets. This task requires suppression of a compelling response altogether. Thus, the stimulus selection, response selection, and response execution tasks each require suppression of information at different proposed stages of attentional processing.

Behavioral data were collected on over 100 children with developmental disorders on these cognitive tasks including patients with autism (Casey et al., 1993), Tourette Syndrome, ADHD (Casey, Castellanos, et al., 1997), childhood onset schizophrenia, and Sydenham chorea (Casey, Vauss, Chused, & Swedo, 1994; Casey, Vauss, & Swedo, 1994; Swedo, Leonard, Schapiro, Casey, et al., 1993). Sydenham chorea is an especially interesting disorder for testing basal ganglia involvement in inhibitory control because of the known pathology of this region in this disorder. Sydenham chorea is a variant of rheumatic fever where antibodies cross-react with the host tissue. In these cases, primarily the brain rather than the heart, is targeted and the region of the brain that is affected is the basal ganglia. These children present with flailing of the limbs and with clinical symptoms. Approximately 75% of these children present with obsessive-

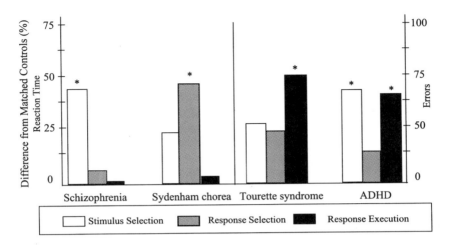

FIG. 14.3. Percent difference in mean reaction times and error rates on the stimulus selection, response selection, and response execution tasks for children with schizophrenia, Sydenham chorea, Tourette syndrome, and ADHD relative to matched controls. Asterisks indicate significant differences between patients and controls in the raw data.

compulsive symptomatology (Swedo et al., 1993) and thus this disorder has been used as a medical model of OCD. All of these disorders (Sydenham chorea, ADHD, OCD, Tourette syndrome, and childhood onset schizophrenia) were assumed to have a deficit in suppression of information at different points in cognitive processing.

Figure 14.3 shows data from four different developmental disorders[1]. Data from seven unmedicated adolescent schizophrenic patients, ages 11 to 16 years, showed deficits in performance of the stimulus selection task (mean reaction times of 1,048 ms vs. 697 ms, $p < .02$), but not the response selection or execution tasks relative to matched controls. Data from 10 unmedicated Sydenham chorea patients, ages 7 to 15 years, showed deficits in performance on the response selection task (mean reac-

---

[1]Our published data from our patients with autism (see Casey et al., 1993) are not summarized with the other four disorders because these individuals and their matched controls were tested as adults. In general, we observed significant deficits for these patients during the stimulus selection task. Specifically, the autistic patients had a difficult time shifting attention from one stimulus attribute to another resulting in significantly slower reaction times ($p < .0001$). It is interesting to note that there was overlap in clinical symptomatology for these two samples (i.e., autistic patients and patients with childhood onset schizophrenia). Autistic patients also had tremendous difficulty on a divided attention task that required the subjects to divide their attention between visual and auditory targets simultaneously (67% vs. 97% mean accuracy, $p < .001$).

tion times of 977 ms vs. 848 ms, $p < .05$), but not the stimulus selection or response execution tasks (Casey, Vauss, & Swedo, 1994). Unmedicated children with Tourette syndrome, ages 7 to 13 years, showed deficits in performance on the response execution task (100% false alarm rate relative to a 20% false alarm rate in normal controls, $p < .01$), but not the response selection or stimulus selection tasks. Finally, data from 20 unmedicated ADHD patients, ages 6 to 16 years, showed poorer performance on the stimulus selection (8% vs. 3%, $p < .05$) and response execution tasks (11% versus 3%, $p < .01$), but not the response selection task relative to matched controls.

In sum, the data show a four-way dissociation in the pattern of performance on these tasks, with the schizophrenic patients showing deficits on the stimulus selection task; the Sydenham chorea patients showing deficits on the response selection task; the children with Tourette syndrome showing deficits on the response execution task; and, the children with ADHD showing deficits on both the stimulus selection and response execution tasks. The pattern of performance for the children with ADHD fits with the distractibility and impulsivity observed in this disorder. These data fit with anatomical and clinical data implicating the involvement of different parallel basal ganglia thalamocortical circuits with distinct clinical disorders (Alexander et al., 1991). Thus, these three marker tasks may be used to assess the integrity of basal ganglia thalmocortical circuits, specifically the two association circuits (dorsolateral and lateral orbital) and the limbic circuit in different developmental populations.

Inhibitory Control: Underlying Circuitry

MRI-based Anatomical Studies. The previously reported behavioral findings suggest that our battery of cognitive tasks map onto distinct frontostriatal circuitry. In an effort to examine this relation more directly, anatomical correlates of inhibitory control as measured by our three cognitive tasks were examined using magnetic resonance imaging (MRI). Based on a sample of 25 children with ADHD and matched controls, task performance correlated only with anatomical measures observed to be abnormal in ADHD (Castellanos et al., 1996). Specifically, size and asymmetry of the right prefrontal cortex, caudate nuclei, and globus pallidum correlated with task perfomance, but not other areas, for example, putamen (Casey, Castellanos, et al., 1997). For significant correlations, tests for parallelism in slopes between groups were performed. The groups differed in slope for the stimulus selection and response execution tasks, but not the response selection task. The behavioral and anatomical measures typically correlated for the normal volunteers. In contrast, behavioral data from the children with ADHD typically did not correlate with anatomical measures, or were in the opposite direction. These results imply that deficits in inhibitory control observed in ADHD may be due to abnormalities of the basal ganglia and related frontostriatal circuitry. These correlational data indicate the ability of our cognitive tasks to access the integrity of frontostriatal circuitry, but more importantly support the hypothesis of the role of these regions in inhibitory control.

Functional MRI Studies. Until recently, the use of functional neuro-imaging techniques in developmental studies has been limited, primarily due to their reliance on harmful radiation and the vulnerability of developmental populations to such exposure. Within the past 6 years, a non-invasive neuroimaging technique, functional magnetic resonance imaging (fMRI), has been developed to examine brain state changes based on changes in blood oxygenation. Perhaps one of the most important aspects of fMRI is its utility in studying human brain development in vivo. This methodology is central to work on the functional circuitry underlying inhibitory control.

We recently completed a number of studies with healthy children (Casey et al., 1995; Casey, Trainor, Orendi, et al., 1997). In one such study, prefrontal activity was examined during the performance of the previously described response execution task (i.e., the go–no-go task) that was modified for the scanner environment (Casey, Trainor, Orendi, et al., 1997). In this version of the task, subjects were instructed to respond to any letter but "X." Letters were presented at a rate of one every 1.5 s and 75% of the trials were target trials to build up a compelling tendency to respond. We hypothesized that performance of the response execution task would activate brain regions of the limbic basal ganglia thalamocortical circuit involved in avoiding or suppressing a response. This prediction was based on our previous behavioral findings summarized in Fig. 14.3 together with evidence from other animal, clinical, and neuroimaging studies. Based on the results from 18 subjects, between the ages of 7 and 24 years (9 adults and 9 children), we found that only activity in the orbito-frontal cortex and right anterior cingulate cortex correlated with behavioral performance ($p < .009$ and $p < .05$, respectively, see Fig. 14.4). In addition, we observed significantly more errors and more overall prefrontal activity ($p < .001$) for children (490 mm^3) relative to adults (182 mm^3). This difference in overall prefrontal activity appeared to be specific to the dorsolateral prefrontal cortex with children activating this region significantly more than adults ($p < .001$). This difference in dorsal prefrontal activity may be due to differences in strategies between groups to perform the task. For example, the adults may have realized that simply remembering conditions specific to avoiding a response (i.e., to not respond to the X) was sufficient to complete the task. According to our working model of inhibitory control, maintaining information about when to approach or avoid an event involves the limbic basal ganglia thalamocortical circuit and thus medial orbitofrontal cortex would be the expected site of activity. Children, on the other hand, may have tried to remember when to approach and avoid a response as well as remember the entire stimulus target set (A–Z, excluding X). According to our model, maintenance of the stimulus set would involve the dorsolateral prefrontal circuit, and maintenance of when to avoid an event would involve the limbic circuit; thus, both dorsolateral and medial orbitofrontal activity would be expected. Alternatively, the activation of both orbital and dorsolateral regions in children may suggest an increased selectivity in representation of the prefrontal cortex with maturation. These interpretations are not mutually exclusive. Nonetheless, consistent with our hypothesis, activity of the limbic circuit (anterior cingulate and orbitofrontal cortex) was observed and directly related to our hypothesis of frontal lobe involvement in inhibitory control.

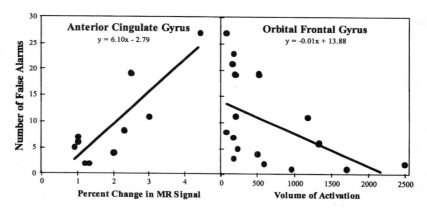

FIG. 14.4.   Correlations between number of false alarms on a go–no-go task and fMRI-based brain activity in the anterior cingulate cortex and orbitofrontal cortex.

In the previous study, our methodology was limited in the number of images we could acquire and so activity in the basal ganglia was not examined. We are currently in the process of completing three developmental studies using our cognitive tasks and whole brain acquisitions that allow full coverage of the basal ganglia thalamocortical circuits.

Inhibitory Control: Disruption in the Development of Inhibitory Circuitry

How and when does disruption in the development of basal ganglia thalamocortical circuitry predispose a child to developmental disabilities? What are the effects of disruption in the development of this circuitry? One period, in particular, is that occurring near birth. For example, a common occurrence of perinatal asphyxia during premature births is a hemorrhage in the region of the basal ganglia. The most metabolically active brain regions are most vulnerable to hemorrhage. In the preterm infant, this region is the germinal matrix, an area within the ventricular wall and adjacent brain regions such as the caudate nuclei (i.e., basal ganglia). This population of premature children is an important one due to the continuing increase in their survival rate with advances in modern medical technology and because they are particularly susceptible to intracranial hemorrhage (IVH), especially in the region of the basal ganglia (Volpe, 1995). Poor long-term outcomes in the most severe cases (e.g., IVH grades III and IV) have been well documented, revealing a greater frequency of mental retardation and neurobehavioral deficits in these children. Milder cases of IVH have not been carefully examined and neither severe nor mild cases have been quantitatively examined with regard to the long-term effects of these insults on later brain development. To date, neuroimaging studies of infants with medical histories of perinatal complications have been limited primarily to the first year of life or to clinical classifications of hemorrhage and ischemic brain damage (Keeney et al., 1991a, 1991b; Lipper et al., 1988; Scher et al., 1989). Although gross abnormalities are easily detected on clinical scans as has been the finding in previous studies, important but subtle abnormalities in size and symmetry of relevant structures may not be detected. We know from recent neuroimaging work (e.g., Casey, Castellanos, et al., 1997; Castellanos et al., 1996) that such measures are critical in highlighting the underlying brain regions involved in serious behavioral problems in the developing child. By identifying the kinds of neuroanatomic lesions associated with poor outcome, we are in a better position to develop more focused rehabilitative interventions.

We recently initiated a study to methodically investigate the long-term effects of mild-to-severe intraventricular hemorrhage (IVH). A multidisciplinary team that assesses their cognitive and neurological development and the development of their brains as shown by quantitative structural MRI and fMRI follows the children enrolled in this program. The objective is to characterize the structural and functional effects of neonatal intraventricular hemorrhage in a systematic manner in comparison to unaffected, matched controls.

Behavioral Studies. Preliminary behavioral results suggest that these children as infants fail to demonstrate the presence of inhibitory attentional processes present in age matched full-term and preterm controls. Specifically, in collaboration with Mark Johnson (Casey, Dobson, et al., 1994), we tested infants on a visual spatial paradigm (Posner, 1980) to examine a phenomenon referred to as inhibition of return. Inhibition of return refers

340   CASEY

to the tendency for individuals to avoid attending to a location or object just
previously attended. It is assumed that this behavior is adaptive and has an
evolutionary basis in that it is more adaptive to scan one's environment for
food or predators than to continue to fixate and focus repeatedly on the
same location or object. This phenomenon is observed behaviorally in
healthy full-term infants by 4 to 6 months of age (Johnson et al., 1991). Presumably, this phenomenon comes online as projections from the frontal
cortex to the basal ganglia develop. The specific basal ganglia thalamocortical circuit assumed to be involved is the oculomotor circuit.

The experimental design for testing this phenomenon is to present two
visual stimuli, a cue and a target, in the same spatial locations with either a
short or long time delay between them (refer to Fig. 14.5). The cue is presented very briefly with little time for the subject to actually make a
saccade in its direction. When there are short delays between the cue and
target, performance is typically facilitated with faster saccades toward the
cued location relative to a noncued location. At longer delays of 600 ms or
more, performance is typically suppressed, with slower saccades toward
the cued location relative to a noncued trial. This slower response latency
to the cued location is what is meant by inhibition of return.

Preliminary data are depicted in Fig. 14.6. The healthy preterm infants
made faster saccades toward the cued location for the short delay and

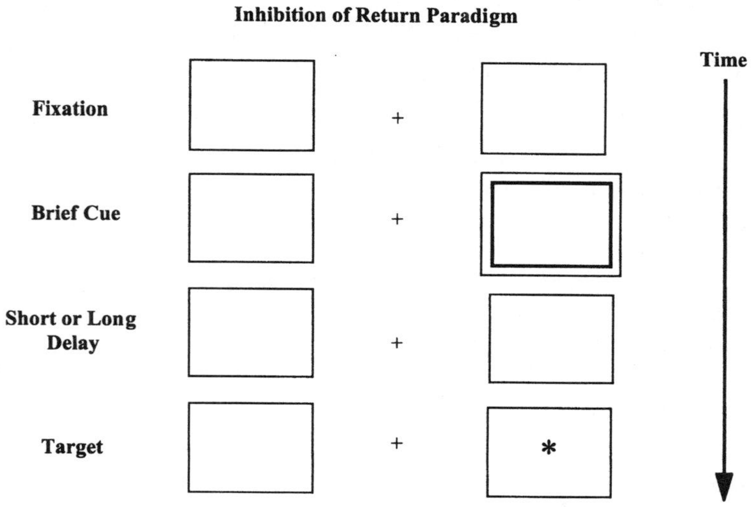

FIG. 14.5.   Experimental design for spatial cueing task used to examine
inhibition of return. Once the subject fixates centrally, a brief cue is presented for 100 msec followed by a short or long delay of 100 to 1200 ms before presentation of the target stimulus. Reaction time and accuracy of saccades to the target are recorded.

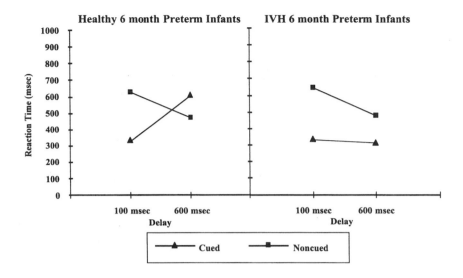

FIG. 14.6. Mean reaction times for healthy preterm infants and preterm infants with IVH during performance of the spatial cueing paradigm. Reaction times are plotted as a function of cue and delay for each group.

slower saccades toward the cued location for the long delay as predicted and as observed in healthy full-term, 6-month-old infants (Johnson et al., 1991). Due to the small sample size, we failed to observe a significant three-way interaction of group by delay by cue, however, our data suggest that the preterm infants with IVH failed to show inhibition of return. They made faster saccades to the cued location regardless of how long the delay between the cue and target were. Because hemorrhages occur most often in the basal ganglia in these infants, we assume such a deficit would be due to disruption of the underlying frontostriatal circuit at the level of the basal ganglia.

At least 17 children between the ages of 6 and 9 years with histories of IVH of Grade II or higher have been tested on our three cognitive tasks. We compared their data to a subset of the previously reported data from our children with ADHD and their matched controls within the same age range. The average age for each group was 7.2 years ($n = 17$), 7.8 years ($n = 13$), and 7.4 years ($n = 13$) for the IVH, ADHD, and control groups, respectively. The results on the three cognitive tasks are summarized in Fig. 14.7. The data are presented as percent differences in errors from the healthy control subjects. The patterns of errors made by the children with IVH are very similar to those observed for our sample of children with ADHD and even more similar to those reported previously for children with Tourette Syndrome (refer back to Fig. 14.3). Specifically, the children

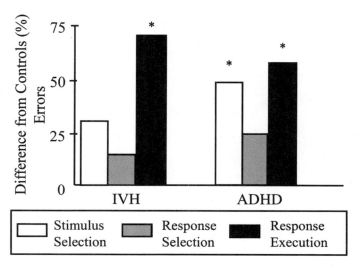

FIG. 14.7.   Percent difference in mean error rates on the stimulus selection, response selection, and response execution tasks for children with IVH and ADHD relative to matched controls. Asterisks indicate significant differences between patients and controls.

with IVH performed worst on the stimulus selection and response execution tasks, but only performance on the response execution task was significantly different from performance of healthy volunteers.

   Clinical Studies.   It is perhaps not surprising that our sample of children with IVH appears similar to children with ADHD and tic disorders in their behavioral performance given the results of structured clinical interviews. Roughly 20% have a psychiatric diagnosis of ADHD (4 times that of the general population) and there appears to be an increased risk for tic disorder and anxiety disorder. Males were more likely than females to have a disorder. These data are consistent with findings recently published by Whitaker and associates (1997) of children with neonatal insults. According to that study, 22% of children with neonatal insults had at least one psychiatric disorder, the most common being ADHD. An increased risk for tic disorders and anxiety disorders was also observed. No effort was made to acquire MRIs on the children in the Whitaker study in order to quantify the extent and location of the brain insult; rather the study relied upon clinical classification of the insult from ultrasound data.

   MRI-Based Anatomical Studies.   As stated previously, neuroimaging studies of infants with medical histories of perinatal complications have been limited primarily to the first year of life or to clinical classifications of

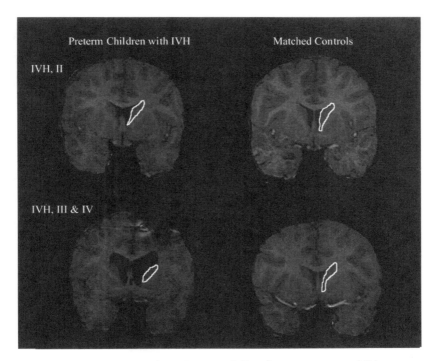

FIG. 14.8.   Representative MRI coronal slices from two preterm children
with IVH of Grade II (upper left) and IVH of Grades III and IV (lower left),
and their respective matched controls (upper and lower right). The caudate
nucleus is outlined in white for each subject. Note how the ventricles are distended at the expense of the caudate nucleus for the children with IVH.

hemorrhage and ischemic brain damage. We have quantified affected
brain regions in children with IVH using volumetric, MRI-based measures
of the basal ganglia (Giedd et al., 1996). The preliminary MRI data have
revealed between 6% to 13% decreases in the size of the caudate nuclei in
children with IVH of Grade II and as much as 20% to 30% in children with
IVH of Grades III and IV when compared to matched controls. Two representative children and their matched controls are depicted in Fig. 14.8.
Note how the ventricles are clearly distended and the caudate nuclei
(traced in white) are smaller in the case of the child with IVH of Grades III
and IV relative to an age matched control. This is less evident in the case
of the child with IVH of Grade II. Only with careful quantitative measures
can we detect subtle but significant differences in the caudate nucleus.

Functional MRI Studies.   We have acquired functional brain imaging
data on 17 of our children with IVH during performance of the response
execution task (go–no-go task). The mean error rate during performance

of this task was 39% compared to 27% for our original sample of healthy children (Casey, Trainor, Orendi, et al., 1997). In our previous study, we observed that activity in the orbitofrontal cortex and right anterior cingulate cortex correlated with the number of false alarms ($r = -.41$, $p <$ .02 and $r = .81$, $p < .0005$, respectively). The greater the volume of activity in mm^3 (i.e., number of significant voxels × voxel size) in the orbital frontal cortex, the fewer the number of false alarms. In our current study, error rate did not correlate with activity in the orbitofrontal gyrus ($r = -.28$) even though more than 70% of subjects had activity in this region (refer to Fig. 14.9). In our previous study, images were acquired covering the prefrontal cortex, but the current study used whole brain echo planar imaging and so we were able to examine patterns of activity in the striatum (caudate and putamen) for all subjects, particularly the caudate nucleus. Only 35% of our children with IVH had striatal activity. As expected, those children with the higher grades of IVH (II–IV) had little to no activity in this region. These data suggest that even though activity is observed reliably in orbitofrontal cortex, disruption at the level of the basal ganglia is sufficient to disrupt performance on inhibitory tasks. Therefore, these data further support our hypothesis of the role of the basal ganglia in inhibitory control.

Overall, our preliminary behavioral, clinical, and neuroimaging data from our children with IVH are consistent with our hypothesis of disruption in inhibitory control at the level of the basal ganglia. First, these children perform poorly on tasks that require them to suppress a salient

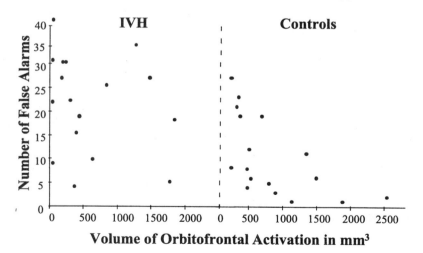

FIG. 14.9. Volume of fMRI-based orbitofrontal activity plotted as a function of number of false alarms for children with IVH and controls during performance of a go–no-go task.

stimulus or compelling response (e.g., sensory selection and response execution tasks). Second, these children are at risk of developing disorders with known inhibitory deficits (e.g., ADHD and Tic Disorders). Third, MRI-based morphometry measures show decreased volume of the basal ganglia, specifically the caudate nucleus, in children with IVH compared to age-matched controls. Fourth, fMRI results showed little to no activity in the caudate nucleus in children with IVH of Grade II or higher and while activity was reliably observed in prefrontal cortex (e.g., orbitofrontal cortex), it was not correlated with behavioral performance. In sum, disruption of the basal ganglia thalamocortical circuits at the level of the basal ganglia appears sufficient to disrupt inhibitory control.

CONCLUSIONS

This chapter presents a mechanistic model of inhibitory control whereby the inhibitory mechanism is the same across circuits (i.e., at the level of the basal ganglia), but the type of information represented by each circuit (i.e., maintained in the frontal cortex) is different. Accordingly, the basal ganglia are involved in suppression of actions while the frontal cortex is involved in representing and maintaining information and conditions to which we respond or act. Developmentally, we propose that the ability to support information in prefrontal cortex against information from competing sources increases with age thereby facilitating inhibitory control. Relevant projections from the prefrontal cortex to the basal ganglia are enhanced while irrelevant projections are eliminated. This organization continues throughout childhood and adolescence as evidenced by the prolonged development of prefrontal regions in synapse elimination and myelination.

More generally, we have taken the position that the basal ganglia thalamocortical circuits underlie inhibitory control and that inhibitory deficits observed across a range of developmental disorders reflect a disruption in the development of these circuits. Four lines of converging evidence for our model were presented including data from cognitive measures, MRI-based morphometry, functional MRI, and lesion studies. First, we reported that children with developmental disorders involving the basal ganglia and prefrontal cortex perform poorly on tasks requiring suppression of attention toward a salient stimulus or competing response choice. Further, a dissociation in the pattern of performance on these tasks for each of four disorders was observed implying the involvement of different basal ganglia thalamocortical circuits for each disorder. Second, MRI-based morphometry measures of the frontal cortex and basal ganglia correlated with performance on inhibitory tasks indirectly supporting our structure–function hypotheses. Third, a more direct line of evidence for the involvement of the prefrontal cortex in inhibitory control was presented based on a functional MRI study. Finally, behavioral, morphometry and fMRI results from our children with neonatal basal ganglia insults showed deficits in inhibitory control and a four- to fivefold increase in developmental disorders with inhibitory deficits (ADHD and tic disorders) in later childhood.

ACKNOWLEDGMENTS

This work was supported in part by a 5-K01 MH01297-03, the Charles A. Dana Foundation, the John D. and Catherine T. MacArthur Foundation, and a John Merck Scholarship in the Biology of Developmental Disabilities.

REFERENCES

Alexander, G. E., Crutcher, M. D., & DeLong, M. R. (1991). Basal ganglia thalamocortical circuits: Parallel substrates for motor, oculomotor, prefrontal and limbic functions. *Progress in Brain Research, 85,* 119–145.

Asarnow, R. F., Brown, W., & Strandburg, R. (1995). Children with a schizophrenic disorder: neurobehavioral studies. *European Archives of Psychiatry and Clinical Neuroscience, 245,* 70–79.

Barkley, R. A. (1994). Impaired delayed responding: A unified theory of attention deficit hyperactivity disorder. In D. K. Routh (Ed.), *Disruptive behavior disorders: Essay in honor of Herbert Quay* (pp. 11–57). New York: Plenum.

Baxter, L. R., Jr., Schwartz, J. M., Mazziotta, J. C., Phelps, M. E., Pahl, J. J., Guze, B. H., & Fairbanks, L. (1988). Cerebral glucose metabolic rates in nondepressed patients with OCD. *American Journal of Psychiatry, 145,* 1560–1563.

Berman, K. F., Illowsky, B. P., & Weinberger, D. R. (1988). Physiological dysfunction of dorsolateral prefrontal cortex in schizophrenia: IV. Further evidence for regional and behavioral specificity. *Archives of General Psychiatry, 45,* 616–622.

Bourgeois, J. P., Goldman-Rakic, P. S., & Rakic, P. (1994). Synaptogenesis in the prefrontal cortex of rhesus monkeys. *Cerebral Cortex, 4*(1), 78–96.

Case, R. (1972). Validation of a neo-Piagetian capacity construct. *Journal of Experimental Child Psychology, 14,* 287–302.

Casey, B. J., Castellanos, F. X., Giedd, J. N., Marsh, W. L., Hamburger, S. D., Schubert, A. B., Vauss, Y. C., Vaituzis, A. C., Dickstein, D. P., Sarfatti, S. E. & Rapoport, J. L. (1997). Implication of right frontostriatal circuitry in response inhibition and attention-deficit/hyperactivity disorder. *Journal of the American Academy of Child and Adolescent Psychiatry, 36*(3), 374–383.

Casey, B. J., Cohen, J. D., Davidson, R., Hu, X., Lowe, M., Nelson, C., Noll, D. C., O'Craven, K., Rosen., B., Savoy, R., Truwitt, C., & Turski, P. (1998). Reproducibility of fMRI results across four institutions using a working memory task. *Neuroimage, 8,* 249–261.

Casey, B. J., Cohen, J. D., Jezzard, P., Turner, R., Noll, D. C., Trainor, R. J., Giedd, J., Kaysen, D., Hertz-Pannier, L., & Rapoport, J. L. (1995). Activation of prefrontal cortex in children during a nonspatial working memory task with functional MRI. *Neuroimage*, 2(3), 221–229.

Casey, B. J., Dobson, V., Scher, M., & Johnson, M. (1994). The effects of intraventricular hemorrhage (IVH) Grades III and IV on visual attention in preterm infants. *Proceedings of the International Conference on Infant Studies.* (abstract) Paris, France.

Casey, B. J., Forman, S. D., Franzen, P., Berkowitz, A., Badgaiyan, R., Braver, T. S., Nystrom, L., Welsh, T., & Noll, D. C. (submitted). A functional MRI study of ventral and dorsal prefrontal activation as a function of target probability.

Casey, B. J., Gordon, C. T., Mannheim, G. B. & Rumsey, J. M. (1993). Dysfunctional attention in calendar-calculating autistic savants. *Journal of Experimental and Clinical Neuropsychology, 15,* 933–946.

Casey, B. J., Trainor, R., Giedd, J., Vauss, Y., Vaituzis, C. K., Hamburger, S., Kozuch, P., & Rapoport, J. L. (1997). The role of the anterior cingulate in automatic and controlled processes: A developmental neuroanatomical study. *Developmental Psychobiology, 30,* 61–69.

Casey, B. J., Trainor, R. J., Orendi, J. L., Schubert, A. B., Nystrom, L. E., Giedd, J. N., Castellanos, F. X., Haxby, J. V., Noll, D. C., Cohen, J. D., Forman, S. D., Dahl, R. E., & Rapoport, J. L. (1997). A developmental functional MRI study of prefrontal activation during performance of a go–no-go task. *Journal of Cognitive Neuroscience, 9*(6), 835–847.

Casey, B. J., Vauss, Y. C., Chused, A., & Swedo, S. E. (1994). Cognitive functioning in Sydenham's chorea: Part 2. Executive Functioning. *Developmental Neuropsychology , 10*(2), 89–96.

Casey, B. J., Vauss, Y. C., & Swedo, S. E. (1994). Cognitive functioning in Sydenham's chorea: Part 1. Attentional Functioning. *Developmental Neuropsychology, 10*(2), 75–88.

Castellanos, F. X., Giedd, J. N., Eckburg, P., Marsh, W. L., King, A. C., Hamburger, S. D., & Rapoport, J. L. (1994). Quantitative morphology of the caudate nucleus in attention-deficit hyperactivity disorder. *American Journal of Psychiatry, 151,* 1791–1796.

Castellanos, F. X., Giedd, J. N., Marsh, W. L., Hamburger, S. D., Vaituzis, A. C., Dickstein, D. P., Sarfatti, S. E., Vauss, Y. C., Lange, N., Kaysen, D., Krain, A. L., Ritchie, G. F., Snell, J. W., Pajapakse, J. C., & Rapoport, J. L. (1996). Quantitative brain magnetic resonance imaging in attention-deficit hyperactivity disorder. *Archives of General Psychiatry, 53*(7), 607–616.

Chugani, H. T., Phelps, M. E., & Mazziotta, J. C. (1987). Positron emission tomography study of human brain functional development. *Annals of Neurology, 22,* 487–497.

Cohen, J. D., & Servan-Schreiber, D. (1992). Context, cortex and dopamine: A connectionist approach to behavior and biology in schizophrenia. *Psychological Review, 99,* 45–77.

Estes, W. K. (1975). *Handbook of Learning and the Cognitive Processes* (Vol. 1). Hillsdale, NJ: Lawrence Erlbaum Associates.

Flavell, J. H., Beach, D. R., & Chinsky, J. M. (1966). Spontaneous verbal rehearsal in a memory task as a function of age. *Child Development, 37,* 283–299.

Frazier, J. A., Giedd, J. N., Hamburger, S. D., Albus, K. E., Kaysen, D., Vaituzis, A. C., Rajapakse, J. C., Lenane, M. C., McKenna, K., Jacobsen, L. K., Gordon, C. T., Breier, A., & Rapoport, J. L. (1996). Brain magnetic resonance imaging in childhood-onset schizophrenia. *Archives of General Psychiatry, 53*(7), 617–624.

Giedd, J. N., Snell, J. W., Lange, N., Rajapakse, J. C., Casey, B. J., Kaysen, D., Vaituzis, A. C., Vauss, Y. C., Hamburger, S. D., Kozuch, P. L., & Rapoport, J. L. (1996). Quantitative magnetic resonance imaging of human brain development: Ages 4–18. *Cerebral Cortex, 6,* 551–560.

Goldman-Rakic, P. S. (1987). Circuitry of primate prefrontal cortex and regulation of behavior by representational memory. In V. B. Mountcastle, F. Plum, & S. R. Geiger (Eds.), *Handbook of physiology, the nervous system: Higher functions of the brain.* Sect. 1, Vol V, Pt 1. (pp. 373–417). Bethesda: American Physiological Society.

Huttenlocher, P. R. (1979). Synaptic density in human frontal cortex - developmental changes and effects of aging. *Brain Research, 163*, 195–205.

Insel, T. R. (1988). Obsessive-compulsive disorder: A neuroethological perspective. *Psychopharmacology Bulletin, 24*, 365–369.

Johnson, M. H., Posner, M. I., & Rothbart, M. K. (1991). *Components of visual orienting in early infancy: Contingency learning, anticipatory looking and disengaging, 3*, 335–344.

Keating, D. P., & Bobbitt, B. L. (1978). Individual and developmental differences in cognitive processing components of mental ability. *Child Development, 49*, 155–167.

Keeney, S. E., Adcock, E. W., & McArdle, C. B. (1991a). Prospective observations of 100 high-risk neonates by high-field (1.5 tesla) magnetic resonance imaging of the central nervous system. I. Intraventricular and extracerebral lesions. *Pediatrics, 87*(4), 421–429.

Keeney, S. E., Adcock, E. W., & McArdle, C. B. (1991b). Prospective observations of 100 high-risk neonates by high-field (1.5 tesla) magnetic resonance imaging of the central nervous system. II. Lesions associated with hypoxic-ischemic encephalopathy. *Pediatrics, 87*(4), 431–438.

Kolb, B., & Whishaw, I. Q. (1990). *Fundamentals of human neuropsychology*, 3rd ed. New York: Freeman.

Leckman, J. F., Price, R. A., Walkup, J. T., Ort, S., Pauls, D. L. & Cohen, D. J. (1987). Nongenetic factors in Gilles de la Tourette's syndrome. *Archives of General Psychiatry, 44*, 100.

Lipper, E. G., Ross, G. S., Heier, L., & Nass, R. (1988). Magnetic resonance imaging in children of very low birth weight with suspected brain abnormalities. *The Journal of Pediatrics, 113*(6), 1046–1049.

Mink, J. W. (1996). The basal ganglia: Focused selection and inhibition of competing motor programs. *Progress in Neurobiology, 50*, 381–425.

Pascual-Leone, J. A. (1970). A mathematical model for transition in Piaget's developmental stages. *Acta Psychologica, 32*, 301–345.

Posner, M. I. (1980). Orienting of attention. *Quarterly Journal of Experimental Psychology, 32*, 3–25.

Rakic, P., Bourgeois, J. P., & Goldman-Rakic, P. S. (1994). Synaptic development of the cerebral cortex: implications for learning, memory, and mental illness. *Progressive Brain Research, 102*, 227–243.

Rosenberg, D. R., Keshevan, M. S., O'Hearn, K. M., Dick, E. L., Bagwell, W. W., Seymour, A. B., Montrose, D. M., Pierri, J. N., & Birmaher, B. (1997). Frontostriatal measurement in treatment-naive children with obsessive–compulsive disorder. *Archives of General Psychiatry, 54*(9), 824–830.

Scher, M. S., Dobson, V., Carpenter, N. A. & Guthrie, R. D. (1989). Visual and neurological outcome of infants with periventricular leukomalacia. *Developmental Medicine and Child Neurology, 31*, 353–365.

Swedo, S. E., Leonard, H. L, Schapiro, M. B., Casey, B. J., Mannheim, M. D., Lenane, M. C., & Rettew, D. C. (1993). The psychological sequelae of Sydenham's chorea. *Pediatrics, 91*, 706–713.

Swedo, S. E., Pietrini, P., Leonard, H. L., Schapiro, M. B., Rettew, D. C., Goldberger, E. L., Rapoport, S. I., Rapoport, J. L., & Grady, C. L. (1989). Cerebral glucose metabolism in childhood-onset obsessive–compulsive disorder. *Archives of General Psychiatry, 49*, 690–694.

Trommer, B. L., Hoeppner, J. A., & Zecker, S. G. (1991). The go–no-go test in attention deficit disorder is sensitive to methylphenidate. *Journal of Child Neurology, 6*, 128–131.

Volpe, J. J. (1995). *Neurology of the Newborn, 3rd Edition*. Philadelphia, PA: W. B. Saunders Company.

Whitaker, A. H., VanRossem, R., Feldman, J. F., Schonfeld, I. S., Pinto-Martin, J. A., Torre, C., Shaffer, D., & Paneth, N. (1997). Psychiatric outcomes on low birth-weight children at age 6 years: Relation to neonatal cranial ultrasound abnormalities. *Archives of General Psychiatry, 54*(9), 847–856.

Wolf, S. S., Jones, D. W., Knable, M. B., Gorey, J. G., Lee, K. S., Hyde, T. M., Coppola, R., & Weinberger, D. R. (1996). Tourette Syndrome: Prediction of phenotypic variation in monozygotic twins by caudate nucleus D2 receptor binding. *Science, 273* (5279): 1225–1227.

Zametkin, A. J., Nordahl, T. E., Gross, M., King, A. C., Semple, W. E., Rumsey, J., Hamburger, S. D., & Cohen, R. M. (1990). Cerebral glucose metabolism in adults with hyperactivity of childhood onset. *New England Journal of Medicine, 323*, 1361–1366.

AUTISM

Dynamic Cortical Systems Subserving Cognition: fMRI Studies With Typical and Atypical Individuals

Patricia A. Carpenter
Marcel Adam Just
Timothy Keller
Vladimir Cherkassky
Jennifer K. Roth
Carnegie Mellon University

Nancy Minshew
University of Pittsburgh

Functional brain imaging brings many new offerings to the table of cognitive neuroscience, particularly offerings that help refine our understanding of the dynamic and adaptive properties of brain function that underpin learning and development. It is a little ironic that a methodology that is classically associated with static, still images, should be informative about dynamics and plasticity. But the association is no longer correct. In the early days of functional neuroimaging, in the late 1980s, the scanners were less sensitive, so that all that was possible was a group average image depicting the areas of brain activation, averaged over several participants, contrasting a small number of experimental conditions. The current instrumentation, namely high-speed fMRI, is far more sensitive yielding enough data to observe reliable effects in single participants in a few minutes per experimental condition. Capturing the dynamics requires only a fast enough shutter speed and a willing participant.

One of the consequences of the new technology of particular relevance here is that it is possible to examine the cognitive system as it adapts to slight differences in the quantitative and qualitative demands of a particular experimental condition. The result is that the brain adaptation is clearly manifest, and this manifestation is one of the key offerings of fMRI: the hardware–software distinction in the analysis of cognition that was previously a useful heuristic scientific strategy obscures one of the main adaptive properties of mind. The hardware (brain tissue) is dynamically recruited to meet the processing needs. The underlying software and the neuralware in which it is implemented are constantly changing. It is likely that this dynamic recruitment scheme underlies performance not only in a changing cognitive task, but also underlies adaptation over a longer time frame in a changing world.

In this chapter, we explore several implications of a new perspective on cognition and its adaptiveness, focusing on examples in the of area of language comprehension. This chapter does not present the details of the theory, which appears elsewhere (Just, Carpenter, & Varma, 1999). Here, we summarize some of the main points of this theoretical perspective, and how they apply to issues of brain plasticity and atypical development. Some of the key hypotheses that are emerging from this perspective include the following:

- Cognition entails physiological work and resource consumption
- A system involves collaboration among multiple neural components, a team
- The team members may have multiple and overlapping functions
- The components are dynamically recruited
- As a behavior becomes more skilled, there is better coordination of its components
- Disturbance, such as cortical injury, can result in rebalancing the work load among the components
- Developmental syndromes, such as autism, may result in an unusual collaboration, with differential amounts of coordination within versus among system components

We first describe evidence for these hypotheses based on fMRI studies of normal young adults in high-level cognitive tasks. Second, we describe some initial fMRI research with adults who have experienced stroke. Last, we briefly describe some preliminary fMRI studies of individuals who are high-functioning intellectually, but who also have autism.

THE CORTICAL SYSTEMS SUPPORTING COGNITION

Cognition, Resource Consumption and Mental Work. Cognitive computations entail physiological work. Moreover, the mapping between brain activation and a theoretical account of the underlying processes depends not just on the type of processing that is occurring, but also on how much work is involved. This hypothesis contrasts the standard assumption that the mapping between brain activation and cognitive process depends only on the qualitative nature of the processes. This hypothesis is supported by neuroimaging studies that quantitatively vary the amount of task demand.

Before describing such a study, it is useful to briefly describe the methodology. Functional Magnetic Resonance Imaging (fMRI) exploits the fact that neuronal cortical activity is accompanied by local increases in the concentration of deoxygenated blood (Kwong et al., 1992; Ogawa et al., 1990). Using a powerful magnet (1.5 or 3.0 T in our studies), fMRI detects the small increases (1–4%) in magnetic susceptibility of the hemoglobin in the microvasculature without any other contrast agent; hence, the method is called Blood Oxygen Level Dependent (BOLD). The typical study described here compares the distribution of activation in several experimental conditions to that of a rest condition.

A Sentence-Comprehension Study. In the first fMRI study to manipulate the quantitative demands imposed by a sentence comprehension task, we contrasted the brain activation in language-related cortical areas for three kinds of sentences, shown in Table 15.1 (Just, Carpenter, Keller, Eddy, & Thulborn, 1996). The three sentence types are superficially similar (each containing two clauses and the same number of content words), but they differ in structural complexity, and consequently in the demand they impose during comprehension, as shown in several behavioral studies (e.g., King & Just, 1991).

The fMRI-measured activation showed systematic increases with sentence complexity in a large scale network of cortical areas: the left posterior superior and middle temporal gyri (roughly, Wernicke's area), the left inferior frontal gyrus (roughly, Broca's area), and to a lesser extent, their right hemisphere homologues. Figure 15.1 shows an example of the data by illustrating the voxels in one slice that were significantly activated above the resting baseline (through the posterior temporal region) for the main conditions for one individual.

Figure 15.2 shows the quantitative results when the voxels are summed for an entire cortical region and averaged across participants. In this particular study, the volume of activated brain tissue in left temporal cortex increased from about .75 cm³ in the processing of the least demanding

TABLE 15.1
Three Sentence Types That Differ in Structural
Complexity (from Just et al., 1996)

1. ACTIVE CONJOINED	*The reporter attacked the senator and admitted the error.*
2. SUBJECT-RELATIVE CLAUSE	*The reporter that attacked the senator admitted the error.*
3. OBJECT-RELATIVE CLAUSE	*The reporter that the senator attacked admitted the error.*

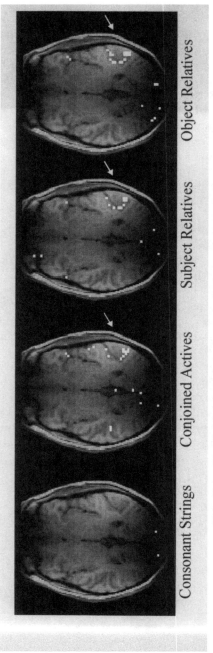

Consonant Strings Conjoined Actives Subject Relatives Object Relatives

FIG. 15.1. Thresholded fMRI activation images (1.5T) (superimposed on structural images) for only the most activated slice through Wernicke's area (indicated by the arrow) from one participant. The number of activated voxels (shown in white) generally increases with sentence complexity. (From "Brain activation modulated by sentence comprehension" by Just, Carpenter, Keller, Eddy, & Thulborn, 1996, *Science, 274,* Figure 2, p. 115. Copyright © 1996 by the American Association for the Advancement of Science. Reprinted with permission).

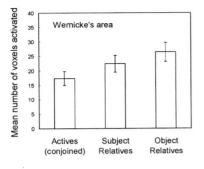
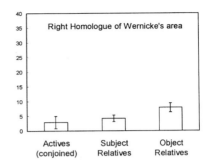
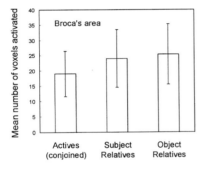
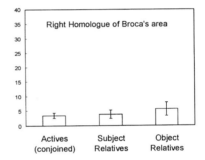

FIG. 15.2. The average number of activated voxels across participants indicates that the processing of more complex sentences leads to an increase in the volume of neural tissue that is highly activated in all four areas. The top panels indicate the average number of activated voxels in the left (Wernicke's area) and right laterosuperior temporal cortex (and standard errors of the means over 15 participants). The bottom panels indicate the average number of activated voxels in the left (Broca's area) and right inferior frontal cortex (and standard errors of the means over only five participants). (From "Brain activation modulated by sentence comprehension" by Just, Carpenter, Keller, Eddy, & Thulborn, 1996, *Science, 274*, Figure 1, p. 115. Copyright © 1996 by the American Association for the Advancement of Science. Reprinted with permission).

sentence to about 1.2 cm³ in the processing of the most demanding sentence. The right hemisphere homologue of this area is barely activated for the least demanding sentences, but is substantially activated for the most demanding ones. Such results demonstrate a dynamic recruitment of additional neural tissue within a region and across regions depending on the amount of computational demand that is imposed by the task. Thus, one cannot specify a particular region of the brain and declare "this is where sentence processing occurs." Sentence comprehension occurs in a dynamically configured network of brain areas whose membership and degree of member participation depends on the comprehension task.

The same type of result has been found in some other domains, such as mental rotation (Carpenter et al., 1999) and short-term list learning (Braver et al., 1997; Grasby et al., 1994). In the research on mental rotation, we found that fMRI-measured activation in the parietal region increased monotonically with the amount of mental rotation required to align two three-dimensional Shepard and Metzler figures. Figure 15.3 shows how the fMRI-measured activation increases monotonically for both the left and right parietal regions; this activation is particularly in and around the intraparietal sulcus.

Both the sentence comprehension and the mental rotation studies assume that the mental workload is being modulated by the independent variable. This assumption is based on several types of evidence. First, behavioral measures indicate that performance indices, such as response

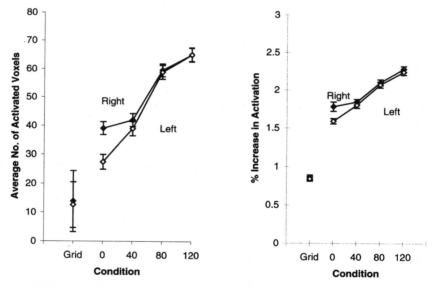

FIG. 15.3.   The average number of activated voxels for the left and right parietal regions increase as a function of the increase in angular disparity (left panel) and so does the mean percentage of increase in MR signal intensity over the fixation condition (right panel) for the 3.0T study. By contrast, the grid condition results in relatively few voxels that are significantly activated and a significantly lower increase in mean activation level. The results for the right hemisphere are shown by filled symbols and for the left, by open symbols. (From "Graded functional activation in the visuospatial system with the amount of task demand" by Carpenter, Just, Keller, Eddy, & Thulborn, 1999, *Journal of Cognitive Neuroscience*, Figure 5, p. 15. Copyright © 1999 by the MIT Press. Reprinted with permission).

times and error rates, increase with the manipulation. Second, computational modeling studies that take account of resource utilization show in precise detail the extra computational work involved in comprehending more complex sentences (Haarmann, Just, & Carpenter, 1997; Just & Carpenter, 1992) or in mentally rotating through a larger rotation angle (Just & Carpenter, 1985). The fMRI studies show how the workload is distributed through the neural infrastructure subserving cognition.

Workload and Time. The workload of a system during the execution of some task can be distinguished from its sheer duration. Response time measures of task performance cannot make the distinction, because tasks can take a relatively long time but consume few resources. For example, in almost any resource-sensitive computational model, mentally adding a long list of single digit numbers $(3 + 2 + 5 + 2 + 3 + 1 + 4 + 6 + 1 + \ldots = ?)$ consumes less activation than solving a complex mental multiplication task (e.g., $63 \times 28 = ?$), although the list length can be manipulated to equate the processing time.

The fMRI experimental paradigms and the data analysis equate the duration of the data acquisition for the less demanding and more demanding trials, so these results indicate that the person is working harder, not simply longer. In addition, in the mental rotation study, we contrasted the effects of cognitive workload versus simply taking more time, by including a condition in which the participant had to scan a fixed grid, making more eye fixations and taking longer than the rotation task. The fMRI-measured activation in the parietal region for that condition, shown in Fig. 15.3 as the "grid" condition, supports the distinction between computations that constitute additional mental work and computations that simply take longer, a distinction not possible with conventional behavioral measures. More generally, these types of studies suggest that the increased activation of nearby cortical tissue is a fairly general mechanism of adaptation to task difficulty. It is also consistent with the hypothesis that during a cognitive task, brain activation reflects cognitive work.

Resource Utilization. Because all physical energy systems require resources, the concept of resource consumption is not novel from the perspective of biology. In fact, many neuroimaging techniques depend on one or another aspect of biological resource consumption. Glucose metabolism underlies PET–FDG; and as explained before, the BOLD–fMRI technique assesses the relative concentration of oxygenated hemoglobin in a region of neural activity. In contrast to biological models, the concept of resources is not well represented in many cognitive science models, particularly computational theories that instead focus on abstract descriptions of cognitive computations and representations. However, the concept of resource consumption in a computational model can bridge from the abstract domain of functional description into the concrete domain of neural implementation (Carpenter & Just, 1999).

An index of the amount of work that a cognitive system performs can be used to map between the computational level and physiological measures of thought (such as brain activation). One such index (adapted from economics) is called *resource utilization*, an index of the rate at which an activity (such as processing information) consumes resources. How should resources be construed in a computational model? In a simulation model of sentence comprehension, resource utilization corresponds to the rate at which different subsystems utilize functional activation (Just & Carpenter, 1992, 1993). The term *activation* has a long history of usage in cognitive science, and typically denotes the availability of a concept in memory. In the comprehension model, each element has an associated activation level that indicates its availability in working memory. In addition, functional activation also plays a role in computation because computations occur through the gradual increase in the activation levels of intermediate and final products, as occurs in connectionist models (McClelland, Rumelhart et al., 1986). For example, more functional activation is used in the course of constructing and maintaining the representations of the more complex sentences with object-relative clauses than for the less complex sentences. Within some dynamic range, the amount of functional activation that is needed to accomplish a task is expected to correlate with the fMRI-measures of brain activation.

Networks and Brain Regions. In contrast to a localist assumption of a one-to-one mapping between cortical regions and cognitive operations, an alternative view is that cognitive task performance is subserved by large-scale cortical networks that consist of spatially separate computational components, each with its own set of relative specializations, that collaborate extensively to accomplish cognitive functions. For example, visual sentence comprehension is subserved by a large-scale network that includes left inferior frontal gyrus (Broca's area), left posterior superior temporal gyrus (Wernicke's area), angular gyrus, extrastriate and primary visual cortex, and in some circumstances, left middle frontal gyrus [left dorsolateral prefrontal cortex, DLPFC] and the right hemisphere homologues of Broca's and Wernicke's areas. For example, Fig. 15.4 shows the thresholded statistical probability maps for 14 slices (each 5 mm thick) that were obtained during a sentence comprehension task and are superimposed on structural images taken of the participant. Note the large number of regions, only some of which are labeled, that show activation. Thus, no single area "does" sentence comprehension. Furthermore, the collaboration among areas is hypothesized to be highly interactive (to be discussed next), making the resulting cognitive computations an emer-

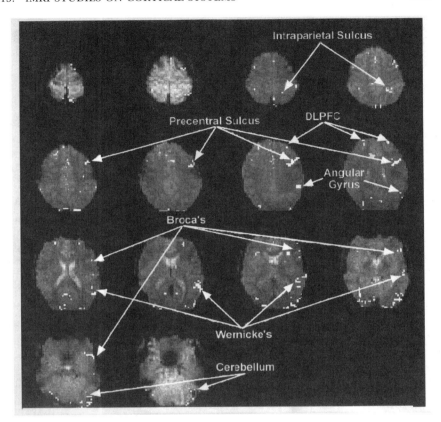

FIG. 15.4.   Statistical probability maps superimposed on structural images for a single individual to illustrate how a cognitive task (like sentence comprehension) elicits activation in multiple cortical loci. The voxels in white are those that are significantly activated above a baseline fixation condition when a normal college student reads a series of sentences. The 14 axial oblique slices were acquired with gradient echo, resonant echo planar MRI at 3.0T. (From "Modeling the mind: Very High-field fMRI-activation during cognition" by Carpenter & Just, 1999, *Topics in magnetic resonance imaging, 10,* Figure 2, p. 18. Copyright © 1999 by Lippincott, Williams, & Wilkins. Reprinted with permission).

gent property of several collaborating team members. This characterization applies not just to visual sentence comprehension, but to any cognitive task of any complexity whatsoever.

Moreover, neuropsychological research on patients with brain damage has produced a compatible view. Even localized lesions produce nonmodular deficits, affecting performance in a variety of tasks (Mesulam, 1990). Further supporting this view is the concept of distributed processing that has been fostered by connectionist models (Elman et al., 1996;

McClelland, Rumelhart et al., 1986). Mesulam's proposal of distributed processing in a large-scale network consisting of multiple brain areas applies the connectionist distribution principle several levels higher.

A second emerging hypothesis is that neural components may participate in supporting more than one function, although a region may preferentially perform a certain type of operation, reflecting a *relative specialization* rather than an absolute and exclusive involvement. In neuroimaging studies, a particular cortical region may be activated in a family of tasks.

The differential effect of various types of computational demand on each member of a large scale cortical network can be informative about the organization of the network. In the domain of sentence processing, we have found that effects of sentence complexity are far more pervasive than expected, and they interact with other types of computational demand. For example, we have recently examined the effects of sentence complexity in conjunction with the effects of lexical frequency (using the same type of paradigm as before) with a 3.0T scanner on 14 separate 5mm slices covering most of the cortex (Keller, Carpenter, & Just, 1999).

As reported previously, the sentence's structural complexity affected the amount of activation in both the left posterior superior/middle temporal region (Wernicke's area) and the left inferior frontal region (Broca's area). In addition, sentence complexity affected the activation in other regions as well, including the left and right ventral striatal pathway, regions that are more associated with visual form processing, as well as the precentral sulcus, the intraparietal sulcus, and DLPFC (dorsolateral prefrontal cortex). Equally importantly, the effect of the sentence's structure typically interacted with the normative frequency of the nouns in the sentence, and this interaction, always taking a similar form, occurred in several areas (Broca's and Wernicke's prominent among them). In general, the sentence's structure had a much a greater effect on brain activation if the nouns in the sentence were infrequent than if they were frequent (an overadditive interaction between the two types of demand). This existence of an interaction within a given brain region indicates that more than one type of process (in this case, syntactic and lexical) involves the region.

The occurrence of the interaction in multiple brain regions indicates the collaboration among areas, such that the creation of a greater workload of one area is propagated (as more work) for another area. The results speak to the multispecialization of areas and the collaboration among areas. Any conclusion about the mapping of a particular region is limited by the spatial resolution of the methodology. For example, the spatial resolution of fMRI is excellent and rapidly improving; nevertheless, it is always possible that within whatever spatial resolution is present, there is some finer subdivision that cannot be detected. However, our viewpoint postulates that cognitive representations and processes are not coded by individual components (be they neurons or small cortical regions), but rather that they are coded by the relations among neural components, which can undoubtedly be characterized at multiple levels. Even methodologies that record from single cells but examine a large number of them are reaching similar conclusions about the dynamic, interdependent nature of the underlying activity (Nicolelis, 1997; Sanes & Donoghue, 1997a, 1997b; Sanes et al., 1995). The localization of function is an old problem in psychology, with fMRI providing a somewhat new type of answer: limited regional equipotentiality with extensive interregional collaboration.

Dynamic Recruitment and Lateralization. The lateralization of activation in the temporal region for sentence comprehension tasks provides one perspective on the flexibility of network configuration. Across a number of reading studies with normal, right-handed college students, we found much more activation in the left posterior superior and middle temporal region than in the right. The degree of lateralization can be quantified by measuring the difference in the activation in the two homologues (Left – Right) relative to the total activation across the hemispheres (Left + Right), an index that ranges from +1 for an entirely left-lateralized pattern, to –1 for an entirely right-lateralized pattern. In a study involving 30 right-handed college students with the sentence comprehension paradigm that was described earlier, the mean of the index was 0.80 (SD = .3) for the posterior superior and middle temporal region, with the most right lateralized value being –0.2 for these right-handed individuals. This indicates that the activation was very left lateralized for this region and task (Keller, Carpenter, & Just, 1999).

The activation in this region is more bilateral when participants read sentences that have a concrete, visuospatial referent that they expect to see (Carpenter et al., 1999). The amount of activation in the right posterior temporal region almost equaled that in the left as participants read sentences like "*It isn't true that the star is above the plus*" before examining a subsequent picture (such as a plus above a star). The fact that the left and right posterior temporal regions were almost equally activated in this instance illustrates how the language networks are dynamically configured. The degree of involvement of the right temporal area reflects in part the content of the task.

Listening comprehension also may be more bilaterally represented in the temporal region than reading comprehension. This conclusion is supported by two studies that contrasted the two modalities. One involved verifying general knowledge questions (such as "*The study of economics concerns the distribution of wealth and goods*"). For the 10 college students who performed the task visually, the laterality index for the superior/middle temporal region was 0.65; by contrast, for the 10 who performed the task auditorily, it was 0.17. (In the latter case, we separately coded and excluded the primary auditory cortex, which is expected to be bilaterally activated.) A second study contrasted reading and listening to active and object-relative sentences followed by questions as in the earlier Just et al. (1996) study (Michaels et al., 2000). That study showed a similar shift in the laterality index, with a value of .63 for reading and .07 for listening. These analyses suggests that there are secondary and higher-level association regions in the right temporal region that are activated by the auditory comprehension task. The apparently greater lateralization in the temporal region for reading than for listening comprehension (for normal, right-handed adults) is itself an interesting phenomenon. These studies suggest that there is not a single language comprehension system, even in the posterior temporal region. The differential lateralization suggests that language comprehension is accomplished by somewhat flexibly configured subsystems that may partially overlap but also may partially diverge, depending on the precise properties of the computational demand. Such data are consistent with the more general argument that the systems underlying cognitive performance are dynamically configured and allocated as a function of both the quantitative and qualitative properties of the task as well as the availability and history of the neural systems themselves.

Another mechanism of dynamic allocation is the recruitment of additional neural components into the network underlying task performance. To take an example from language processing, the participation of a goal-monitoring/executive system (often associated with the prefrontal cortex) may depend on whether the sentence comprehension requires a substantial amount of reasoning. If a typical college student is given even a complex sentence to understand, this processing usually activates a set of cortical regions that does *not* include DLPFC. By contrast, the executive system should be much more likely to enter into the network if the task involves complex reasoning, such as in understanding "*Brothers and sisters have I none, but this man's father is my father's son.*" (The problem is to determine the family relation between the speaker of this sentence and the person depicted in the portrait to which the speaker is referring, (Casey, 1993). Although system components may enter or exit a network, equally often the components simply participate to a greater or lesser degree. For example, normal college students who are trained in a sentence comprehension task show relatively little activation (less than 1 voxel, on average) of DLPFC; if the same task is embedded in a dual task situation, the amount of DLPFC activation doubles.

System Combinatorics.   A systems approach to cognition raises the issue of what constitutes a cognitive system. Several recent taxonomies can be defined by task domain. For example, Gardner's (1983), theory of multiple intelligences, which grew out of his considerations of both neuropsychological and developmental data, suggested seven systems—linguistic, spatial, quantitative, music, kinesthetic, interpersonal, and intrapersonal intelligences. Moreover, a meta-analysis of a very large compendium of psychometric data suggested an overlapping set of candidate systems: general reasoning, verbal language comprehension, rote memory, visualization, auditory perception (including music), creativity and speed on low-level tasks (Carroll, 1993). A third proposal, based on the deficits of patients with focal lesions, distinguished spatial attention, rote memory, and language systems, without an attempt to be exhaustive (Mesulam, 1990). These taxonomies usefully capture some empirical observations concerning the partially dissociable effect of lesions in adults and the patterns of individual differences.

Although useful, such taxonomies lack the generative and combinatoric properties that are needed to capture the spectrum of cognitve skills that humans acquire, including skills that did not exist during the development of Nature's original toolkit for human intelligence. The notion of combinatorics is central if we are to account for the human ability to learn and adapt to such a wide spectrum of environments, from programming in Java, or flying a lunar module, to proving a theorem in topology. The combinatoric approach proposes that larger systems reflect the recruitment of smaller subsystems, perhaps with different weights and orders. However, the recruitment is not simply some linear concatenation because the components themselves are affected by the other components with which they collaborate.

Developmental Implications. Although we have discussed recruitment over in the context of short time spans, dynamic recruitment may be the appropriate characterization for discussing learning and development as well. During development, the cortical systems are recruited as described before, but the relative specializations and degree of collaboration are probably much less well established. The developmental account must explain how an adult "end" state can arise from some initial state. Elman and colleagues (1996) suggested that the cortex in the infant gradually differentiates into regions containing cells that are differentially proficient in various types of processing. Repeated exposure to different types of processing can result in some areas gradually becoming more proficient (and hence specialized) in particular types of processing, namely those types of processing for which the given cell types are most suited. The fMRI contribution here is that the cycles of learning in this account require a dynamic recruitment of multiple possible areas in performing a task.

As a task is performed repeatedly during development, there is a differential increase in proficiency of the best suited area for a given computation. Moreover, this learning regimen can also lead to some redundancy of function across areas, such that inherently less proficient areas could also accrue some degree of proficiency. For example, cells in more than one brain area may initially attempt to process speech sounds, but the area containing cells that are particularly sensitive to the fine timing distinctions that differentiate phonemes may eventually become specialized for speech processing. Another area may also have initially attempted to process speech, and eventually lost the competition for the specialization, but may nevertheless have retained a residual capability to process speech, albeit less efficiently. Thus, there might be some overlap in function between areas, such as between the left and right homologues of the language network.

There is much more to cognitive development than recruitment of brain areas, but the flexibility of the brain response to task variation in adults provides a possible clue to the development of large scale cortical networks. In a developing child, any new task might evoke a network of brain areas which might initially constitute a larger or different set than in an adult, by virtue of the task properties evoking even remotely relevant areas. With practice and learning, the best-suited areas come to be favored, and probably collaborate increasingly with each other. This same account applies equally to learning in adults, where the increased collaboration between areas with learning (measured as increased synchrony in fMRI activation) has been demonstrated (Buchel et al., 1999).

RECRUITMENT AND PLASTICITY: EVIDENCE FROM INDIVIDUALS WHO HAVE HAD A STROKE

The dynamic recruitment and overlap of function across areas underlies not only development, but also adaptation to brain damage. One clue can be found in a recent set of fMRI studies of adults who have had left hemisphere strokes that affected their language comprehension ability. The first study examined two individuals soon after their strokes, and the results suggest that there can be relatively rapid recruitment of the right hemisphere homologues of the classic language cortical regions. Two adults were studied using a sentence comprehension paradigm shortly after the acute phase of their strokes and a few months thereafter, while they showed spontaneous language recovery (Thulborn, Carpenter, & Just, 1999). One individual had been studied prior to his stroke, and so constitutes a rare before–after case study. Of the two patients, one had a left anterior stroke (B) and the other, a stroke in the left posterior region (W). For both patients, the fMRI data obtained shortly after the stroke showed that there was more right hemisphere activation in the area that was homologous to the stroke than the normal controls. Moreover, the relative amount of activation in the right homologue increased over time, suggesting a redistribution of the workload in the language network. For both individuals, this redistribution was accompanied by a relatively rapid improvement in language skill, eventually resolving in mild language deficits. The redistribution is consistent with other PET neuroimaging studies that were performed years after the patients' strokes and indicated considerable right hemisphere involvement (Engelien et al., 1995; Weiller et al., 1995).

An fMRI Study of Young Children.   The redistribution of language processing to the homologous hemisphere was also found in a fMRI study of school age children who had early left hemisphere focal lesions (Booth et al., in press). These children participated in a fMRI written sentence-comprehension study that was essentially identical to the Just et al. (1996) study with college students. The children with left hemisphere damage showed activation in the typical language processing areas, but the activation was strongly right lateralized. These data suggest that when a lesion has damaged the preferred mechanisms, then the participation of the right homologue is increasingly recruited. Such an interpretation is also consistent with the behavioral data from a similar population of children who showed delayed language production, but more normal language skills by the school age years (Stiles & Thal, 1993). The general plasticity pattern here is to recruit areas to a task that were secondary in proficiency, but nevertheless had the potential for the task or had actually participated in the task performance previously. In effect, this mechanism recruits an existing member of a large scale network to perform the function of a damaged component, a function that the intact member may have previously performed in a secondary (less proficient) role.

An fMRI Study Showing Prefrontal Recruitment With Language Therapy.   Another mechanism of plasticity in response to brain damage may implicate the recruitment of the prefrontal regions in conjunction with the learning of a new strategy. The learning study involved two adults with aphasia, long past their focal left hemisphere stroke who underwent an intensive language therapy program (Just, Carpenter, & McNeil, 1999). Both individuals had difficulty understanding sentences at the beginning of the therapy, but by the end, showed significant behavioral improvement and generalization to other sentences. The therapy was a version of semantic mapping therapy (Schwartz et al., 1994); it involved the auditory presentation of a variety of sentences and the participant had to identify the semantic roles of the individuals mentioned in the sentence.

Both before and after the therapy, the patients participated in a fMRI study that contrasted the visual comprehension of structurally simpler and more complex sentences. Before the therapy, the patients could not understand any of the more complex sentences (i.e., they could not answer simple comprehension questions about sentences that they had just read), and correspondingly, they had little activation in the language regions (Broca's and Wernicke's areas) for those conditions. However, when simpler sentences were presented, these patients were able to understand them and their brain activation looked approximately normal. After the therapy, however, the patients were able to understand complex sentences with reasonable accuracy. At that point, there was activation in the language regions for all three types of sentences presented, with increasing amounts related to the increasing complexity of the sentence. In addition, however, posttherapy there was pronounced activation in the dorsolateral prefrontal region, a region that typically shows relatively little activation for college students in this paradigm. For the patient with an intact left DLPFC, the new activation, which increased with sentence complexity, occurred in left DLPFC. For the patient in whom left DLPFC had been damaged, the new activation occurred in right DLPFC.

The suggestion of these results is that a new member was recruited into the network; the recruited region is associated with planning, problem solving, and sequential analyses. Our interpretation of these results is that the behavioral therapy regimen instilled a new set of strategic processes for sentence comprehension, and this new component of processing was supported in least in part by the involvement of the prefrontal regions. Of equal interest is the fact that the new DLPFC involvement was associated with renewed involvement of the language regions during the processing of complex sentences. The DLPFC role was collaborative, entailing activation of other members of the network that had previously been absent.

This result cumulates with developmental studies that have used EEG (e.g., Case, 1992; Thatcher, 1992) or have analyzed the effects on children who have had early lesions in this region. Those studies indicated that there may be prefrontal involvement throughout development, but there is some suggestion that this area is particularly important during adolescence and the child's increasing ability to do more abstract thinking. Such data suggest that the prefrontal regions may be particularly implicated in learning and development, although not exclusively. Rather, the role of the prefrontal region may be dynamic; it may be particularly useful in assembling and coordinating subsystems in the early stages of learning, and its role may actually decrease with learning and automaticity.

SPECULATIONS ON THE DEVELOPMENT
OF COGNITIVE SYSTEMS

Mechanisms of Dynamic Recruitment.   The recruitment of cortical systems is dynamic through the course of a child's development, as it is the adult, although the recruitment latitude is much greater in the child. The qualitative processing characteristics of the various cognitive systems that emerge surely reflect both the child's biological propensities as well as his or her socio-cultural experiences. Piaget postulated that the sensory–motor reflexes of the very young infant constitute the scaffolds of cognition. A specific instantiation of this shaping process can be observed in the development of visual attention (Johnson, 1997). Johnson has studied the orienting of young infants toward head-like visual patterns, and he also has studied how young birds imprint on patterns that resemble mother hens. Based on both behavioral, animal, and physiological studies, he argued that early processing systems shape the infant (or baby bird) to attend to certain features of the environment (i.e., head-like patterns). This selection mechanism then biases what is available to the other organizing systems of the developing infant. Hence, the cortical systems that are phylogenetically shaped for certain broad types of information (say visual lines, patterns, and motion in conjunction with human sounds), then receive correlated information that corresponds to an event like "Mother picking up, cuddling, and talking to the baby." Thus, the early sensory–motor reflexes biases the developing cortical and subcortical systems by filtering what otherwise could be a "booming buzzing confusion."

Dynamic recruitment during development is also consistent with the studies of individuals who are either congenitally blind or congenitally deaf (Neville, 1993). Those studies suggest, for example, that there is auditory processing in the occipital region of congenitally blind individuals, suggesting both plasticity and dynamic recruitment of that region. This view is also consistent with the general point of *Rethinking Innateness* (Elman et al., 1996), which argues that the existence of highly organized cortical systems in the adult does not imply that such systems are prewired in the infant.

Another component of this view is that the neural systems that support cognition are not internally closed systems, but rather they are shaped by the activities of the person in the world. The idea that the person and environment form a system is one way to incorporate the constructivist views espoused by numerous developmentalists. It is also consistent with the concept of affordances, from Gibson's perceptual theory (Gibson, 1979), which stresses that the environment is coded in terms of what it affords perceptually and motorically. All of these perspectives point to the inherent role of adaptation in development and learning, adaptation that is shaped both by the environment as well as the individual.

The Qualitative Properties of Developmental Systems. Piaget suggested that the sensory–motor schemas are the foundations for the next systems to develop, which he proposed, are the schemas of the permanent object, space, time and causality (Piaget & Inhelder, 1969). Of course, these schemas are interdependent because the permanent object is the argument that undergoes the transformations through time and space, and is linked into events through causality. An extrapolation of the Piagetian approach is to propose that these systems, plus others, contribute to the genesis of other cognitive systems. To illustrate how this might occur, we can look at language development. The acquisition of sign language by congenitally deaf infants parallels the acquisition of spoken language by hearing infants in both the nature and sequence of the content and timing of particular developments. These parallels have led at least one researcher (Pettito, 1996) to suggest that the unifying dimension underlying signed and spoken language is temporal organization. (This is not to deny the role that space can play as a communicative tool as well as a topic, but simply that the invariant code may be temporal). Time may be one of the features, along with social communication, that gives rise to the elements that become organized. The emerging system may build on the initial sensitivity to temporal information (as well as other sensitivities), and the emerging language systems have their own unique properties.

Coordination and Collaboration in System Development.   Complex systems, whether biological or social, involve the coordination of activity across components. In the case of cognitive skill, this coordination occurs across a variety of cortical and subcortical components. One mechanism of coordination may relate to the time course of activity across units. This hypothesis is a generalization of Hebb's (1949) proposal that coactivation at the neural level is a mechanism of learning, specifically, a mechanism for constructing the cell assembly. The current proposal extends this insight to suggest that coordinated activity (not necessarily just activation) among larger units might lead to the organization of still larger units, up through networks and large scale cortical networks systems.

A second aspect of this issue may be more appropriately labeled *collaboration*, the sharing of cognitive computations. Collaboration among components may be important in cognitive development because it may allow for greater specialization of individual components, analogous to the way in which collaborations among individuals in a work environment allows for individuals to develop deeper skills. It is important to keep in mind that specialization and collaboration (in the sense of frequent communication of relevant information) are mutually compatible. The other side of the coin is that collaboration may foster some variation in how computations are performed; such variation may provide the backup that can be drawn from if the environment becomes more demanding or the individual changes. A component that works closely with another component may learn a secondary specialization, one that can be recruited when the primary component is occupied or otherwise unavailable. For example, the conjoint modulation of the inferior frontal and posterior superior temporal region in language processing is indicative of such collaborative processing. The particular primary computations of these regions appear to be so collaborative that the two areas are similarly affected by many, but not all, language factors (Just, Carpenter, & Varma, 1999).

These speculations about system collaboration may be particularly important in the case of neural development. Impediments to the normal development of any individual component, to their coordination or specific collaboration may affect a number of different types of cognitive tasks. Any such impediment is also likely to affect the development of other components that are without a primary problem, because the entire network must adapt to the unusual coordination or collaboration. This point was noted by Parks et al. (1988) in the context of a theory of network efficiency, that dysfunctional neurochemical systems or developmental inadequacies in particular neural structures could result in poor feedback. This poor feedback could lead to the inadequate recruitment and poor organization of processing elements, resulting in the inefficient communication within topographically separated assemblies of neuronal networks.

IMPAIRED AND PRESERVED SYSTEMS
IN AUTISTIC INDIVIDUALS

Autism is the prototype of a developmental disorder that has a wide-ranging impact on both social and cognitive behaviors. Individuals with autism often show a lack of reciprocity in social give-and-take. As children, they may lack interest in playing with toys, and as adults, they typically have a restricted range of interests, although they may develop great focus and expertise in the domains that do draw their interest. Of particular relevance here is the fact that individuals with autism often show deficit language skills and difficulty with problem solving and self-initiated conceptual behavior, sometimes characterized as problems with the executive system (Hughes, 1996; Hughes, Russell, & Robbins, 1994; Minshew, 1996; Minshew, Goldstein, & Siegel, 1995; Ozonoff, Pennington, & Rogers, 1991). Executive deficits have been documented for preschool children, older children, adolescents and adults with autism (Ozonoff et al., 1991; Ozonoff et al., 1994; Rumsey, 1985; Rumsey & Hamburger, 1988) and longitudinally (Ozonoff & McEvoi, 1994).

That autism is not some global retardation is indicated by the fact that these individuals have an uneven profile of abilities, with remarkable gifts as well as limitations. In a large sample of individuals with near-normal or above-normal IQs, the more pattern-oriented skills, such as word decoding and visuospatial processing, of the individuals with autism, exceeded the skill levels exhibited by age- and IQ-matched controls (Siegel et al.,

1996). Even more dramatic are the prodigious talents of artists such as Nadia (Selfe, 1977), the incredibly complex buildings designed by Temple Grandin (Grandin, 1995; Grandin & Scariano, 1986), and skills of calendar calculators and musical prodigies (Howe, 1989). Again, these skills tend to involve patterns, either visuospatial, musical or numeric, that can be the basis of complex achievements that are beyond the capability of normal controls. The skills tend not to be in linguistic, abstract reasoning or social domains. The unevenness of the profile of individuals with autism is important for theories of autism as well as theories of cognitive development in general.

Our fMRI research on autism is a preliminary progress report of a project to map between cognitive performance (based on behavioral studies) and cortical function (using fMRI) in the language and executive processing domains for individuals who are high-functioning (IQs in the normal range) and who have autism. These individuals have all the criterial characteristics of autism, but its less severe form allows a more normal developmental environment. In addition, the characterization of cognition in high-functioning autistic patients that has been developed by one of us (Minshew, Goldstein, & Siegel, 1995, 1997) seems particularly compatible with the systems approach that has emerged from the fMRI studies of young adults and patients. The general framework suggests that cognitive deficits in autistic individuals are due, in large part, to difficulties with higher-order abstraction (Minshew, Goldstein, & Siegel, 1995, 1997; Ozonoff, 1995). This preliminary report should not be overinterpreted because it is based on a small number of participants and an incomplete analysis of the data. Nevertheless, we have discovered some regularities, which are compatible with and make concrete the notion of a deficit in higher-order abstraction.

Psychometric test results suggest that the language deficits of individuals with autism appear primarily when the processing demand is high. In fact, in a battery of psychoeducational tests, the largest and most reliable difference between a sample of high-functioning (IQ > 70) individuals with autism and age- and IQ-matched controls occurred with a test called the Detroit Test of Oral Directions (Goldstein, Minshew, & Siegel, 1994). The Detroit Test is a type of language working-memory test. It consists of a series of picture sets of several common objects (e.g., a pail, baseball, hat, and bird). With each set, the participant hears a set of directions, such as *draw a circle around the baseball, draw a line through the hat, circle the container, and draw a line to the bird without touching the other items.* Because the participant is not allowed to begin any action until all of the directions are heard, the individual must process and simultaneously store a sequence of commands. The complexity of the test increases through 3 main manipulations: (a) the successive sets increase in the number of items and number of instructions, (b) the phrase used to denote a picture can be more abstract rather than concrete (e.g., *container* vs. *pail*), (c) several cognitive operations may be embedded in one direction (e.g., *cross out the triangle* vs. *cross out the biggest number that is in a square*). All three factors can be viewed as ways to increase the demand on the language working-memory system that must both process and store linguistic information; hence, the deficits were associated with a highly demanding working memory task. These findings are especially noteworthy precisely because the groups were matched for IQ. A likely explanation is that the difference in language processing between autistic and control individuals arises if the language task makes a substantial demand on executive processes to coordinate the memory and processing demands.

Multiple Deficits Hypotheses. One type of hypothesis concerning autism is that it is a syndrome of multiple primary deficits (e.g., Goodman, 1989, 1994), and two such proposals are particularly relevant to the current context. One proposal links the deficits in executive processing to the frontal system (Ozonoff, Pennington, & Rogers, 1991). There is evidence of metabolic abnormality in the frontal system in the cell membranes that are vital to synaptic function. A hypermetabolic energy state (abnormally decreased levels of phosphocreatine) was found in the dorsal prefrontal cortex of 11 high-functioning autistic adolescent and young adults, using NMR spectroscopy (Minshew, Goldstein, Dombrowski, Panchalingam, & Pettegrew, 1993). Also, the degree of metabolic abnormality correlated with performance in complex language and problem-solving tasks. The fact that the correlation occurs with complex comprehension suggests that the dysfunction plays a role in more domains than those that are explicitly labeled "executive." Note that this frontal hypothesis does not exclude the possibility of other sites of dysfunction, including, for example, the rather widely documented abnormalities in the cerebellum (Courchesnse, 1997). Indeed, such a possibility is consistent the data reported in this volume (Galaburda & Rosen, this volume) that showed that cortical lesions in a developing animal can cause the malformations in a subcortical region. Such data suggest that neural development is interactive, and it is the degree and nature of that interdependence that needs to be elucidated.

A second and not mutually exclusive hypothesis arises from the interdependence of various components in a network. This second hypothesis is that autism affects the interconnectivity among and within various cortical systems. A relative lack of interconnectivity might explain why the difficulties of high-functioning individuals with autism span a variety of complex information-processing tasks. At present, our research gives no indication of whether such a problem might be traced to a single or multiple neurobiological mechanisms. Rather, our goal is to produce a more precise description of the functional characteristics and neural correlates of the relatively preserved and compromised systems.

An fMRI Sentence Comprehension Study. To examine these hypotheses in the area of language processing, we used a simplified version of the sentence-comprehension paradigm that was described previously. Currently, the study involves approximately 11 individuals with autism (depending on the particular study) and half that number of control subjects who are matched in age, gender and IQ. The comparisons between the individuals with autism and the controls concerned several key issues:

- Do the same areas activate during the task?
- Do the different areas activate to the same relative degree?
- Is the activation between and within activated areas equally coordinated?

We presented affirmative active and passive sentences containing high-frequency words, which we knew to be easily comprehensible for this group. The sentences were constructed so that either noun was equally likely to perform the action, for example, *The boy told the farmer* and *The doctor was splashed by the artist.* The task was to read the sentence, press a button to terminate the sentence, and initiate the presentation of a question (such as or *Who did the splashing?—the doctor or the artist*), and press one of two buttons to signal the answer. There were 35 active and 35 passive sentences; five sentences of the same type were presented successively in a single epoch. In addition, there was a baseline condition in which the participant fixated a fixation point. The participants were run through two studies with the same design, although different sentences. We present the data only for individuals who had acceptably low head-motion.

Distribution of Activation. The same major cortical regions are involved in sentence comprehension for the adults with autism as for the control individuals. The major activation is in the posterior superior and middle temporal region, the inferior frontal region, with additional activation in the DLPFC, the more posterior part of the middle frontal gyrus, and of course, in the visual areas including the primary visual cortex, the ventral extrastriate and inferior temporal regions, and the motor regions (partially associated with eye fixations). Thus, there are no differences between individuals with autism and controls with respect to which areas activate during sentence comprehension.

Most but not all of the individuals with autism showed the left lateralization of activation that we typically find in right-handed adults. For two individuals, the language activation was right lateralized, and the analysis included these data; thus, the conclusions apply to the language dominant hemisphere. (While the proportion of right-lateralized processing among this sample seems high, the sample size itself is so small it should not be over interpreted.)

Activation Within the Language Network.   The second key issue concerns the distribution of activation within the language network. We analyzed the relative amount of activation across three main areas (the dominant inferior frontal, posterior superior and middle temporal regions, and dorsolateral prefrontal region) after setting a threshold for each individual such that there was a total of 35 voxels activated in these three areas. The distribution of these 35 voxels turned out to differ for the two groups. Specifically, the individuals with autism had significantly more activation in the temporal region than the controls and less in the inferior frontal (and the activation was similar for DLPFC for both groups for most but not all comparisons). The results were consistent across the two types of sentences (active and passive), and for the two studies. In this limited sample, there is a reliable difference in the relative degree of activation of the inferior frontal and posterior temporal areas. It is possible that the individuals with autism do less semantic processing in the inferior frontal regions, and instead, rely more on the temporal regions.

The difference in the relative amount of activation in inferior frontal gyrus is associated with a difference in the precise location of the activation. There is a suggestion that the autistic individuals show less activation than controls in the anterior, inferior portion of the inferior frontal gyrus (the pars triangularis), but a similar amount in the superior portion. We are still refining and testing the reliability of this subtle difference in the locus of the activation, as well testing various hypotheses about the possible functional significance of the difference.

Functional Connectivity.   The third key issue concerns the functional connectivity among these regions. Our hypothesis was that there might be less coordination between major regions, such as the frontal and temporal regions, for the individuals with autism than for the control individuals. This hypothesis is motivated by the repeated findings that high-functioning autistic individuals show deficits in tasks requiring high-level abstraction, but not in more concrete tasks. We have assessed functional connectivity by examining the correlations in the time course of the fMRI-measured activation of the voxels both within and between the major regions of interest.

To be conservative, we have analyzed the correlations only within sentence-processing epochs, eliminating the data from the fixation epochs. In these analyses, and average time course is computed for one area, and the time course of each voxel of another area is correlated with that average time course. Functional connectivity can be measured as the number of voxels in the second area that are highly correlated (say with, $r > .4$) with the average time course, or the mean of the correlations between the voxels in the second area and the average time course computed from the first area. This analysis yields a relatively consistent pattern, regardless of which measure of connectivity is used. As predicted, the functional connectivity between regions is systematically lower for the individuals with autism. Additionally, some of the areas show higher within-area correlations (each voxel is better correlated with the area's average time course) for the individuals with autism than for controls. These results also generalize to a problem-solving task involving a different set of activated areas. In summary, the analysis suggests a different pattern of functional connectivity for the two groups, with lower interarea connectivity for the networks. This pattern suggests a deficit in the coordination of processing, a deficit that is consistent with the known cognitive deficits of high-functioning individuals with autism.

The theoretical framework we initially presented helped to frame the issues and studies in high-function autism, and the preliminary findings relate directly to the framework: There is no noticeable difference in the constituency of the large scale cortical networks between the high-functioning autistic individuals and the control group. There are differences between the high-functioning autistic individuals and the control group in the way that the workload is distributed among the members of the large scale cortical network. There also appears to be a lower degree of coordination among the members of the network in the high-functioning autistic individuals than in the control group.

How can such brain activation characteristics be related to psychological processes? The finding of lower coordination may be the key. The results show (and everyone knows) that there is more than one way to produce a particular cognitive outcome. In this case, the outcome is the understanding of a sentence. The tasks were constructed to allow comparable cognitive outcomes to occur, and comprehension accuracy was indeed comparable. However, the results show that systematically different patterns of brain activation led to the comparable outcomes. We can refer to these different patterns as different strategies but this labeling, in itself, does not explain the nature or the possible sources of the difference between the groups. We tentatively propose that one key to understanding the difference in the underlying mechanisms lies in the finding of lower functional connectivity in the high-functioning autistic individuals than in the control group.

It may be that some biological factor prevents the interarea collaboration from developing normally, although the individual areas may initially function normally. During the dynamic recruitment of areas to perform new tasks in the course of development, the competition for "most proficient" or "most specialized" may favor the area that is already favored or specialized for performing some other facet of that task, and disfavor another area whose inclusion in the task performance would require interarea coordination. This scheme would gradually lead to extreme intraregion specialization and unusually low interarea coordination. This is a large extrapolation from limited fMRI data, but it is consistent with many documented facets of high-functioning autism, some of which we describe next.

Nonuniform Cognitive Profiles or Relative Sparings in Autism. High-functioning adults with autism often show particular asymmetries in cognitive skills. For example, they have better word level decoding skills and certain visuospatial skills than age-matched and IQ-matched controls along with poorer linguistic and abstract reasoning skills (Minshew, Goldstein, & Siegel, 1997). This asymmetry echoes the striking asymmetry of the profiles reported for the estimated 10% of the autistic population who are savants (Rimland & Fein, 1988). These individuals have truly outstanding talents that are primarily in the domains of music, rote memory, art, word decoding or spelling, math, and mechanics, talents that coexist with marked deficits in abstraction and linguistic comprehension (Aram & Healy, 1988; Huttenlocher & Huttenlocher, 1973; Rimland & Fein, 1988; Siegel, 1994). The music, art, and mnemonic skills are also grounded in the concrete. For example, their artistic skills may involve drawing concrete objects rather than the abstract, conceptual art. The invited inference is that there may be a trade-off between the more concrete skills that reflect the relative over development of some system and the less developed abstraction skills, that may reflect a relative lack of coordination among subsystems, perhaps in addition to other problems.

Developmental disorders, such as autism, provide an important window on neural development by virtue of the developmental process itself. What makes the domain both so challenging and so central is that it highlights the interdependence of various systems, not simply during adult performance, but pervading development itself. The essential characteristic of development is the imperative to grow. It is quite possible that the hyper-development that occurs in individuals with autism is one such outcome of development. If a system cannot develop normally, it may clear the way for some other system to become "overly" developed. We noted that the asymmetry in the profile is marked in what has been estimated as 10% of the low-functioning children with autism who are savants; the asymmetry is found in high-functioning individuals who are compared to age- and IQ-matched controls. Thus, the possibility exists that the asymmetry is common to autism, but simply difficult to identify given that there is such large general variation among individuals in the profile of cognitive and social skills. Thus, the suggestion is that the uneven profiles associated with autism may be an exaggeration of the variation in profiles that mark everyone, an exaggeration that is partially a consequence of the developmental growth process as well as their experiences.

Although our results are still at an early stage, the neuroimaging techniques, in combination with simulation modeling and further experimentation, provide a more precise, testable, and therefore potentially useful, analysis of the developmental processes that underlie autism. It may allow us to understand some of the unique strengths that these individuals possess, as well as point toward additional ways to shore up weaknesses.

Summary.   In this chapter, we have described, in broad outlines, a theoretical picture that is emerging from the functional imaging of the neural systems that subserve high-level cognitive processes. One of the overarching themes to emerge is that the systems are adaptive both over the short term, as a task's demands change, and over the longer term, as the individual learns and develops. Indeed, the adaptive processes in the case of human cognition, include biological evolution and embryology, as well as learning and development. The centrality of adaptation in the development of complex systems is not an insight that is unique to developmental cognitive neuroscience; the same insight is emerging from research in a variety of disciplines, such as genetics and physics. Indeed, several theorists have proposed that at an abstract level, the development of complex systems shows similarities across all domains, from the physical ones to the social ones (Holland, 1992, 1995; Kauffman, 1993).

Drawing closer to our home discipline, it is interesting to consider how the current account refines and builds upon the insights of earlier neurologists, including Luria (1966), and a contemporary neurologist, Mesulam (1990), who eloquently elucidated some similar insights. In both cases, much of the insights came from their own meticulous studies with neuropsychological patients, as well as the experimental data of various researchers in the field. Both accounts, and the earlier ones on which they build, stress the interactivity of various neural systems. Additionally, Luria draws considerable attention to the adaptiveness of physiological systems in general and neural systems in particular.

The fMRI results quantify and further refine these insights. In addition to noting the adaptiveness of a cognitive system, we can observe and measure the dynamic recruitment of cortical areas that underlies the adaptiveness. We can measure changes in activation volume with a particular change in computational demand. We can observe a shift of activation from a damaged area to its contralateral homologue. More generally, fMRI permits a progression from the observation of adaptiveness to the characterization of the underlying mechanism, operating at the level of activation of cortical regions. The early fMRI results begin to bring the mechanisms into relief, but major challenges remain in characterizing the mechanisms of adaptiveness, the dynamics of brain recruitment, in more detail. These mechanisms are likely to be central to development, learning, and adaptation.

Finally, we note Luria's argument that some of the theoretical confusions in neuroscience have roots that may be partially due to ambiguity in the constructs. Specifically, Luria notes that *function* in psychology does not mean the activity of a particular organ. Rather, function means an organism's complex adaptive activity, directed toward the performance of some physiological or psychological task; moreover, the activity may be performed in various ways, with the requirements of the organism determining the means of execution. "Functional systems such as these, complex in composition, plastic in the variability of their elements, and possessing the property of dynamic autoregulation, are apparently the rule in human activity" (Luria, 1966). The contribution of fMRI is to give additional substance to these and other abstract ideas about cognition, and to provide an additional opportunity to quantify these features and to study the operating characteristics of the underlying mechanisms.

ACKNOWLEDGMENTS

The research was partially supported by the National Institute of Child Health and Human Development 1 P01-HD35469, and by the National Institute of Mental Health Research Scientist Awards MH00661 and MH00662.

Address correspondence to Patricia Carpenter, Center for Cognitive Brain Imaging, Department of Psychology, Carnegie Mellon University, Pittsburgh PA 15213 or Carpenter+@cmu.edu.

REFERENCES

Aram, D. M., & Healy, J. M. (1988). Hyperlexia: A review of extraordinary word recognition. In L. K. Obler & D. Fein (Eds.), *The exceptional brain: Neuropsychology of talent and special abilities* (pp. 70–102). New York: Guilford.

Booth, J. R., MacWhinney, B., Thulborn, K. R., Sacco, K., Voyvodic, J. T., & Feldman, H. M. (in press). Developmental and lesion effects in brain activation during sentence comprehension and mental rotation. *Developmental Neuropsychology*.

Braver, T., Cohen, J. D., Jonides, J., Smith, E. E., & Noll, D. C. (1997). A parametric study of prefrontal cortex involvement in human working memory. *NeuroImage, 5*, 49–62.

Buchel, C., Coull, J. T., & Friston, K. J. (1999). The predictive value of changes in effective connectivity for human learning. *Science, 283,* 1538–1541.

Carpenter, P. A., & Just, M. A. (1999). Modeling the mind: Very high-field functional magnetic resonance imaging activation during cognition. In K. R. Thulborn (Issue Ed.), *Topics in magnetic resonance imaging* (Vol. 10, pp. 16–36). Philadelphia, PA: Lippincott, Williams & Wilkins.

Carpenter, P. A., Just, M. A., Keller, T., Eddy, W. F., & Thulborn, K. R. (1999). Graded functional activation in the visuospatial system with the amount of task demand. *Journal of Cognitive Neuroscience, 11,* 9–24.

Carroll, J. B. (1993). *Human cognitive abilities: A survey of factor-analysis studies.* New York: Cambridge University Press.

Case, R. (1992). The role of the frontal lobes in the regulation of human development. *Brain and Cognition, 20,* 51–73.

Casey, P. J. (1993). "That man's father is my father's son": The roles of structure, strategy, and working memory in solving convoluted verbal problems. *Memory & Cognition, 21,* 506–518.

Courchesne, E. (1997). Brainstem, cerebellar and limbic neuroanatomical abnormalities in autism. *Current Opinions in Neurobiology, 7,* 269–278.

Elman, J. A., Bates, E. A., Johnson, M. H., Karmiloff-Smith, A., Parisi, D., & Plunkett, K. (1996). *Rethinking innateness: A connectionist perspective on development.* Cambridge, MA: MIT Press.

Engelien, A., Silbersweig, D., Stern, E., Huber, W., Doring, W., Frith, C. D., & Frackowiak, R. J. S. (1995). The functional anatomy of recovery from auditory agnosia: A PET study of sound categoriazation in a neurological patient and normal controls. *Brain, 118,* 1395–1409.

Gardner, H. (1983). *Frames of mind: The theory of multiple intelligence.* New York: Basic Books, Inc.

Gibson, J. J. (1979). *The ecological approach to visual perception.* Boston: Houghton Mifflin.

Goldstein, G., Minshew, N., & Siegel, D. (1994). Age differences in academic achievement in high-functional autistic individuals. *Journal of Clinical and Experimental Neuropsychology, 16,* 671–680.

Goodman, R. (1989). Infantile autism: A syndrome of multiple primary deficits? *Journal of Autism and Developmental Disorders, 19,* 409–424.

Goodman, R. (1994). Brain disorders. In M. Rutter, E. Taylor, & L. Hersov (Eds.), *Child and adolescent psychiatry: Modern approaches* (3rd ed., pp. 569–593). London: Blackwell.

Grandin, T. (1995). *Thinking in pictures: And other reports from my life with autism.* New York: Doubleday.

Grandin, T., & Scariano, M. (1986). *Emergence: Labeled autistic.* Novato, CA: Arena Press.

Grasby, P. M., Frith, C. D., Friston, K. J., Simpson, J., Fletcher, P. C., Frackowiak, R. S. J., & Dolan, R. J. (1994). A graded task approach to the functional mapping of brain areas implicated in auditory–verbal memory. *Brain, 117,* 1271–1282.

Haarmann, H. J., Just, M. A., & Carpenter, P. A. (1997). Aphasic sentence comprehension as a resource deficit: A computational approach. *Brain and Language, 59,* 76–120.

Hebb, D. O. (1949). *The organization of behavior.* New York: Wiley.

Holland, J. H. (1992). *Adaptation in natural and atificial systems: An introductory analysis with applications to biology, control, and artificial intelligence.* Cambridge, MA: MIT Press.

Holland, J. H. (1995). *Hidden order: How adaptation builds complexity.* Reading, MA: Addison-Wesley.

Howe, M. J. A. (1989). *Fragments of genius: The strange feats of Idiots Savantas.* New York: Routledge.

Hughes, C. (1996). Brief report: Planning problems in autism at the level of motor control. *Journal of Autism and Developmental Disorders, 26,* 99–107.

Hughes, C., Russell, J., & Robbins, T. W. (1994). Evidence for executive dysfunction in autism. *Neuropsychologia, 32*, 477–492.

Huttenlocher, P. R., & Huttenlocher, J. (1973). A study of children with hyperlexia. *Neurology, 23*, 1107–1116.

Johnson, M. H. (1997). *Developmental cognitive neuroscience: An introduction.* Cambridge, MA: Blackwell.

Just, M. A., & Carpenter, P. A. (1985). Cognitive coordinate systems: Accounts of mental rotation and individual differences in spatial ability. *Psychological Review, 92*, 137–172.

Just, M. A., & Carpenter, P. A. (1992). A capacity theory of comprehension: Individual differences in working memory. *Psychological Review, 99*, 122–149.

Just, M. A., & Carpenter, P. A. (1993). The intensity of thought: Pupillometric indices of sentence processing. *Canadian Journal of Experimental Psychology, 47*, 310–339.

Just, M. A., Carpenter, P. A., Keller, T. A., Eddy, W. F., & Thulborn, K. R. (1996). Brain activation modulated by sentence comprehension. *Science, 274*, 114–116.

Just, M. A., Carpenter, P. A., & McNeil, M. (1999). Neurocognitive remediation of aphasic language understanding. Unpublished manuscript, Carnegie Mellon University, Center for Cognitive Brain Imaging, Pittsburgh, PA.

Just, M. A., Carpenter, P. A., & Varma, S. (1999). Computational modeling of high-level cognition and brain function. *Human Brain Mapping.*

Kauffman, S. A. (1993). *The origins of order: Self-organization and selectin in evolution.* New York: Oxford University Press.

Keller, T. A., Carpenter, P. A., & Just, M. A. (1999). The neural basis of sentence comprehension: fMRI examination of syntactic and lexical processing. Unpublished manuscript, Carnegie Mellon University, Center for Cognitive Brain Imaging, Pittsburgh, PA.

King, J., & Just, M. A. (1991). Individual differences in syntactic processing: The role of working memory. *Journal of Memory and Language, 30*, 580–602.

Kwong, K. K., Belliveau, J. W., Chesler, D. A., Goldberg, E. I., Weisskoff, R. M., Poncelet, B. P., Kennedy, D. N., Hoppel, B. E., Cohen, M. S., Turner, R., Cheng, H. M., Brady, T. J., & Rosen, B. R. (1992). Dynamic magnetic resonance imaging of human brain activity during primary sensory stimulation. *Proceedings of the National Academy of Science, USA, 89*, 5675–5679.

Luria, A. R. (1966). *Higher cortical functions in man: Second Edition Revised and Expanded.* New York: Basic Books.

McClelland, J. L., Rumelhart, D. E., & The PDP Research Group. (1986). *Parallel distributed processing: Explorations in the microstructures of cognition Volume 2: Psychological and biological models.* Cambridge, MA: MIT Press.

Mesulam, M.-M. (1990). Large-scale neurocognitive networks and distributed processing for attention, language and memory. *Annals of Neurology, 28*, 597–613.

Michaels, E., Just, M. A., Keller, T. A., & Carpenter, P. A. (1999). An fMRI comparison of visual and auditory sentence comprehension. Unpublished manuscript, Carnegie Mellon University, Center for Cognitive Brain Imaging, Pittsburgh, PA.

Minshew, N. J. (1996). Autism. In B. O. Berg (Ed.), *Principles of child neurology* (pp. 1713–1729). New York: McGraw-Hill.

Minshew, N. J., Goldstein, G., Dombrowski, S. M., Panchalingam, K., & Pettegrew, J. W. (1993). A preliminary ^{31}P MRS study of autism: Evidence for undersynthesis and increased degradation of brain membranes. *Biological Psychiatry, 33*, 762–773.

Minshew, N. J., Goldstein, G., & Siegel, D. J. (1995). Speech and language in high-functioning autistic individuals. *Neuropsychology, 9*, 255–261.

Minshew, N. J., Goldstein, G., & Siegel, D. J. (1997). Neuropsychologic functioning in autism: Profile of a complex information processing disorder. *Journal of the International Neuropsychological Society, 3*, 303–316.

Neville, H. J. (1993). Neurobiology of cognitive and language processing: Effects of early experience. In M. H. Johnson (Ed.), *Brain development and cognition: A reader* (pp. 424–448). Cambridge, MA: Blackwell.

Nicolelis, M. A. L. (1997). Dynamic and distributed somatosensory representations as the substrate for cortical and subcortical plasticity. *Seminars in Neurobiology, 9*, 24–33.

Ogawa, S., Lee, T. M., Kay, A. R., & Tank, D. W. (1990). Brain magnetic resonance imaging with contrast dependent on blood oxygenation. *Proceedings of the National Academy of Science, USA, 87*, 9868–9872.

Ozonoff, S. (1995). Executive functions in autism. In E. Schopler & G. B. Mesibov (Eds.), *Learning and cognition in autism* (pp. 199–219). New York: Plenum.

Ozonoff, S., & McEvoi, R. E. (1994). A longitudinal study of executive function and theory of mind: Development in autism. *Development and Psychopathology, 6*, 415–431.

Ozonoff, S., Pennington, B. F., & Rogers, S. J. (1991). Executive function deficits in high-functioning autistic individuals: Relationship to theory of mind. *Journal of Child Psychology and Psychiatry, 32*, 1081–1105.

Ozonoff, S., Strayer, D. L., McMahon, W. M., & Filloux, F. (1994). Executive function abilities in autism and Tourette syndrome: An information processing approach. *Journal of Child Psychology and Psychiatry, 35*, 1015–1032.

Parks, R. W., Lowenstein, D. A., Dodrill, K. L., Barker, W. W., Yoshii, F., Chang, J. Y., Emran, A., Apicella, A., Sheramata, W. A., & Duara, R. (1988). Cerebral metabolic effects of a verbal fluency test: A PET scan study. *Journal of Clinical and Experimental Neuropsychology, 10* , 565–575.

Petitto, L. A. (1996). In the beginning: On the genetic and environmental factors that make early language acquisition possible. In M. Gopnik & S. Davis (Eds.), *The genetic basis of language* (pp. 46–71). Hillsdale, NJ: Lawrence Erlbaum Associates.

Piaget, J., & Inhelder, B. (1969). *The psychology of the child.* New York: Basic Books.

Rimland, B., & Fein, D. (1988). Special talents of autistic Savants. In L. K. Obler & D. Fein (Eds.), *The exceptional brain: Neuropsychology of talent and special abilities* (pp. 474–492). New York: Guilford.

Rumsey, J. M. (1985). Conceptual problem-solving in highly verbal, nonretarded autistic men. *Journal of Autism and Developmental Disorders, 15*, 23–36.

Rumsey, J. M., & Hamburger, S. D. (1988). Neuropsychological findings in high-functioning autistic men with infantile autism, residual state. *Journal of Clinical and Experimental Neuropsychology, 10*, 201–221.

Sanes, J. N., & Donoghue, J. P. (1997a). Dynamic motor cortical organization. *The Neuroscientist, 3*, 158–165.

Sanes, J. N., & Donoghue, J. P. (1997b). Static and dynamic organization of motor cortex. *Advances in Neurology, 73*, 277–296.

Sanes, J. N., Donoghue, J. P. T. V., Edelman, R. R., & Warach, S. (1995). Shared neural substrates controlling hand movements in human motor cortex. *Science, 268*, 1775–1777.

Schwartz, M. F., Saffran, E. M., Fink, R. B., Myers, J. L., & Marin, N. (1994). Mapping therapy: A treatment program for agrammatism. *Aphasiology, 8*, 19–54.

Selfe, L. (1977). *Nadia: A case of extraordinary drawing ability in an autistic child.* London, UK: Academic Press.

Siegel, D. J., Minshew, N. J., & Goldstein, G. (1996). Wechsler IQ profiles in diagnosis of high-functioning autism. *Journal of Autism and Developmental Disorders, 26*, 389–406.

Siegel, L. S. (1994). The modularity of reading and spelling: Evidence from hyperlexia. In G. D. A. Brown & N. C. Ellis (Eds.), *Handbook of spelling: Theory, process and intervention* (pp. 227–248). Sussex, UK: Wiley.

Stiles, J., & Thal, D. (1993). Linguistic and spatial cognitive development following early focal brain injury: Patterns of deficit and recovery. In M. H. Johnson (Ed.), *Brain development and cognition: A reader* (pp. 643–664). Cambridge, MA: Blackwell.

Thatcher, R. W. (1992). Cyclic cortical reorganization during early childhood. *Brain and Cognition, 20,* 24–50.

Thulborn, K. R., Carpenter, P. A., & Just, M. A. (1999). Plasticity of language-related brain function during recovery from stroke. *Stroke, 30,* 749–754.

Weiller, C., Isensee, C., Rijntjes, M., Huber, W., Muller, S., Bier, D., Dutschka, K., Woods, R. P., Noth, J., & Diener, H. C. (1995). Recovery from Wernicke's aphasia: A positron emission tomographic study. *Annals of Neurology, 37,* 723–732.

CONCLUDING COMMENTARY

Chapter 16

Educating the Human Brain:
A Commentary

Michael I. Posner
Sackler Institute

Two major themes emerged from this meeting. The first is that cognitive development might benefit from new ideas from neuroscience. The second is that taken together, cognitive development and neuroscience might tell us something about both helping normal children with education and aiding children with special problems. Before dealing with these issues directly, it is important to recognize that there is nothing really new about efforts to use current psychology to reform education and child raising.

History

The psychology of learning certainly had influence on education in the United States early in this century. For example, there was an important discussion about the role of formal classical education between those who believed that learning was primarily about developing a general capacity of mind and those who argued for domain specific knowledge. Thorndike's studies (Thorndike & Woodworth, 1901), that demonstrated the importance of common elements in order for learning to transfer between domains, was one basis for the development of elective routes through schooling.

This same issue of domain specificity has been examined in more contemporary terms. Efforts to produce models of mental processes in the form of computer programs such as the general problem solver (Newell,

Shaw, & Simon, 1958), showed that relatively few general principles could be used to provide a program that would solve a broad range of problems. Many followers of this effort felt that humans would need to acquire only a few general principles to perform a very wide range of complex problem solving. Research in artificial intelligence, like the studies of Thorndike, revealed quite the opposite. Expertise in computer systems required a great deal of knowledge about the specific domain in which the problem is posed. In addition, research on human chess masters led Herbert Simon to conclude that their skills were based almost entirely on domain-specific knowledge about chess acquired in thousands of hours of practice.

Until 30 years ago, psychology was still dominated by an interest in learning. Changes in the behavior of rats and humans as they mastered mazes, guessed which light would occur next or anticipated the response in paired-associate learning, were dominant paradigms. Some principles affected educational practice. For example, B. F. Skinner held that it was important to program learning so that the learner would not make errors and lose motivation. The teaching machine movement was one outcome of these experiments and direct instruction of reading was another.

Cognitive psychologists (for example, Jerome Bruner) argued for the importance of a sense of discovery as an element in the teaching and motivation of the student. This tension between education as drill and as discovery remains present in the current effort to apply cognition and neuroscience to education.

After the cognitive revolution, the interest in learning generally subsided and was replaced with an effort to understand the structures of memory. This history perhaps explains some of the complaints heard here by developmental psychologists (e.g., Siegler) that studies of the process of change have made relatively little progress.

For more than forty years, the ideas of Jean Piaget held sway in developmental psychology. Piaget presented a sophisticated treatment describing how the initial genetic endowment of human infants matured in a series of stages into the formal operations of the thinking adult. Piaget placed relatively little emphasis on the ability of education to alter or speed these changes. Thus, his views supported the idea that schooling ought to introduce subjects only when the child's stage of mental development is ready. In the late 1970s, Jerome Bruner and others took a different view. They argued—based on a cognitive perspective of the type—that properly analyzed, any subject can be introduced to the students at any age.

Arguments of psychologists like Skinner and Simon that influenced education did not concern any detailed model of the brain. In fact, most psychologists, while regarding brain activity as basic to learning in some theoretical sense, believed that we could acquire the principles of psychology needed for understanding learning without any reference to brain activity. While B. F. Skinner was the leading figure in American behaviorism and Herbert Simon was the founder of Cognitive Science, they were in agreement with the belief that by studying the learners' environment and by knowing the relatively simple principles by which organisms learn, it was possible to guide the educational process.

Perhaps the popular acceptance of phrenology as brain theory at the turn of the century had inoculated psychologists against the acceptance of any particular view of how this mysterious organ, the human brain, actually worked. Roger Sperry's Nobel prize-winning work improved understanding of the human brain by studying people whose cerebral hemisphere had been separated to stop otherwise uncontrolled seizures. This work demonstrated the different complex capabilities of each separated hemisphere and led to renewed interest in the problem of localization of function. The split-brain research has had a profound influence on a generation of psychologists whose studies tended to examine the specialized processes that might be present in the right and left hemisphere of the normal person. There emerged theories of human intelligence based on consideration of the organization of the brain and the effects of brain injury (e.g., Gardner, 1991), and renewed interest in the role of laterality in the acquisition and execution of language and other high level human skills.

As in the case of phrenology, there was wide public interest that often overdramatized the separate contributions of the two hemispheres in normal function. Training models for the right and left hemisphere became fashionable, and it was common to describe people or institutions as either left or right hemisphere. Despite these extremes, there is little question that the split-brain studies have done a great deal to rekindle interest in the neurology of higher mental processes and hopes that there could be an understanding of how cognitive and emotional processes were carried out by the brain. However, in normal persons, there is always very strong interaction between the hemispheres and the relatively few split-brain patients had intractable cases of epilepsy in their childhood, making it difficult to base firm conclusions about normal function. These problems made it difficult to establish an empirical agenda that could counter very simplified views of hemispheric function that became the basis of popular understanding.

Methodology

Research about what happens in the normal brain when it carries out high level skills had to await new methods that could open up the normal human brain to careful scrutiny. With this history in mind, we might ask: Are we better equipped to speak to these critical issues now than we were 10 or 100 years ago?

Up until the last 10 years, neuroscience tended to deny the importance of specific learning experience in shaping brain structures and circuits. There was no way to examine such changes in normal humans and thus no way except changes in behavior to gage the effectiveness of rehabilitation and teaching methods. These factors have now changed, as is clear from the papers in this volume. Distinguished neuroscientists pursue issues of brain plasticity thanks heavily to the contribution of Michael Merzenich. Those interested in the development of the human brain and its adult state have available several techniques that allow the analysis of activity in ever smaller areas of the human brain (Casey, Just, Haier, Johnson, Neville).

The various chapters introduce imaging of metabolism or blood flow with PET (Haier) and changes in blood oxygenation with fMRI (Casey, Just, Neville). These methods reflect the close ties between neural activity and changes in the vascular system. They are becoming more accurate in spatial resolution so that it is now possible to resolve differences in the area of the brain activated by two different conditions within about a millimeter. Recent developments have also allowed trial-by-trial analysis of evoked blood flow so that designs can be used that mix different trial types which are later averaged. However, any measure based on hemodynamic responses will lag behind neuronal events by several hundred to thousands at millisecs. We know from chronometric studies that mental operations that underlie cognitive processes often take place in the range of ten to hundreds of ms.

Most of the tasks studied in this volume involve five or more active brain areas. Tracing the fine temporal dynamics of when each of these brain areas becomes active while people carry out high-level skills, has required combining hemodynamic methods with those of temporal precision such as electrical or magnetic recording.

For example, obtaining the use for a stimulus noun, once sensory and motor processes are subtracted, shows four unique active brain areas. These are the anterior cingulate, left frontal and posterior cortex, and the right cerebellum (Posner & Raichle, 1994). Recordings from scalp electrodes (Abdullaev & Posner, 1998) show that the anterior cingulate (attentional processing) starts at about 170 ms, with the left frontal (lexical semantics) at about 220 ms and the left posterior only at about 500 ms. Although it is assumed that the cerebellar activity occurs in conjunction with the left frontal area, it has not been possible to record a scalp signature of this activity.

In order to use neuroimaging and other methods to trace change in the developing brain, it is necessary to obtain observations at the moment change is occurring. This point is made very forcefully by chapters on the microgenetic method (Kuhn, Siegler). By careful analysis of when change might occur, and by the introduction of methods to hasten its occurrence, it is possible to observe the moment of change in the child's verbal responses, in their gestures (Goldin-Meadow), or in their eye movements (see Johnson). Even when that change is not consciously available to the child, Goldin-Meadow has shown how it may be expressed by gesture. Indeed, 30 years of work with mental chronometry in adults (Posner, 1978) has provided us with a variety of methods in which unconscious information may be expressed in the person's responses to probe events. In addition, the microgenetic method itself benefits from a long period of study of the analysis of verbal protocols (Ericsson & Simon, 1991) and nonverbal (e.g., eye movement) protocols.

The use of chronometric and/or neuroimaging methods can allow examination of events even during the pauses in verbal behavior that often occur at the moment of insight (such as in Siegler's analysis of his patient KBK). These observations can occur whether or not the subject is yet conscious of what will eventually be the product of thought. Semantic priming and tip of the tongue studies have shown clearly that associations still below consciousness can be observed by their influence on probe events.

These methods taken as a whole provide a rich tool kit for examining change at the moment it occurs. How are these changes likely to be manifested in terms of brain circuitry? We turn now to the analysis of the question of the mechanisms of plasticity.

PLASTICITY

In neuroscience, plasticity has been examined primarily as a change at a given synapse (e.g. the Hebb synapse) for example, as discussed in the McClelland chapter. In order to examine the many forms of change in behavior that are discussed in this volume, it is going to be necessary to greatly enlarge the types of brain activity examined. Some behavior change occurs very rapidly as a person shifts from one strategy to another; other changes may involve many years of development. Table 16.1 consid-

TABLE 16.1
Mechanisms of Plasticity

Time	Name	Mechanisms	Chapters
(1) milliseconds	Attention	Strategies	Siegler, Kuhn
(2) seconds to minutes	Priming	Automation	Haier
(3) days to weeks	Practice	Connections	McClelland
(4) weeks to months	Learning	Maps	Merzenich
(5) months to years	Development	New Networks	Thelen, Johnson, Neville

ers the relation between various time scales in which behavior changes and the most likely form of brain plasticity that might underlie each of them.

The importance of strategies in the development of cognitive skills was discussed by Siegler and by Kuhn. What is a strategy in the brain? Many studies have shown that attention can increase blood flow (neuronal activity) within areas of the brain that are performing a particular computation. For example, attending to visual motion boosts activity in the region of the mid-temporal lobe devoted to processing moving stimuli.

Recent work shows that attention can reprogram the temporal organization of the anatomical areas that perform computations during thought (Posner & Pavese, 1998; Posner & Raichle, 1994). For example, if subjects are asked to press a key indicating the lexical category of a word (e.g., manufactured), there is activity in the left frontal area. Processing whether a word fits into a sentence frame activates a left posterior area. When subjects are asked to press a key *"if the word is manufactured and fits into the sentence frame,"* the left frontal area is increased in electrical activity early in processing and the left posterior, late. It is as if priority is first given to the lexical classification and then to the sentence. However, this time course is changed if subjects are asked to determine *"if the word fits the sentence and is manufactured"* (Posner & Pavese, 1998). Now, the increase in left posterior activity occurs earlier. It is as though the instruction rapidly reprograms the priority of the computations creating a new circuit of information flow that fits the new strategy.

Haier discussed the studies of Raichle showing how just a few minutes of practice with generating uses for a list of words can change the circuit involved. In this example, the use of each noun is well known to the subject even before the task starts. Practice works by priming the particular responses used by the person. At the start of practice it took about 1100 ms to generate the use of a familiar noun (e.g., pound to hammer), but after practice, the reaction time was down by about half. The brain areas involved in word generation dropped out and instead the anterior insula, which is very active in reading words aloud and is suppressed below reading while generating uses, returns to the same level of activity as in reading. Thus, a few minutes of practice has changed the circuitry involved in the task from one related to thinking and heavily involving prefrontal cortex and attention to one involved when a task is highly automated, such as reading words aloud. This study shows clearly how the brain adapts to automation by developing a new circuitry. For many years, neurologists have known that patients unable to perform a task deliberately can do so when the situation calls for an automatic response. Now, we have a good idea of how the brain performs this interesting trick, and it turns out to be part of normal brain machinery. There have been many neuroimaging studies of priming conducted in a wide variety of tasks. The universal finding is that priming tunes the brain so that less activity is required to carry out the task.

Row 3 of Table 16.1 involves learning of new associations, such as involved when connectionist models acquire a new stimulus–response pairing (McClelland). In one study of this process, undergraduates learned the meaning of 40 lexical items in a new artificial language (McCandliss, Posner, & Givon, 1997). After over 50 hours of practice, the new items showed the same "word superiority effect" in physical matching tasks as were found for the native language. McCandliss found that the new items activated left hemisphere brain areas that appeared to be related to the meaning of the word in a way that closely resembled the activity found for their native English. Connections between reading the word silently and the frontal and posterior areas related to meaning were established by the practice. However, the study showed no evidence that the visual system of subjects were automatically packing the letters into a visual word form in the way in which they would with English. New connections between brain areas seem to be formed at a rate that may be much faster than the structures represented by activity within one brain area were altered.

Merzenich said that monkeys could alter sensory maps by appropriate training. Because monkeys are kept in the laboratory working many hours a day, high levels of training can be achieved rather rapidly. However, human studies most often distribute learning across rather short sessions, so that it may be many weeks of practice before the structure related to the learned task actually changes. For example, practice on a repetitive motor task led to an expansion of the size of motor cortex devoted to the learned sequence after about 3 weeks of training (Karni et al., 1998). Of course, this time course may differ greatly depending upon what sort of map is being studied. As mentioned previously, McCandliss failed to observe changes in the visual word form area even after 50 hours of practice in a lexical acquisition tasks that was sufficient to produce a normal word superiority effect. Much more will need to be done to understand the time course necessary to achieve skills and to observe how each form of plasticity might change with age and experience. Observation of brain areas related to semantic categories (Martin et al., 1995; Spitzer et al., 1996) makes such studies of great importance as we try to understand how learning of new concepts takes place.

The final row of Table 16.1 refers to the initial acquisition of skills in the process of human development. Presumably a skill such as reaching (Thelen), face recognition (Johnson) or language processing (Neville) begins by developing both the brain structures and their connections necessary to perform the task. Perhaps both continue to develop over many years before reaching anything approaching their adult form. In this regard a recent study by McCandliss related to the visual word form system

(Posner & Raichle, 1994) may be of particular importance. In adults, the visual word form system seems to involve posterior structures that respond to orthographically regular letters strings, but do not respond to consonant strings. The data suggest the response of this system is not based on familiarity since highly familiar words and very unfamiliar but orthographically regular strings seem to have the same impact. Moreover, adult studies suggest that 50 hours of practice with new strings had no impact on this system. However, in his work with 10-year-old children, McCandliss showed that the system responded to familiar, but not unfamiliar words (Posner & McCandliss, 1999). It appears as though this system begins by storing individual exemplars, but as neighborhoods become dense, computes what appear to be orthographic rules. If this is a correct result, it suggests that studies of the acquisition of skills in children may be particularly revealing of the initial forms of computation. For example, the number comparison task that Case describes in his chapter from the work of Dehaene (1996), seems to be nearly in its adult form in a group of high-performing 5-year-olds (Temple & Posner, 1998). This result is very much unlike that of McCandliss for the visual word form in reading and suggests that it will be necessary to explore the development of individual skills before coming to any strong conclusions about differences between child and adult learning mechanisms.

CONTROL

The issue of high level (conscious) control over the mechanisms of cognition and emotion is raised very forcefully in the beautiful chapter on the control of reaching (Thelen). She says

> "coordination of the parts happens without a central executive giving commands, but rather the interaction of components, such that the systems are truly self-organizing."

This view contrasts strongly with the need for executive control of those authors dealing with strategies (e.g., Kuhn, Siegler). I believe both approaches have merit. Cognitive researchers who have proposed a central executive may have left the impression that the central executive was fully in control of our behavior. Thelen shows that behavior can emerge without central control and studies of priming also show that even high-level processes may interact and influence each other without guidance or executive control. However, both cognitive models and PET and other neuroimaging studies suggest that there may be attentional networks whose purpose is to select among and influence, specific computational systems related to particular tasks. Attention has both been shown to influence neuronal activity within widely separated brain anatomy, and also to arise in specific brain networks (Posner & DiGirolamo, 1998).

The executive control exercised by attention appears to arise in some of the striatal–frontal networks discussed by Casey in her chapter. Tasks that involve resolving conflict between stimuli dealing with novel events or monitoring errors all produce activity in the anterior cingulate gyrus or closely related other midfrontal structures. Studies of the Stroop effect are the most prominent. They seem to produce strong activity in the midfrontal region associated with the act of selecting the ink color rather than the competing color name. Studies of the Stroop effect with high-density electrodes (Posner & Rothbart, 1998) show activity in this midline area, where trials with a colored word segregate for noncolor words (neutral trials) by 270 ms. It is assumed that in mixed block trials, it is necessary for both congruent and incongruent trials to amplify information from the extrastriate color errors to reduce the color name competition. Later on, incongruent trials are particularly difficult because their competing responses often lead to error. Both response competition and error detection produce additional cingulate activity (Casey, this volume; Dehaene, 1996). Thus, attention serves both to select among competing stimuli and among competing responses.

Why should that anterior cingulate, which is the outflow of the limbic system and thus clearly related to emotion, be involved in the monitoring and control of purely cognitive activity? One important observation is that this same brain area seems to be involved in response to pain (Rainville et al., 1997). We suggest (Posner & Rothbart, 1998) that the need to control distress in early infancy involves regulation of areas involved in distress (e.g., the amygdala) by the anterior cingulate (Posner & Rothbart, 1998). This interaction produces soothing by distraction, which can occur either through the mediation of a caregiver (Harman et al., 1997), or eventually through the child's own attention system. When it becomes necessary to resolve conflict in purely cognitive tasks, by selecting among competing stimuli or responses, we speculate that the same brain system, already developed for the regulation of emotion, is used for this purpose. Casey's studies of the striatal–frontal circuitry outlined several different networks that may be involved in control by inhibition and it remains an important task for students of executive control to understand the functions of each of them.

Remediation

The chapters here include study of a large number of groups with difficulties in cognitive and emotional processing, either due to specific brain injury or to development. These include autism (Just), Attention Deficit Disorder (Casey), closed head injury (Stiegler), Dyslexia (Galaburda), language difficulties (Merzenich), and others. As pointed out by Haier, it has been difficult to study individual differences using the PET method. There is a severe limit to the amount of data that can be collected on any individual with PET and there are difficulties in the registration of anatomical with functional images. Functional MRI has greatly aided the study of individual differences both by increasing the data collection and using the same basic method to collect both structural and functional images. There still remain problems in comparing normal controls with patients because subject designs, with small number of subjects, can make for serious problems of subject selection. If the brains of the two groups should behave differently during a task, it is not possible to know if the brain abnormality was the cause of the disorder or resulted from difficulty in doing the task induced by a disorder whose cause could be rather remote.

One of the most striking aspects of this volume is that many of the authors included a remediation program as a part of the testing of their theory (Case, Galaburda, Just, McClelland, Merzenich, Siegler). This is a new development of great potential practical significance and, of course, it also allows the subjects to serve as their own control. Most of the chapters presented some evidence of a behavioral improvement from the therapy. Some of these were documented in very great detail (Case, Merzenich), while others were more impressionistic. Even in the chapters where detailed documentation of the behavior was provided, it was not always easy to figure out what aspect of the therapy led to what improvement. Two systematic efforts solve this difficulty. One of these was in McClelland's use of a simulation that allowed testing of the degree to which stimulus enhancement and feedback, two of the many variables used in the Merzenich–Tallal program, may have contributed to learning the specific phonemic contrast he studied. Galaburda described an animal model that allowed one to study the interaction between brain areas in development that might have led to difficulties in resolving small temporal differences.

The most obvious way to observe a change in the brain with remediation is to use neuroimaging methods to study the same subjects prior to and following therapy. Because many of the changes induced by therapy may lie in the circuits involved, a combination of fMRI and event-related magnetic or electrical potentials might be especially useful in tracing the effects of the therapy. However, except for the single aphasic patient studied by Just, we did not have any completed examples of this strategy, though several of the authors mentioned efforts to carry out this kind of experiment. The results of the Just experiment were quite interesting. The frontal lesion produced a loss in the ability of Wernicke's area to respond normally during the processing of sentences. Both behavioral results and the operation of Wernicke's area were normalized by the therapy and additional activity was seen in the neighborhood of the frontal lesion. If we knew the organization of this activation in real time, we would have even a stronger set of clues as to what the therapy might be doing. Although a change in behavior is an important result to document from therapy, the analysis of changes in brain circuitry is also important. Klahr asked how we might go from guessing to knowing. The convergence of a performance and a brain change has this function. This kind of convergence produces an explanatory model for the observed change that is more persuasive in convincing people and institutions of the importance of therapy. In addition, images of the change can provide a rapid means of guiding changes in therapy that may make it more effective.

It is still very early in the strategy of imaging brain changes resulting from therapy, to evaluate its theoretical and practical significance. Moreover, there is even less data on the use of neuroimaging to evaluate the effects on the brain of educational methods with normal subjects (e.g., the study being performed by Case). However, judging from the very preliminary results described here and from my own limited experience in working on research projects with reading in normal subjects (McCandliss, Posner & Givon, 1997), and in closed head injury patients (Sohlberg et al., in press), I believe it will be a major aspect of the study of brain development in the coming years. The practical opportunities to assist with developmental and acquired disorders and at the same time test and advance theory, make the mating of educational and therapeutic methods with neuroimaging almost irresistible.

CONCLUSIONS

The organizers of this volume pose four questions as the major issues to which it is addressed. We have had some brilliant flashes of the kind of answers likely to emerge from research.

Why, they asked, do cognitive abilities emerge when they do? Thelen's chapter presents a beautiful analysis of why the child reaches. Remember the lesson that McClelland teaches; if you arrange events right, perhaps a new ability will emerge even at a much older age.

What are the sources of individual differences in behavior and of developmental abnormalities? Casey finds many abnormalities related to fronto–striatal circuitry and a picture of the role of these structures in normal development may be emerging from attentional research described in the control section of this commentary.

What happens when people learn? I offer the material summarized in Table 16.1 from the various chapters to provide an overview of the many types of brain changes that accompany normal learning.

How can experience be ordered to optimize learning? McClelland's chapter provides an interesting empirical strategy for discovering what parts of an intervention make a difference. Perhaps a combination of neuroimaging with behavioral assessment may improve interventions. However, we should remember a caution raised in discussion by Klahr; it is often difficult, if not impossible, to optimize, and perhaps we will have to be content with a merely satisfactory solution (what Simon called satisficing). If this leads to improvement in the lives and potential of those undergoing education or therapy, that in itself is a great gain.

Even if neuroimaging methods do not, in the end, allow us to greatly improve our interventions, there may be another gain. In a recent study of closed head injury patients, we found that our placebo (teaching our patients a course in the neuropsychology of brain function) improved their feelings about themselves and their social functioning (Sohlberg, et al., in press). We should have known in advance that it helps human beings to understand their condition and to have a way of connecting it to general information and to the fate of others with similar problems. We should not forget that knowledge has a certain pleasure and power for its own sake. On that score, this volume and the meeting that led to it, has been a great success.

REFERENCES

Abdullaev, Y. G., & Posner, M. I. (1998). Event-related brain potential imaging of semantic encoding during processing single words. *Neuroimage, 7*, 1–13.

Bruer, J. T. (1997) Education and the brain: A bridge too far. *Educational Researcher, 26*(8), 4–16.

Dehaene, S. (1996). The organization of brain activations in number comparison: event related potentials and additive factors method. *J. Cognitive Neuroscience, 8*, 47–86.

Dehaene, S., Posner, M. I., & Tucker, D. M. (1994). Localization of a neural system for error detection and compensation. *Psychological Science, 5*, 303–305.

Ericsson, K. A., & Simon, H. A. (1991). *Protocol analysis.* Cambridge MA: MIT Press.

Gardner, H. (1991). *Multiple intelligences.* New York: Basic Books.

Harman, C., Rothbart, M. K., & Posner, M. I. (1997). Distress and attention interactions in early infancy. *Motivation and Emotion, 21*, 27–43.

Karni, A., Meyer, G., Rey-Hipolito, C., Jezzard, P., Adams, M. M., Turner, R., & Ungerleider, L. G. (1998). The acquisition of skilled motor performance: Fast and slow experience-driven changes in primary motor cortex. *Proc. U.S. Nat'l Acad of Sci, 95*, 861–68.

Martin, A., Haxby, J. V., Lalonde, F. M., Wiggs, C. L., & Ungerleider, L. G. (1995). Discrete cortical regions associated with knowledge of color and of action. *Science, 270*, 102–105.

McCandliss, B. D., Posner, M. I., & Givon, T. (1997). Brain plasticity in learning visual words. *Cognitive Psychology, 33*, 88–110.

Newell, A., Shaw, J. C., & Simon H. A. (1958). Elements of a theory of human problem solving. *Psychological Review, 65*, 151–165.

Posner, M. I. (1978). *Chronometric explorations of mind.* Hillsdale, NJ: Lawrence Erlbaum Associates.

Posner, M. I., & DiGirolamo, G. J. (1998). Executive attention: Conflict, target detection and cognitive control. In R. Parasumaman (Ed.), *The attentive brain* (pp. 401–423). Cambridge, MA: MIT Press.

Posner, M. I., & McCandliss, B. D. (1999). Brain circuitry during reading. In R. Klein & P. McMullen (Eds.), *Converging methods for understanding reading and dyslexia* (pp. 305–337). Cambridge, MA: MIT Press.

Posner, M. I., & Pavese, A. (1998). Anatomy of word and sentence meaning. *Proceedings of the National Academy of Sciences, USA, 95*, 899–905.

Posner, M. I., & Raichle, M. E. (1994). *Images of mind.* New York: Scientific American Books.

Posner, M. I., & Rothbart, M. K. (1998). Attention, self-regulation and consciousness. *Transactions of the Philosophical Society of London B, 353*, 1915–1927.

Rainville, P., Duncan, G. H., Price D. D., Carrier, B., & Bushness, M. C. (1997). Pain affect encoded in human anterior cingulate but not somatosensory cortex. *Science, 277*, 968–970.

Sohlberg, M. M., McLaughlin, K. A., Pavese, A., Heidrich, A., & Posner, M. I. (in press). Evaluation of attention process therapy training in persons with acquired brain injury. *J of Exp. and Clinical Neuropsych.*

Spitzer, M., Bellemann, M. E., Kammer, T., Friedemann, G., Kichka, U., Maier, S., Schwartz, A., & Brix, G. (1996). Functional MR imaging of semantic information processing and learning-related effects using psychometrically controlled stimulation paradigms. *Cognitive Brain Research, 4*, 149–161.

Temple, E., & Posner, M. I. (1998). Brain mechanisms of quantity are similar in 5-year olds and adults. *Proceedings of the National Academy of Sciences, USA, 95*, 7836–7841.

Thorndike, E. L., & Woodworth, R. S. (1901). The influence of improvement in one mental function upon the efficiency of other functions. *Psychological Review, 8*, 247–261.

Author Index

A

Abdullaev, Y. G., 390, *398*
Abel, L., 126, 136, *143*
Abeles, M., 76, *91*
Aboitiz, F., 311, *321*
Adams, C., 222, *249*
Adams, M. M., 138, *143*, 393, *398*
Adcock, E. W., 339, *348*
Ahern, S., 127, *141*
Ahissar, E., 76, *91*
Ahissar, M., 74, 76, 77, 88, 89, *91*
Ahmed, S. A., 319, *320*
Ahn, W., 228, *248*
Akahane-Yamada, R. A., 115, 118, *119*
Albright, T. D., 213, *216*
Albus, K. E., 328, *347*
Alegria, J., 308, *322*
Alexander, D. W., 114, *120*
Alexander, G. E., 328, 333, 336, *346*
Alibali, M. W., 6, 8, 9, 12, 13, 17, 19, 20, 21, 22, 26, *28, 29*
Alkire, M. T., 124, 129, 131, 132, 139, 140, *141, 142, 143*
Allard, T. A., 68, 69, 74, *91, 92, 93*, 284, *286*
Allard, T. T., 69, 76, *93*
Alliquant, B., 81, *92*
Allison, T., 257, 260, *267, 270*
Almassy, N., 167, *182*
Almeida, G. L., 177, *183*
Almli, C. R., 33, *62*
Amsel, E., 229, *248*
Andersen, C., 6, *30*, 222, 229, 230, 240, *248*

Andersen, R. A., 98, *120, 121*
Anderson, C., 228, *249*
Anderson, C. T., 124, 139, *142*
Anderson, K. C., 309, *320*
Andreasen, N. C., 134, 138, *141*
Anglade, P., 81, *92*
Anker, S., 276, *285*
Apicella, A., 126, *144*, 371, *382*
Aram, D. M., 377, *379*
Arieli, A., 76, *91*
Arissian, K., 78, *93*
Armstrong, B. A., 274, *285, 286*
Arndt, S., 134, 138, *141*
Arterberry, M. E., 163, *183*
Arzi, M., 164, *182*
Asanuma, H., 78, *93*
Asarnow, R. F., 327, *346*
Asgari, M., 257, *270*
Ashford, J. W., 126, *144*
Ashmead, D. H., 166, *182*
Atkinson, J., 276, *285*
Ayman-Nolley, S., 26, *29*

B

Bacharach, S., 129, *143*
Bachevalier, J., 262, *266*
Baddeley, A., 89, *91*, 104, 105, *120*
Badgaiyan, R., *347*
Bagwell, W. W., 328, *348*
Bahro, M., 135, 138, *144*
Bailenson, J., 228, *248*
Bakker, P. E., 294, *300*
Baldwin, J. M., 188, *215*
Ballard, D. H., 98, *120*
Barba, D. B., 259, *270*

401

Subject Index